Compliantly-Driven Robots with Application to Exoskeletons

Xiang Li · Haoyong Yu

Compliantly-Driven Robots with Application to Exoskeletons

Mechanical Design, Interaction Control, and Interactive Learning

Springer

Xiang Li
Department of Automation
Tsinghua University
Beijing, China

Haoyong Yu
Department of Biomedical Engineering
National University of Singapore
Singapore, Singapore

ISBN 978-981-96-9543-0 ISBN 978-981-96-9544-7 (eBook)
https://doi.org/10.1007/978-981-96-9544-7

This Springer imprint is published by the registered company Springer Nature Singapore Pte Ltd.
The registered company address is: 152 Beach Road, #21-01/04 Gateway East, Singapore 189721, Singapore

Competing Interests The authors have no competing interests to declare that are relevant to the content of this manuscript.

Contents

Notation and Definition

q	Robot joint angles
θ	Rotor angles of actuation motors
$M(q)$	Inertia matrix of the robot
$C(\dot{q}, q)$	Matrix of the Coriolis and Centripetal forces
D_q, D_θ	Damping matrices
$g(q)$	Gravitational torques
K	Stiffness matrix of the compliant actuator
τ_e	Interaction torques
B	Inertia matrix of the actuator
u	Control input exerted on the actuator
I_n	Identity matrix
$\lambda_{\min}$	Minimal eigenvalue
$\lambda_{\max}$	Maximal eigenvalue
M_d, C_d, K_d	Desired impedance parameters

Chapter 1
Introduction to Compliant Actuators

The demand for service robots, as well as assistive and rehabilitation robots, has been growing rapidly in both domestic and hospital settings. This trend is driven largely by the aging populations in most developed nations, which require advanced robotic solutions to support healthcare and daily living activities. As robots are increasingly expected to interact directly with humans, particularly in sensitive environments such as homes and hospitals, ensuring safety has become the most critical concern. The ability of robots to operate safely around humans without causing harm has become a key focus in the design of robotic systems, prompting significant research into the development of technologies that prioritize safety and human-friendliness.

Compliant actuators have emerged as a groundbreaking innovation in this context, offering a new paradigm in the design and functionality of robotic systems. Unlike traditional rigid actuators, which are designed to be stiff and resist external forces, compliant actuators introduce a degree of flexibility and adaptability that closely mimics the natural movements observed in biological organisms. This flexibility allows compliant actuators to absorb and respond to external forces in a way that rigid actuators cannot, making them particularly suitable for tasks requiring delicate interaction with humans or unpredictable environments. As a result, compliant actuators have become central to the development of robots that are capable of safe and effective human-robot collaboration, soft robotics applications, and adaptive control systems that require subtle responses to dynamic conditions.

A representative compliant actuator is the series elastic actuator (SEA). This chapter will focus on the principles of SEA and a novel SEA with two-stage stiffness, addressing the position control problem of this complex system by introducing a continuous tracking algorithm. The stability and convergence of the closed-loop system are analyzed, and experimental results are presented to demonstrate the effectiveness of the proposed control scheme.

X. Li and H. Yu, *Compliantly-Driven Robots with Application to Exoskeletons*,
https://doi.org/10.1007/978-981-96-9544-7_1

1.1 Principle and Structure of SEA

In human-robot interaction scenarios, compliant actuators are particularly valued for their ability to achieve precise force control, impedance control, and back-drivability. These characteristics are crucial for ensuring that robots can safely interact with humans by adjusting their movements based on the forces they encounter. Among the various designs of compliant actuators developed for human-friendly robotic applications [1, 2], the SEA has become the most commonly adopted.

The first SEA design was proposed by Pratt and Williamson in [3], where an elastic element was placed in a series with the motor and gear transmission. The basic components of SEAs proposed in [3] consist of a motor, a gear transmission, a set of linear springs, and a sensor measuring the deflection of the spring. The force output of SEA is calculated using Hooke's law. Later, Robinson carried out further studies on the modeling and control analysis of the SEA in [4].

By introducing an elastic element between the load and the geared motor, the inertia and non-linear frictions of the motor and the transmission are decoupled from the load, and external impacts and shocks are isolated from the gear transmission. It also puts the force sensing and control problem into a position sensing and control problem. Therefore, in human-friendly robotic applications, SEAs are known to offer a range of advantages over stiff actuators, such as high force/torque controllability and fidelity, low output impedance, back-drivability, tolerance to shock and impacts. The principle of SEA is shown in Fig. 1.1. In the presence of physical interaction, the interaction torques are exerted on the robot side. In addition, random disturbances may arise due to unmodeled dynamics and uncertain human motion intention (when the SEA-driven robot interacts with a human subject).

Traditional SEAs are limited to a fixed stiffness. To widen its applications requiring different stiffness, a new compact compliant actuator was first proposed in [5]. The key novelty is that the combination of a low-stiffness spring and a high-stiffness spring is employed instead of a single spring, where the two springs are utilized in the low-force range and the high-force range of the compliant actuator, respectively. Based on the two-stage stiffness, the compliant actuator in [5] is able to generate high-force fidelity and low output impedance and also extend the force range and bandwidth. Therefore, the performance of SEA is significantly improved.

The aforementioned SEA with a two-stage stiffness mechanism is illustrated in Fig. 1.2. This actuator integrates several key components: a servomotor, two rotary encoders, a torsional spring, a set of linear springs, a ball screw, and a potentiometer. Each component plays a crucial role in enhancing the actuator's functionality and performance. The motion from the motor is initially transmitted to the ball screw

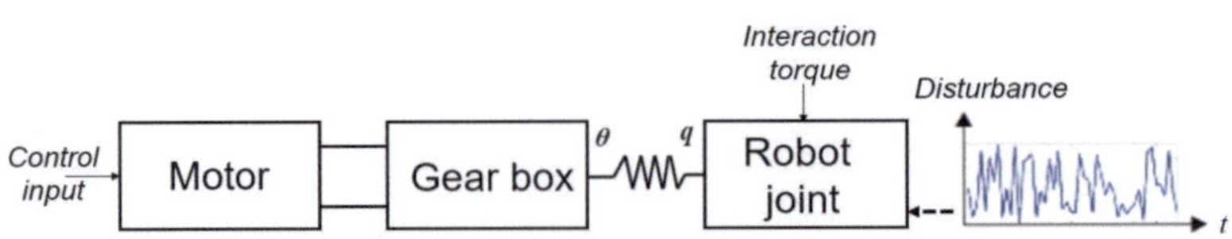

Fig. 1.1 The basic principle of SEA

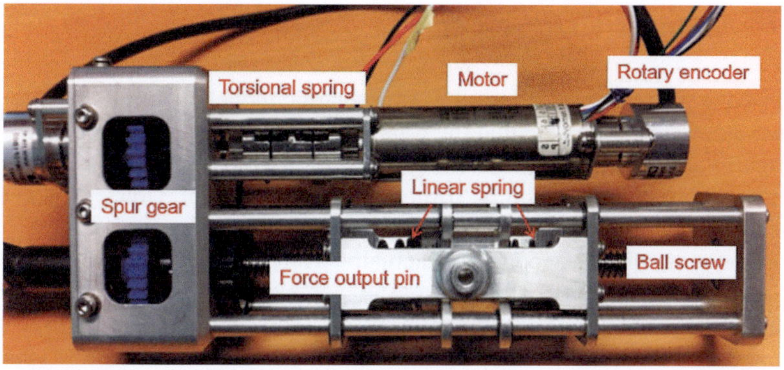

Fig. 1.2 Mechanical structure of the compliant actuator: **a** Principle of the actuator design (CAD model); **b** A prototype of the novel compliant actuator

through the torsional spring and spur gear, which converts the rotational motion of the shaft into the linear motion of the ball screw nut. This linear motion is then transferred to the output carriage via the linear springs. An encoder is mounted on the motor to measure the angular displacement of both the motor and the torsional spring. A second encoder measures the angular displacement of the ball screw. Additionally, a linear potentiometer is used to measure the displacement of the linear springs.

In this design (see Fig. 1.2a), the stiffness of the linear spring is selected to achieve the desired average output force, while the linear spring is designed to be soft, compact, and lightweight. A very small torsional spring is placed at the output end of the motor, with an effective spring stiffness over 100 times greater than that of the linear spring. Due to this significant difference in stiffness, the torsional spring effectively acts as a rigid link when the linear spring is dominant. Conversely, when the torsional spring is engaged, the soft linear spring is nearly saturated, and force control relies primarily on the torsional spring. This approach enables us to achieve a substantially smaller physical size compared to traditional SEA designs.

Figure 1.2b shows the prototype of the actuator. The actuator is designed to be able to provide up to 55 Nm output torque. A Maxon DC brushless motor (EC 4-pole 120 Watt 36 V) is employed due to its lightweight (0.175 kg) and low moment of inertia. The ball screw selected from Eichenberger Gewinde AG has a pitch of 2 mm/rev and can output over 1100 N force. The stiffness of linear springs is 24 N/mm while the working stroke is 10 mm. The linear springs provide a maximum output force of 240 N. The stiffness of the torsional spring is 0.29 Nm/rad, and the deflection range is 72 degrees. The resolution of the incremental rotary encoder is 1024 lines/rev, and the total mass of the actuator is less than 0.85 kg.

The ball screw used in the actuator has a high precision with a pitch of 2 mm/rev. Therefore, the actuator exhibits excellent reversibility, extremely low friction, and high efficiency, compared with conventional gearboxes. In this sense, the actuator is backdrivable even if the power is turned off. Also, in such a design, the ball screw is fixed at both ends and only rotates to make the nut move in both directions. As the overall friction is very low, there is no noticeable difference in terms of friction and damping in the two directions. The backdrivability of the actuator has been tested in [6].

1.2 Dynamic Model

The subsequent development describes the dynamic model of the proposed actuator with two-stage stiffness. Note that the rotary elements of the compliant actuator are converted to equivalent translational elements without loss of generality. Then, the soft linear spring and the torsional spring are referred to as the low-stiffness spring and the high-stiffness spring, respectively.

Next, the principle of the compliant actuator with two-stage stiffness is illustrated in Fig. 1.3, where m_1 denotes the mass of the motor, m_2 denotes the mass of the ball screw and spur gear, k_1 and k_2 are the high stiffness and the low stiffness, respectively, x_1, x_2, and x_3 denote the positions of the motor, the ball screw and the spur gear, and the load respectively, b_m, b_1, and b_2 denote the damping coefficients for the motor, the high-stiffness spring, and the low-stiffness spring, respectively, and u is the control input, which is also the exerted force.

The dynamic equation for the compliant actuator can be described using Newton's law as:

$$m_1\ddot{x}_1+b_m\dot{x}_1+b_1(\dot{x}_1-\dot{x}_2)+k_1(x_1-x_2)=u, \tag{1.1}$$

$$m_2\ddot{x}_2+b_2(\dot{x}_2-\dot{x}_3)+k_2(x_2-x_3)=b_1(\dot{x}_1-\dot{x}_2)+k_1(x_1-x_2). \tag{1.2}$$

where (1.1) describes the dynamic model of the motor, and (1.2) describes the dynamic model of the ball screw and the spur gear. The two subsystems are linked by the forces $k_1(x_1-x_2)+b_1(\dot{x}_1-\dot{x}_2)$.

In the compliant actuator, the low-stiffness spring works in the low-force range, and it is saturated when the force exerted on the spring reaches its limit. Then, the high-stiffness spring is activated to work in a high-force range. The different modes of the compliant actuator are summarized as follows.

(1) *Low-stiffness mode*: when the low-stiffness spring plays a dominant role, the dynamic model can be illustrated in Fig. 1.4. That is, the high-stiffness is not activated, and thus $x_1 \approx x_2$. The output force of the actuator is then obtained as: $F_o=k_2(x_2-x_3)+b_2(\dot{x}_2-\dot{x}_3)$ [5].
(2) *High-stiffness mode*: when the high-stiffness spring plays a dominant role, the dynamic model is illustrated in Fig. 1.5. That is, the low-stiffness spring is almost

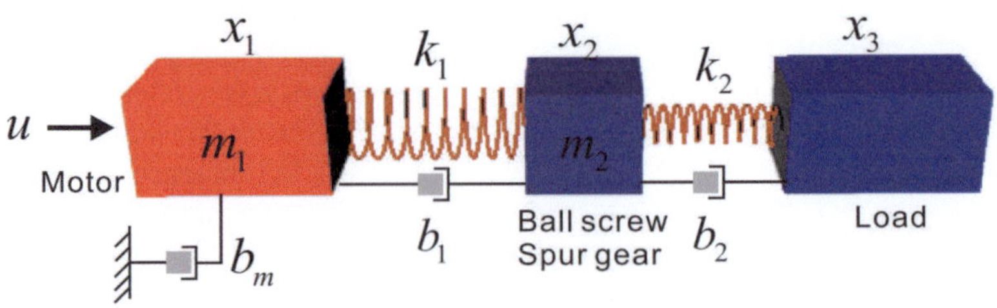

Fig. 1.3 The general case of the compliant actuator

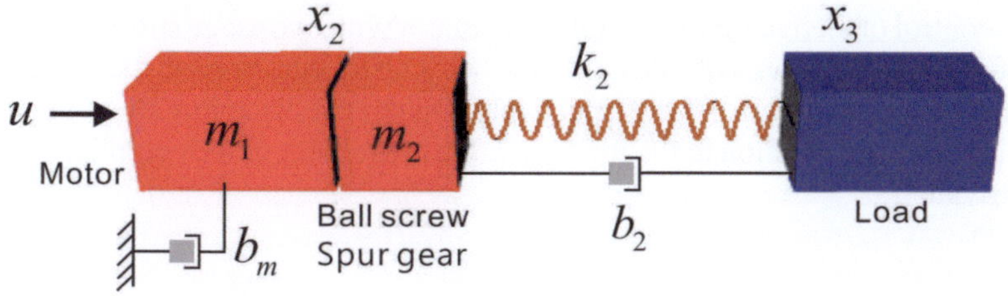

Fig. 1.4 The *low-stiffness mode* of the actuator

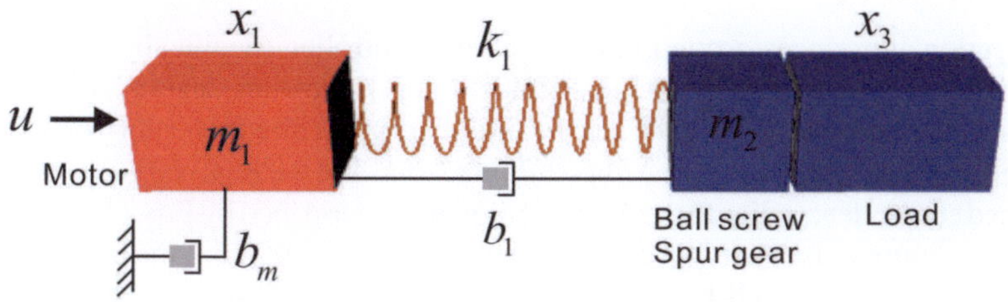

Fig. 1.5 The *high-stiffness mode* of the actuator

Fig. 1.6 The compliant actuator with two-stage stiffness: (a) the overall stiffness, where the low stiffness is given as k_2, and the high stiffness is given as k_1; (b) the output force

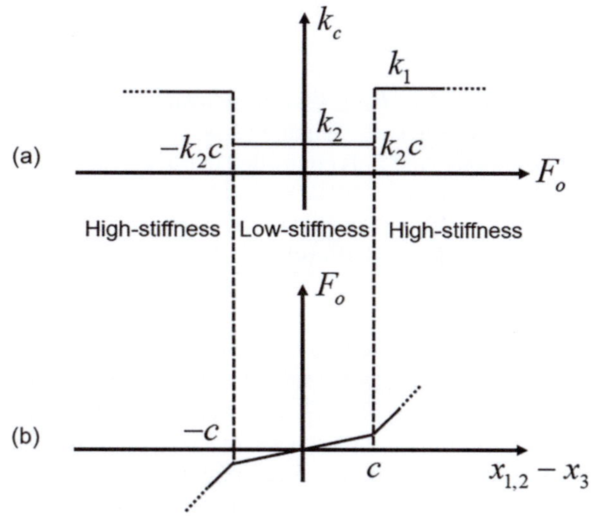

saturated such that $|x_2 - x_3| \approx c$ where c is a positive constant. Hence, we have $k_2|x_2 - x_3| \approx k_2 c$, and $b_2(\dot{x}_2 - \dot{x}_3) \approx 0$, and the output force is obtained as: $F_o = k_1(x_1 - x_3) + b_1(\dot{x}_1 - \dot{x}_3)$ [5].

Due to the hard switching of two springs, the overall stiffness of the compliant actuator is discontinuous, as illustrated in Fig. 1.6, which is specified as:

$$k_c(F_o) = \begin{cases} k_1, & |F_o| > k_2 c, \\ k_2, & |F_o| \le k_2 c. \end{cases} \tag{1.3}$$

The control objective for the compliant actuator with two-stage stiffness is to drive the state variable x_2 or x_1 to the desired time-varying trajectory x_d, in the *low-stiffness mode* or the *high-stiffness mode* respectively. Due to the discontinuity of the overall stiffness, finding a solution to the tracking control problem of the compliant actuator is nontrivial.

1.3 Trajectory Tracking Control

This section introduces a continuous tracking controller specifically designed for compliant actuators with two-stage stiffness. This controller enables the two state variables, x_1 and x_2, to smoothly track the time-varying desired trajectory x_d, ensuring a seamless transition between the *low-stiffness mode* and the *high-stiffness mode*. The proposed controller applies to the actuator dynamics described by Eqs. (1.1) and (1.2). When the compliant actuator is controlled to drive the robot joint, the load dynamics can be modeled as the well-known rigid robot dynamics [7], and the overall system is a high-order system consisting of both the robot dynamics and the actuator dynamics.

First, a force region is formulated to monitor the variation of the output force as:

$$f(F_o) = F_o^2 - (k_2 c)^2 \leq 0. \tag{1.4}$$

When $f(F_o) \leq 0$, $|F_o| \leq k_2 c$, the actuator works in the *low-stiffness mode*, and x_2 is controlled to track x_d. When $f(F_o) > 0$, $|F_o| > k_2 c$, the actuator works in the *high-stiffness mode*, and x_1 is controlled to track x_d. Then, a weight factor $w(F_o)$ is defined as:

$$w(F_o) = 1 - \frac{(\min(0, (\min(0, f(F_o)))^4 - ((\kappa \cdot k_2 \cdot c)^2 - (k_2 \cdot c)^2)^4))^4}{((\kappa \cdot k_2 \cdot c)^2 - (k_2 \cdot c)^2)^{16}}, \tag{1.5}$$

where $0 < \kappa < 1$ is a constant which determines the gradient of the weight between 0 and 1. The weight factor $w(F_o)$ is second-order continuous, and an illustration of $w(F_o)$ is shown in Fig. 1.7. The weight factor equals to 0 when the output force is outside the force region where $f(F_o) > 0$, and it smoothly increases to 1 when the output force is inside the force region where $f(F_o) \leq 0$. Note that the force F_o is obtained by measuring the displacements x_1, x_2, and x_3, and thus the force sensor is not required.

Next, two reference variables x_{o1} and x_{o2} are specified as:

$$\begin{aligned} x_{o1} &= (1 - w(F_o))x_d + w(F_o)x_1, \\ x_{o2} &= w(F_o)x_d + (1 - w(F_o))x_2. \end{aligned} \tag{1.6}$$

From Eqs. (1.5) and (1.6), when the actuator works in the *low-stiffness mode*, $w(F_o) = 1$. Then, the reference variables in Eq. (1.6) can be written as: $x_{o1} = x_1$, $x_{o2} = x_d$. When the actuator works in the *high-stiffness mode*, $w(F_o) = 0$. Then, the reference variables become: $x_{o1} = x_d$, $x_{o2} = x_2$.

Two sliding variables are proposed for different modes of the actuator as:

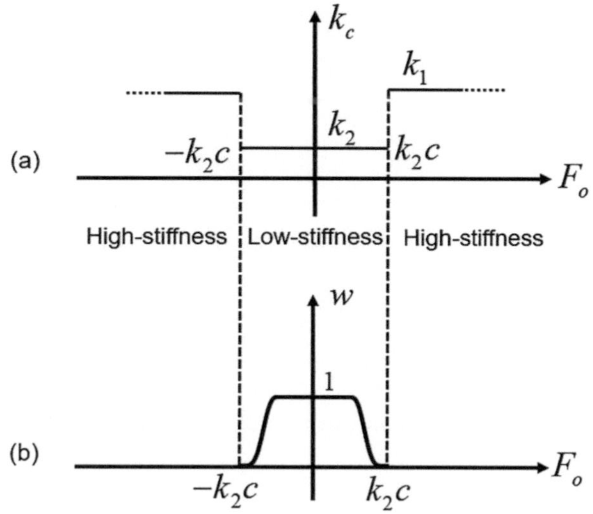

Fig. 1.7 An illustration of the relationship between the stiffness and the weight factor: **a** the two-stage stiffness k_c; **b** the weight factor $w(F_o)$

$$
\begin{aligned}
s_1 &= \dot{x}_1 - \dot{x}_{r1} \\
&= \dot{x}_1 - \dot{x}_{o1} + \alpha_1(x_1 - x_{o1}), \\
s_2 &= \dot{x}_2 - \dot{x}_{r2} \\
&= \dot{x}_2 - \dot{x}_{o2} + \alpha_2(x_2 - x_{o2}),
\end{aligned} \tag{1.7}
$$

where

$$
\begin{aligned}
\dot{x}_{r1} &= \dot{x}_{o1} - \alpha_1(x_1 - x_{o1}) \\
&= \dot{x}_{o1} - \alpha_1 \Delta x_1, \\
\dot{x}_{r2} &= \dot{x}_{o2} - \alpha_2(x_2 - x_{o2}) \\
&= \dot{x}_{o2} - \alpha_2 \Delta x_2,
\end{aligned} \tag{1.8}
$$

are auxiliary variables, α_1 and α_2 are positive constants, and $\Delta x_1 = x_1 - x_{o1}$ and $\Delta x_2 = x_2 - x_{o2}$. When the actuator works in the *low-stiffness mode*, $x_{o1} = x_1, x_{o2} = x_d$, and thus s_1=0, $s_2 \neq 0$. That is, the sliding variable s_2 works while s_1 remains zero. When the actuator works in the *high-stiffness mode*, $x_{o1} = x_d, x_{o2} = x_2$, and thus $s_1 \neq 0$, s_2=0. That is, the sliding variable s_1 works while s_2 reduces to zero. Therefore, the two sliding variables s_1 and s_2 are employed in different control modes to drive the state variables x_1 or x_2 to the desired trajectory x_d, respectively.

Using the sliding variables s_1 and s_2, the dynamic model described by (1.1) and (1.2) can be written as:

Table 1.1 Variations of control terms

	$w(F_o)$	x_{o1}	x_{o2}	s_1	s_2
Low-stiffness	1	x_1	x_d	0	$\Delta\dot{x}_2+\alpha_2\Delta x_2$
High-stiffness	0	x_d	x_2	$\Delta\dot{x}_1+\alpha_1\Delta x_1$	0

$$\begin{aligned} &m_1\dot{s}_1+m_1\ddot{x}_{r1}+b_m s_1+b_m\dot{x}_{r1} \\ &+ b_1(\dot{x}_1-\dot{x}_2)+k_1(x_1-x_2) = u, \end{aligned} \tag{1.9}$$

$$\begin{aligned} &m_2\dot{s}_2+m_2\ddot{x}_{r2}+b_2 s_2+b_2\dot{x}_{r2}-b_2\dot{x}_3+k_2(x_2-x_3) \\ =&b_1(\dot{x}_1-\dot{x}_2)+k_1(x_1-x_2). \end{aligned} \tag{1.10}$$

The tracking control scheme for the compliant actuator with two-stage stiffness is now proposed as:

$$\begin{aligned} u =& - k_{s1}s_1 - k_{s2}s_2-b_2\dot{x}_3+k_2(x_2-x_3) \\ &+ m_1\ddot{x}_{r1}+b_m\dot{x}_{r1}+m_2\ddot{x}_{r2}+b_2\dot{x}_{r2}, \end{aligned} \tag{1.11}$$

where k_{s1} and k_{s2} are positive constants. The first two terms in Eq. (1.11) denote the sliding control in the *high-stiffness mode* and the *low-stiffness mode* respectively, and the other terms denote the dynamic compensation. The key novelty of the proposed controller (1.11) is that the new reference variables x_{o1} and x_{o2} are employed instead of the conventional desired trajectories. The reference variables can be treated as weighted sums between the desired trajectory and the state variables. The feedback control is activated when the corresponding reference variable is specified as the desired trajectory, and it is suspended when the reference variable is just the state variable. The variations of the control terms in different modes are summarized in Table 1.1.

The choice of the control parameters should be balanced between the maximum motor input force and the tracking performance. Then, α_i and k_{si} $(i=1, 2)$ are fine-tuned together to achieve the desired transient and steady-state error performance. Short rise time and smaller steady-state error can be achieved by tuning α_i up, and better velocity tracking can be done by tuning k_{si} up.

Substituting the control input (1.11) into (1.9), the closed-loop equation is obtained as:

$$\begin{aligned} &m_1\dot{s}_1+(b_m+k_{s1})s_1-m_2\ddot{x}_{r2}-b_2\dot{x}_{r2}+k_{s2}s_2+b_2\dot{x}_3 \\ &- k_2(x_2-x_3)+b_1(\dot{x}_1-\dot{x}_2)+k_1(x_1-x_2)=0, \end{aligned} \tag{1.12}$$

which also implies that

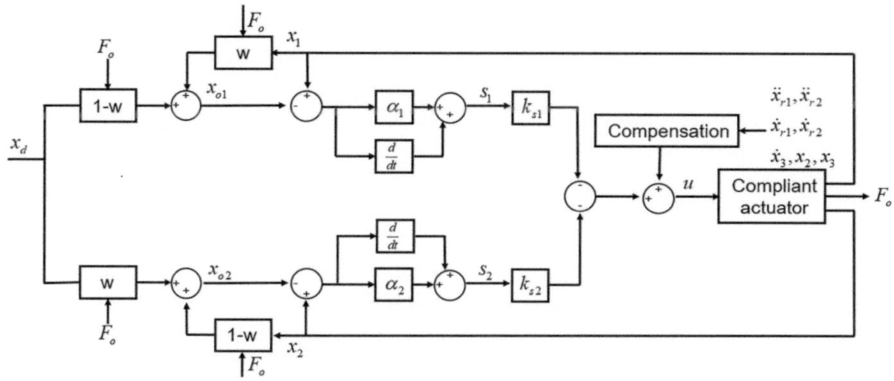

Fig. 1.8 Block diagram of closed-loop system

$$
\begin{aligned}
&b_1(\dot{x}_1-\dot{x}_2)+k_1(x_1-x_2)\\
&=-m_1\dot{s}_1-(b_m+k_{s1})s_1+m_2\ddot{x}_{r2}+b_2\dot{x}_{r2}\\
&\quad-k_{s2}s_2-b_2\dot{x}_3+k_2(x_2-x_3).
\end{aligned}
\tag{1.13}
$$

Substituting (1.13) into the right side of (1.10) yields:

$$
m_1\dot{s}_1+(b_m+k_{s1})s_1+m_2\dot{s}_2+(b_2+k_{s2})s_2=0. \tag{1.14}
$$

The block diagram of the closed-loop system is shown in Fig. 1.8.

We can now state the following theorem.

Theorem 1.3.1 *The proposed tracking controller (1.11) for the compliant actuator system with two-stage stiffness described by (1.14) gives rise to the convergence of tracking errors in the low-stiffness mode and the high-stiffness mode respectively.*

Proof When the compliant actuator works in the *low-stiffness mode*, $s_1 \to 0$, and Eq. (1.14) is simplified as:

$$
m_2\dot{s}_2+(b_2+k_{s2})s_2=0, \tag{1.15}
$$

which implies that $s_2 \to 0$. Since $s_2=\Delta\dot{x}_2+\alpha_2\Delta x_2$, the convergence of $s_2 \to 0$ implies that $\Delta\dot{x}_2 \to 0$ and $\Delta x_2 \to 0$. Since $w(F_o)=1$ in the *low-stiffness mode*, $x_{o2}=x_d$ from (1.6), and thus we have $\dot{x}_2 \to \dot{x}_d$ and $x_2 \to x_d$ as $t \to \infty$. That is, the state variable x_2 tracks the desired trajectory when the actuator works in *low-stiffness mode*.

When the compliant actuator works in the *high-stiffness mode*, $s_2 \to 0$, and hence Eq. (1.14) is simplified as:

$$
m_1\dot{s}_1+(b_m+k_{s1})s_1=0, \tag{1.16}
$$

which then implies that $s_1 \to 0$. Since $s_1 = \Delta\dot{x}_1 + \alpha_1 \Delta x_1$, the convergence of $s_1 \to 0$ implies that $\Delta\dot{x}_1 \to 0$ and $\Delta x_1 \to 0$. Since $w(F_o) = 0$ in the *high-stiffness mode*, $x_{o1} = x_d$ from (1.6), and thus we have $\dot{x}_1 \to \dot{x}_d$ and $x_1 \to x_d$ as $t \to \infty$. That is, the state variable x_1 tracks the desired trajectory when the actuator works in the *high-stiffness mode*. □

During the transition between the two control modes, we have the following theorem to state the convergence.

Theorem 1.3.2 *The proposed tracking controller (1.11) guarantees the convergence of tracking errors when the compliant actuator transits between the low-stiffness mode and the high-stiffness mode, if the control gains α_1, α_2, k_{s1}, and k_{s2} are chosen such that:*

$$\begin{aligned} \alpha_1 &= \alpha_2 \triangleq \alpha, \\ k_{s1} &= k_{s2} \triangleq k_s, \end{aligned} \tag{1.17}$$

where k_s is set sufficiently large.

Proof It has been shown in Theorem 1.3.1 that the control gains α_1, α_2, k_{s1}, and k_{s2} can be chosen as any positive values to ensure the convergence of tracking errors when the actuator works in either the *low-stiffness mode* or the *high-stiffness mode*. We now consider the transition stage where $0<(F_o)<1$. During the transition stage, the high-stiffness spring is not activated such that $x_1 \approx x_2$. Then, from (1.6) and (1.7), we have:

$$\begin{aligned} s_1 &= \dot{x}_1 - \dot{x}_{o1} + \alpha_1 (x_1 - x_{o1}) \\ &= (1 - w(F_o))(\dot{x}_1 - \dot{x}_d) + (\alpha_1 (1 - w(F_o)) - \dot{w}(F_o))(x_1 - x_d), \end{aligned} \tag{1.18}$$

and

$$\begin{aligned} s_2 &= \dot{x}_2 - \dot{x}_{o2} + \alpha_2 (x_2 - x_{o2}) \\ &= w(F_o)(\dot{x}_2 - \dot{x}_d) + (\alpha_2 w(F_o) + \dot{w}(F_o))(x_2 - x_d). \end{aligned} \tag{1.19}$$

Note that $x_1 \approx x_2$, and also $\alpha_1 = \alpha_2 = \alpha$ when condition (1.17) is satisfied. Adding (1.18) with (1.19) yields:

$$s_1 + s_2 = (\dot{x}_2 - \dot{x}_d) + \alpha (x_2 - x_d). \tag{1.20}$$

Since $k_{s1} = k_{s2} = k_s$, Eq. (1.14) can be written as:

$$\frac{m_1}{k_s}\dot{s}_1 + \left(\frac{b_m}{k_s} + 1\right) s_1 + \frac{m_2}{k_s}\dot{s}_2 + \left(\frac{b_s}{k_s} + 1\right) s_2 = 0. \tag{1.21}$$

When k_s is chosen sufficiently large such that $\frac{m_1}{k_s}, \frac{b_m}{k_s}, \frac{m_2}{k_s}, \frac{b_2}{k_s} \to 0$, Eq. (1.21) reduces to:

$$s_1 + s_2 = 0. \tag{1.22}$$

From Eq. (1.20), it is obtained that

$$(\dot{x}_2 - \dot{x}_d) + \alpha(x_2 - x_d) = 0, \tag{1.23}$$

which implies that $x_2 \to x_d$, and $\dot{x}_2 \to \dot{x}_d$ as $t \to \infty$. □

1.4 Model-Based Observer

Note that the acceleration information $\ddot{x}_1$ and $\ddot{x}_2$ is required in the derivatives of reference variables $\ddot{x}_{o1}$ and $\ddot{x}_{o2}$, which is thus required in the control input u (1.11). To eliminate the requirement of the acceleration information, observers are developed to estimate $\dot{x}_{o1}$ and $\dot{x}_{o2}$ respectively.

First, the dynamic model of the motor is reproduced as:

$$m_1\ddot{x}_1 + b_m\dot{x}_1 + b_1(\dot{x}_1 - \dot{x}_2) + k_1(x_1 - x_2) = u. \tag{1.24}$$

Then, an observer can be developed for the estimated reference variable $\hat{x}_{o1}$ as:

$$\begin{aligned} \dot{\hat{x}}_{o1} &= \eta_1 + \beta_1 e_1, \\ \dot{\eta}_1 &= \frac{1}{m_1}(u - (b_m + b_1)\eta_1 + b_1\dot{x}_2 - k_1(x_1 - x_2) + k_{e1}e_1), \end{aligned} \tag{1.25}$$

where $\hat{x}_{o1}$ denotes the estimated signal of x_{o1}, $e_1 = x_{o1} - \hat{x}_{o1}$ is the observation error, η_1 is an auxiliary variable, and β_1 and k_{e1} are positive constants. By following the similar development in [8], it can be shown that the observer (1.25) ensures $e_1 \to 0$. That is, $\hat{x}_{o1} \to x_{o1}$ as $t \to \infty$.

Then, the dynamic model of the ball screw and the spur gear is reproduced as:

$$m_2\ddot{x}_2 + b_2(\dot{x}_2 - \dot{x}_3) + k_2(x_2 - x_3) = b_1(\dot{x}_1 - \dot{x}_2) + k_1(x_1 - x_2). \tag{1.26}$$

Then, another observer can be developed for the estimated reference variable $\hat{x}_{o2}$ as:

$$\begin{aligned} \dot{\hat{x}}_{o2} &= \eta_2 + \beta_2 e_2, \\ \dot{\eta}_2 &= \frac{1}{m_2}(-(b_1 + b_2)\eta_2 + b_1\dot{x}_1 + k_1(x_1 - x_2) \\ &\quad + b_2\dot{x}_3 - k_2(x_2 - x_3) + k_{e2}e_2), \end{aligned} \tag{1.27}$$

where $\hat{x}_{o2}$ denotes the estimated signal of x_{o2}, $e_2 = x_{o2} - \hat{x}_{o2}$ is the observation error, and η_2 is also an auxiliary variable, and β_2 and k_{e2} are positive constants. Similarly, it can be shown that $\hat{x}_{o2} \to x_{o2}$ as $t \to \infty$.

By using the estimated reference variables $\hat{x}_{o1}$ and $\hat{x}_{o2}$, the two sliding variables are revised as:

$$\begin{aligned}\hat{s}_1 &= \dot{x}_1 - \dot{\hat{x}}_{o1} + \alpha_1(x_1 - \hat{x}_{o1}),\\ \hat{s}_2 &= \dot{x}_2 - \dot{\hat{x}}_{o2} + \alpha_2(x_2 - \hat{x}_{o2}).\end{aligned} \tag{1.28}$$

Therefore, the requirement for acceleration information and its derivative is eliminated.

1.5 Results

To better demonstrate the continuous tracking control scheme introduced in this chapter, we deployed it on the aforementioned compliant actuator system with two-stage stiffness to test its effectiveness, as shown in Fig. 1.9. The system comprises an actuator mounted on a testing jig, a real-time control system, a motor driver, and a host PC. The motor driver is the Elmo Harmonica 5/60, capable of delivering a maximum power output of 200 W and a continuous output current of 5 A. The control algorithm is implemented within the NI CompactRIO 9074 embedded control system (National Instruments). Feedback data is acquired through the NI data acquisition system, which includes the NI 9215 (analog input module), NI 9263 (analog output module), and NI 9516 (encoder module). The system operates with a sampling frequency of 2 kHz.

The switching point of the two springs illustrated in Fig. 1.6 is calibrated as: F_o=240 N. Therefore, the low-stiffness spring works at $-240 \leq F_o \leq 240$ N, while the high-stiffness spring works at: F_o>240 N, or $F_o < -240$ N. In experiments, the position of the load is fixed and calibrated as: $x_3 \equiv 0$, for the testing purpose. The dynamic parameters in (1.1) and (1.2) are obtained by referring to [5]. A series of time-varying desired trajectories were specified to illustrate the performance of the proposed controller in the *low-stiffness mode*, the *high-stiffness mode*, and during the

Fig. 1.9 Experimental setup

Table 1.2 Parameter settings

	κ	k_{s1}, k_{s2}	α_1, α_2	k_{e1}, k_{e2}	β_1, β_2
Value	0.9	400	0.06	100	60

transition between two modes. The corresponding parameter settings are illustrated in Table 1.2.

In the first experiment, the desired trajectory was defined as

$$x_d = 0.012 \cdot \sin(2\pi t) \text{ m}. \tag{1.29}$$

The compliant actuator worked in the *low-stiffness mode*, such that only the low-stiffness spring was activated and the weight factor $w(F_o)$ remained 1. Therefore, the reference variables $x_{o1}=x_1$, $x_{o2}=x_d$. The path of the position variable x_2 and the reference variable x_{o2} is shown in Fig. 1.10a. The corresponding tracking error is shown in Fig. 1.10b, which are less than 0.003 m. The observation error, i.e., $x_{o2} - \hat{x}_{o2}$, is shown in Fig. 1.10c, which is around 0.002 m. The bounded control input u is shown in Fig. 1.10d.

In the second experiment, the desired trajectory was given as:

$$x_d = 0.08 + 0.012 \cdot \sin(2\pi t) \text{ m}. \tag{1.30}$$

The compliant actuator worked in the *high-stiffness mode* such that only the high-stiffness spring was activated and the weight factor $w(F_o)$ remained zero. Therefore, the reference variables $x_{o1}=x_d$, $x_{o2}=x_2$. The path of the position variable x_1 and the reference variable x_{o1} is shown in Fig. 1.11a. The corresponding tracking error is shown in Fig. 1.11b, which is less than 0.005 m. The observation errors, i.e., $x_{o1} - \hat{x}_{o1}$, is shown in Fig. 1.11c, which is around 0.001 m. The control input u is shown in Fig. 1.11d, which is also bounded.

In the third experiment, the desired trajectory was given as

$$x_d = 0.06 + 0.012 \cdot \sin(2\pi \cdot t) \text{ m}. \tag{1.31}$$

The compliant actuator transited between the *low-stiffness mode* and the *high-stiffness mode*. The path of the position variable x_2 and the reference variable x_{o2} is shown in Fig. 1.12a, and the path of the position variable x_1 and the reference variable x_{o1} is shown in Fig. 1.12b. It can be seen that when the actuator works in the *low-stiffness mode*, $x_{o1}=x_1$, and $x_{o2}=x_d$. i.e., x_1 is controlled to track itself while x_2 is controlled to track x_d. On the other hand, when actuator works in the *high-stiffness mode*, $x_{o1}=x_d$, $x_{o2}=x_2$. Since the weight factor $w(F_o)$ is continuous, the reference variables are also continuous. The tracking error $x_2 - x_{o2}$ is shown in Fig. 1.12c which is less than 0.003 m, and the tracking error $x_1 - x_{o1}$ is shown in Fig. 1.12d which is less than 0.005 m. Note that the tracking error $x_2 - x_{o2}$ is

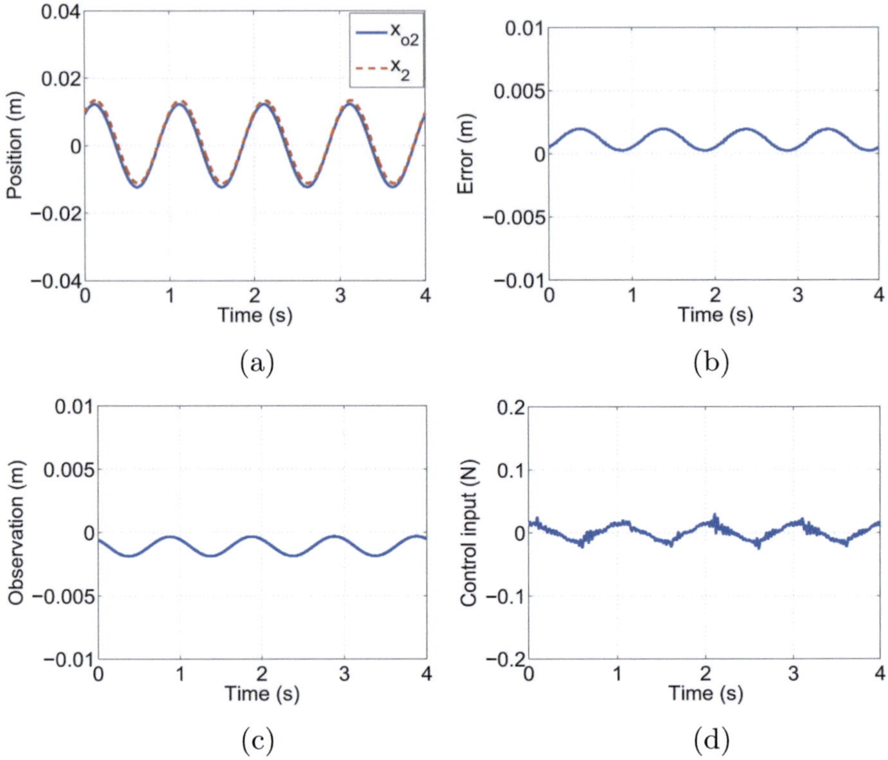

Fig. 1.10 Experiment 1: *Low-stiffness mode*: **a** path of x_2 and x_{o2}; **b** tracking error $x_2 - x_{o2}$; **c** observation error $x_{o2} - \hat{x}_{o2}$; **d** control input u

larger when $w(F_o) = 1$ than those when $w(F_o) = 0$, which is because x_2 is controlled to track x_d when $w(F_o) = 1$, and track itself when $w(F_o) = 0$. On the other hand, the tracking error $x_1 - x_{o1}$ is smaller when $w(F_o) = 1$ than those when $w(F_o) = 0$. The observation error $x_{o2} - \hat{x}_{o2}$ is shown in Fig. 1.12e, which is around 0.001 m, and the observation error $x_{o1} - \hat{x}_{o1}$ is shown in Fig. 1.12f which is around 0.002 m. The experimental results show that the proposed control scheme has been successfully implemented in the compliant actuator with two-stage stiffness. By using the proposed controller, the actuator is able to track the desired trajectory throughout the *high-stiffness mode*, the *low-stiffness mode*, and also the transition between two modes, with high tracking accuracy. The transition is smooth and stable, and the overall control input is continuous.

In the fourth experiment, a disturbance resulting from external forces on the actuator was added in trajectory tracking, and the results are shown in Fig. 1.13. As seen in Fig. 1.13b, d, the external disturbance, occurring at around $t \approx 3.8$ s, resulted in an increase of tracking error, but didn't compromise the stability of a closed-loop system, i.e., the boundedness of tracking errors. In the fifth experiment, the desired

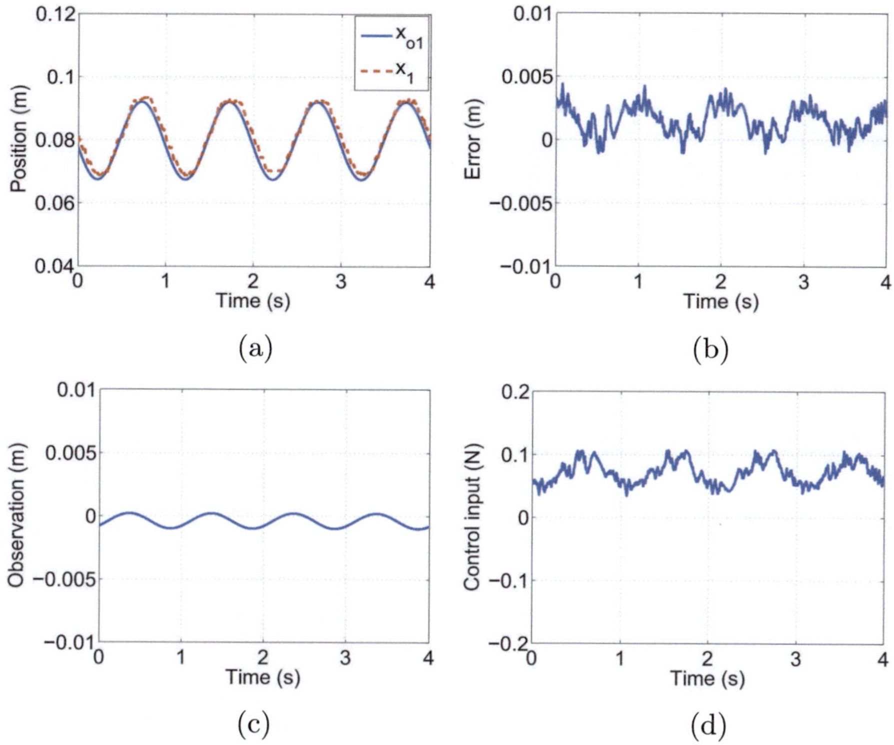

Fig. 1.11 Experiment 2: *High-stiffness mode*: **a** path of x_1 and x_{o1}; **b** tracking error $x_1 - x_{o1}$; **c** observation error $x_{o1} - \hat{x}_{o1}$; **d** control input u

trajectory was specified as a ramp function with the frequency of 0.5 Hz, and the controller also transited between the *high-stiffness mode* and *low-stiffness mode*. The results are shown in Fig. 1.14, where the tracking errors are less than 0.006 m (see Fig. 1.14c, d), the observation errors (see Fig. 1.14e, f) are less than 0.01 m. The results demonstrate the good tracking performance of the proposed controller for different trajectories. A comparison was also carried out between the proposed controller and the hard-switching controller. To implement the hard-switching controller, the weight factor $w(F_o)$ was set as $w(F_o) = 0$ when $F_o > 240$ N, and $w(F_o) = 1$ when $F_o \leq 240$ N. The results are shown in Fig. 1.15, where the hard-switching controller results in more significant chattering at the switching point, compared with the proposed continuous tracking controller.

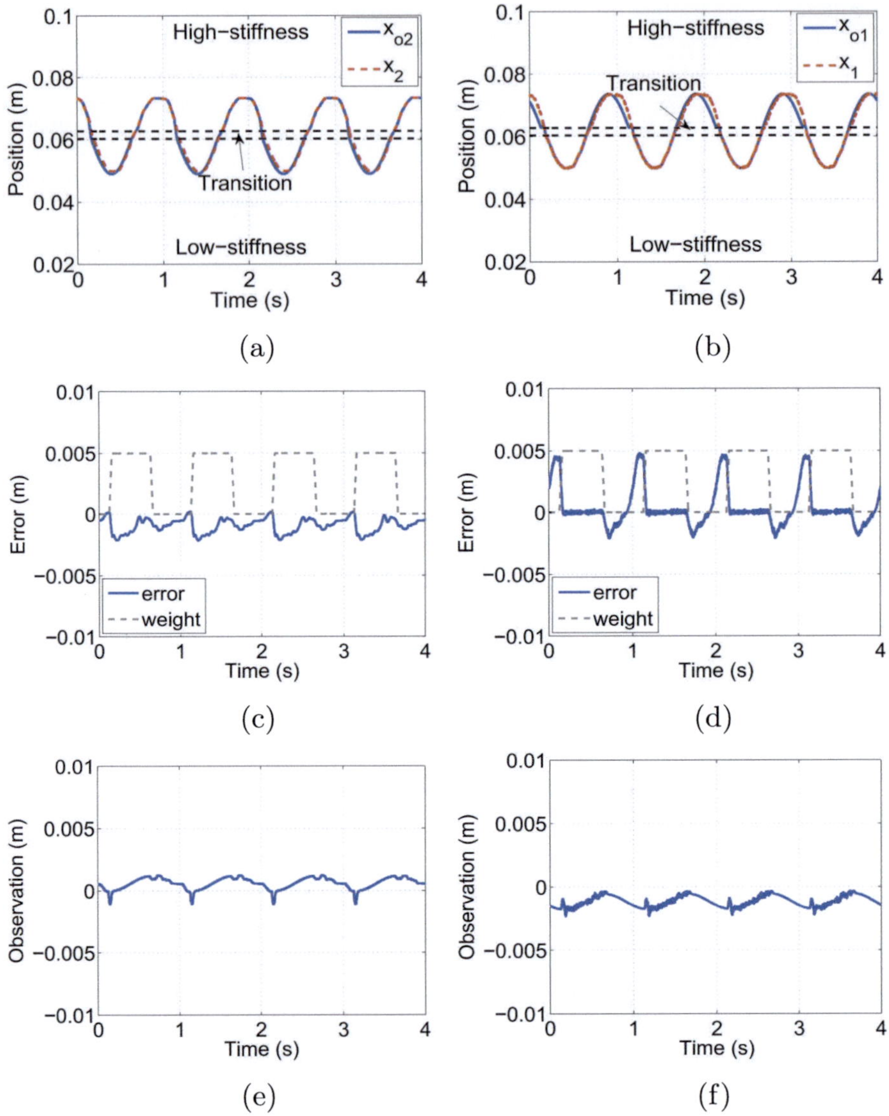

Fig. 1.12 Experiment 3: Transition between two modes: **a** path of x_2 and x_{o2}; **b** path of x_1 and x_{o1}; **c** tracking error $x_2 - x_{o2}$; **d** tracking error $x_1 - x_{o1}$; **e** observation error $x_{o2} - \hat{x}_{o2}$; **f** observation error $x_{o1} - \hat{x}_{o1}$

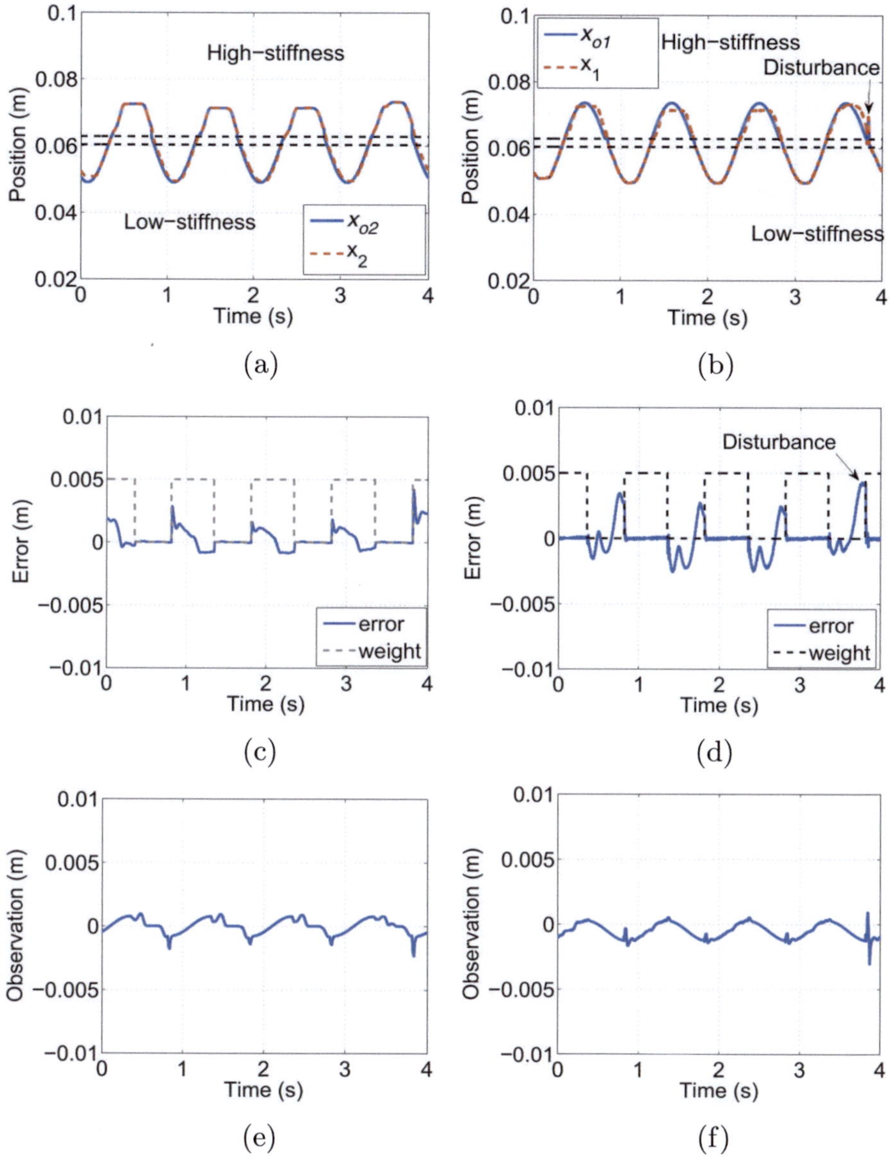

Fig. 1.13 Experiment 4: Tracking control subject to disturbance: **a** path of x_2 and x_{o2}; **b** path of x_1 and x_{o1}; **c** tracking error $x_2 - x_{o2}$; **d** tracking error $x_1 - x_{o1}$; **e** observation error $x_{o2} - \hat{x}_{o2}$; **f** observation error $x_{o1} - \hat{x}_{o1}$

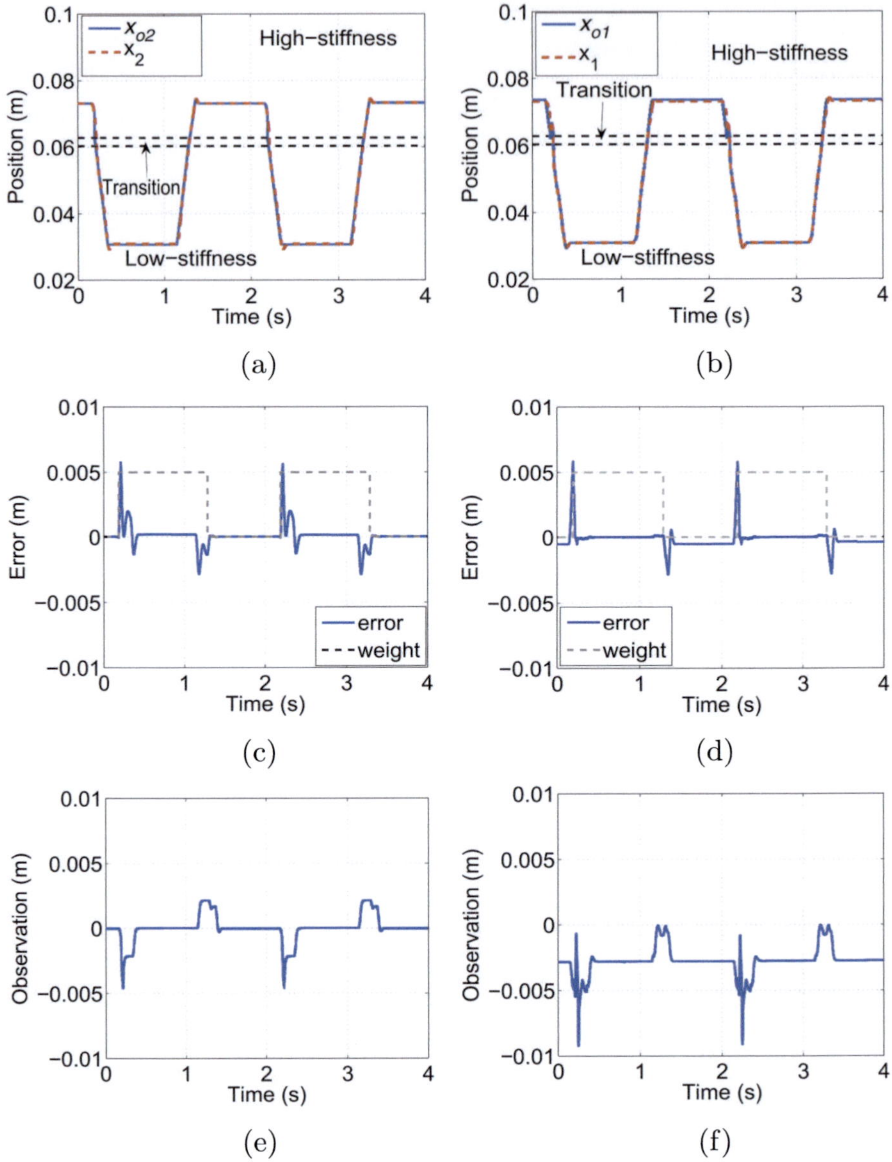

Fig. 1.14 Experiment 5: Tracking control for ramp functions: **a** path of x_2 and x_{o2}; **b** path of x_1 and x_{o1}; **c** tracking error $x_2 - x_{o2}$; **d** tracking error $x_1 - x_{o1}$; **e** observation error $x_{o2} - \hat{x}_{o2}$; **f** observation error $x_{o1} - \hat{x}_{o1}$

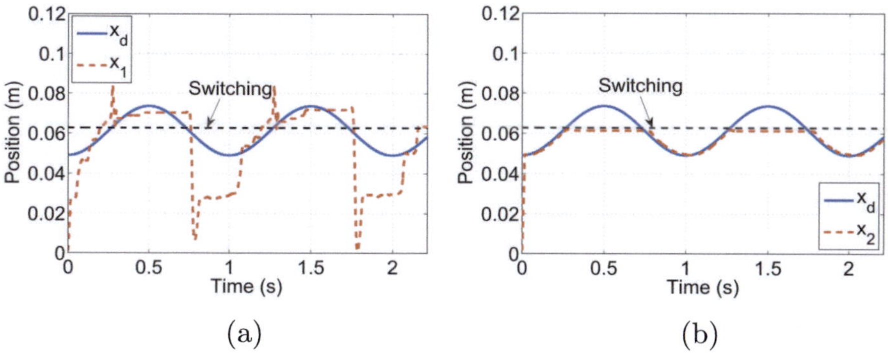

Fig. 1.15 Experiment 6: Hard-switching control: **a** path of x_1; **b** path of x_2

1.6 Summary and Notes

Inspired by the original work [3, 4], much progress has been achieved in the development and applications of SEAs [9–12]. The SEA was implemented in the RobotKnee system [9] to achieve low impedance and high force fidelity. In [10], a Bowden-cable-based SEA was developed for the LOPES robotic exoskeleton system. A rotary SEA was designed in [11] to provide partial support to knee flexion/extension during overground walking. A compact and lightweight prismatic SEA was reported in [12], which is based on a piston-style ball screw support mechanism and a concentric compliant element. However, existing SEAs commonly face a fundamental limitation, i.e., the fixed stiffness of the elastic element as discussed by Pratt et al. in [4, 13]. The performance of SEAs largely depends on the selection of the stiffness of a single spring [13]. Soft spring produces high fidelity, low output impedance, and reduces stiction, but also limits the force range and the bandwidth. On the other hand, a stiff spring increases the bandwidth but reduces force fidelity, leading to low intrinsic compliance and back-drivability and bulky and heavy systems.

To deal with the aforementioned limitations, this chapter provides a detailed explanation of the structural design and dynamics of the compliant actuator with two-stage stiffness. Such a design can combine the advantages of both the high stiffness and the low stiffness. Due to the discontinuous overall stiffness of the compliant actuator caused by the hard switching between two springs, achieving precise position control over this type of actuator presents a significant challenge. Although various control methods [5, 14–18] have been proposed for SEAs, the issue of discontinuity remains unresolved with existing techniques.

Accordingly, this chapter proposes a new trajectory-tracking control scheme for the compliant actuator with two-stage stiffness. While the overall stiffness is discontinuous, the proposed control method ensures a continuous and automatic transition between the *low-stiffness mode* and the *high-stiffness mode*. The stability and convergence of the closed-loop system have been analyzed. The proposed control method has been successfully implemented in the compliant actuator system, and experimental results show that the actuator is able to track different time-varying trajectories in different modes.

References

1. R. Ham, T.G. Sugar, B. Vanderborght, K.W. Hollander, D. Lefeber, Compliant actuator designs. IEEE Robot. Autom. Mag. **16**, 81–94 (2009)
2. R. Schiavi, G. Grioli, S. Sen, A. Bicchi, Vsa-ii: A novel prototype of variable stiffness actuator for safe and performing robots interacting with humans, in *IEEE International Conference on Robotics and Automation* (2008), pp. 2171–2176
3. G.A. Pratt, M.M. Williamson, Series elastic actuators, in *Proceedings 1995 IEEE/RSJ International Conference on Intelligent Robots and Systems. Human Robot Interaction and Cooperative Robots*, vol. 1 (IEEE, 1995), pp. 399–406
4. D.W. Robinson, *Design and analysis of series elasticity in closed-loop actuator force control.* Ph.D. thesis, Massachusetts Institute of Technology (2000)
5. H. Yu, S. Huang, G. Chen, N. Thakor, Control design of a novel compliant actuator for rehabilitation robots. Mechatronics **23**, 1072–1083 (2013)
6. H. Yu, S. Huang, G. Chen, Y. Pan, Z. Guo, Human-robot interaction control of rehabilitation robots with series elastic actuators. IEEE Trans. Robot. **31**(5), 1089–1100 (2015)
7. S. Arimoto, *Control Theory of Non-Linear Mechanical Systems: A Passivity-Based and Circuit-Theoretic Approach* (Oxford University Press, London, U.K., 1996)
8. X. Li, C.C. Cheah, S. Hu, D. Sun, Dynamic trapping and manipulation of biological cells with optical tweezers. Automatica **49**(6), 1614–1625 (2013)
9. J.E. Pratt, B.T. Krupp, C.J. Morse, S.H. Collins, The roboknee: an exoskeleton for enhancing strength and endurance during walking, in *IEEE International Conference on Robotics and Automation* (2004), pp. 2430–2435
10. J.F. Veneman, R. Ekkelenkamp, R. Kruidhof, F.C.T.V.D. Helm, H.V.D. Kooij, A series elastic- and bowden-cable-based actuation system for use as torque actuator in exoskeleton-type robots. Int. J. Robot. Res. **25**(3), 261–281 (2006)
11. F. Sergi, D. Accoto, G. Carpino, N.L. Tagliamonte, E. Guglielmelli, Design and characterization of a compact rotary series elastic actuator for knee assistance during overground walking, in *IEEE RAS & EMBS International Conference on Biomedical Robotics and Biomechatronics* (2012), pp. 1931–1936
12. N. Paine, S. Oh, L. Sentis, Design and control considerations for high-performance series elastic actuators. IEEE/ASME Trans. Mechatron. **19**(3), 1080–1091 (2014)
13. D.W. Robinson, J.E. Pratt, D.J. Paluska, G.A. Pratt, Series elastic actuator development for a biomimetic walking robot, in *IEEE/ASME International Conference on Advanced Intelligent Mechatronics* (1999), pp. 561–568
14. D.W. Robinson, G.A. Pratt, Force controllable hydro-elastic actuator, in *IEEE International Conference on Robotics and Automation* (2000), pp. 1321–1327
15. H. Vallery, J. Veneman, E.V. Asseldonk, R. Ekkelenkamp, M. Buss, H.V.D. Kooij, Compliant actuation of rehabilitation robots - benefits and limitations of series elastic actuators. IEEE Robot. Autom. Mag. **15**, 60–69 (2008)
16. G. Wyeth, Control issues for velocity sourced series elastic actuators, in *Australasian Conference on Robotics and Automation* (2006)
17. H. Vallery, R. Ekkelenkamp, H.V.D. Kooij, M. Buss, Passive and accurate torque control of series elastic actuators, in *IEEE/RSJ International Conference on Intelligent Robots and Systems* (2007), pp. 3534–3538
18. K. Kong, J. Bae, M. Tomizuka, Control of rotary series elastic actuator for ideal force-mode actuation in human-robot interaction applications. IEEE/ASME Trans. Mechatron. **14**, 105–118 (2009)

Chapter 2
Compliantly-Actuated Robots

As shown in the previous chapter, SEAs offer a level of compliance that enhances safety, particularly in unpredictable environments where human contact is frequent. This is mainly because of its inherent flexibility, which allows for smoother motion and reduces the risk of injury during collisions or accidental impacts.

While various SEA designs have been developed and successfully implemented in physical human-robot interaction systems, relatively few control strategies have been proposed to address the challenges that arise from the dynamic stability and uncertainties inherent in SEA-driven robotic systems. Moreover, the crucial issue of ensuring safety in these systems, particularly in scenarios where there may be conflicting motions between the human and the robot, has not yet been systematically explored or resolved. These gaps in control strategies can potentially limit the effectiveness and adoption of SEA-based systems in real-world applications, especially when precise, stable, and safe interactions are required.

In this chapter, we delve into two novel control algorithms specifically designed for SEA-driven robots engaged in human-robot interactions. This first controller is capable of detecting the human motion intention and smoothly transiting between the robot-in-charge mode (where the robot plays a dominant role in following a desired trajectory) and the human-in-charge mode (where the human plays a dominant role and the robot follows the human). The second controller specifies the control objective as a desired impedance model and achieves it in an iterative manner by exploring the repeating nature of rehabilitation training. These methods not only account for the dynamic behavior of the system but also ensure the stability of the closed-loop control through rigorous theoretical analysis using the Lyapunov method. Additionally, the practical effectiveness of these control strategies is validated through a series of physical experiments, demonstrating their potential to enhance both the performance and safety of SEA-driven robots in dynamic, human-centered environments.

X. Li and H. Yu, *Compliantly-Driven Robots with Application to Exoskeletons*,
https://doi.org/10.1007/978-981-96-9544-7_2

2.1 Backstepping Approach

This chapter considers the robot driven by a specially designed compact SEA, as shown in Fig. 2.1. This SEA (Series Elastic Actuator) is composed of several key components: a servomotor equipped with a rotary encoder, a set of linear springs, a ball screw mechanism, and two potentiometers. Here's how it works: the motor's motion is transmitted to the ball screw through a coupler, which transforms the rotary motion of the motor shaft into linear movement of the ball screw nut. This linear movement is then passed on to the output carriage via the linear springs, and the carriage, in turn, moves a robot joint using a pair of cables. For accurate feedback, the system is equipped with an encoder on the motor to measure its angular displacement, a linear potentiometer to track the displacement of the springs, and a rotary potentiometer at the robot joint to measure its angle. The design of this SEA is able to generate high fidelity of force control, low output impedance, and extend the force range and bandwidth. This allows for smooth and responsive robot joint movements, which is ideal for applications requiring delicate human-robot interaction.

The dynamic model of SEA-driven robots, with the consideration of human-robot interaction, can be given as [1]:

$$\boldsymbol{M}(\boldsymbol{q})\ddot{\boldsymbol{q}}+\boldsymbol{C}(\dot{\boldsymbol{q}},\boldsymbol{q})\dot{\boldsymbol{q}}+\boldsymbol{D}_q\dot{\boldsymbol{q}}+\boldsymbol{g}(\boldsymbol{q})+\boldsymbol{K}(\boldsymbol{q}-\boldsymbol{\theta}) =\boldsymbol{\tau}_e, \tag{2.1}$$

$$\boldsymbol{B}\ddot{\boldsymbol{\theta}}+\boldsymbol{D}_\theta\dot{\boldsymbol{\theta}}+\boldsymbol{K}(\boldsymbol{\theta}-\boldsymbol{q}) =\boldsymbol{u}, \tag{2.2}$$

where $\boldsymbol{q}\in\Re^n$ denote the joint configurations, $\boldsymbol{\theta}\in\Re^n$ is the vector of motor rotor shaft positions, $\boldsymbol{M}(\boldsymbol{q})\in\Re^{n\times n}$, $\boldsymbol{C}(\dot{\boldsymbol{q}},\boldsymbol{q})\dot{\boldsymbol{q}}\in\Re^n$, and $\boldsymbol{g}(\boldsymbol{q})\in\Re^n$ denote the inertia matrix, the vector of centripetal and Coriolis torque, and the vector of gravitational torque of the robot respectively, $\boldsymbol{K}\in\Re^{n\times n}$ is a diagonal stiffness matrix of SEA, $\boldsymbol{\tau}_e=\boldsymbol{J}^T(\boldsymbol{q})\boldsymbol{f}_e$ represents a vector of interaction torque, $\boldsymbol{J}^T(\boldsymbol{q})\in\Re^{n\times m}$ is the transpose of $\boldsymbol{J}(\boldsymbol{q})$, and $\boldsymbol{J}(\boldsymbol{q})\in\Re^{m\times n}$ is the Jacobian matrix from joint space to task space, $\boldsymbol{f}_e\in\Re^m$ is a vector of interaction force which is also continuous, $\boldsymbol{B}\in\Re^{n\times n}$ is a diagonal and constant inertia matrix of actuator, $\boldsymbol{D}_q\dot{\boldsymbol{q}}\in\Re^n$ and $\boldsymbol{D}_\theta\dot{\boldsymbol{\theta}}\in\Re^n$ represent the viscous frictions for the robot and the actuator, respectively, and $\boldsymbol{u}\in\Re^n$ is a vector of the input torque exerted on the actuator. Equation (2.1) describes the well-known rigid body dynamics, and Eq. (2.2) describes the dynamic model of SEA. The output torque is given as

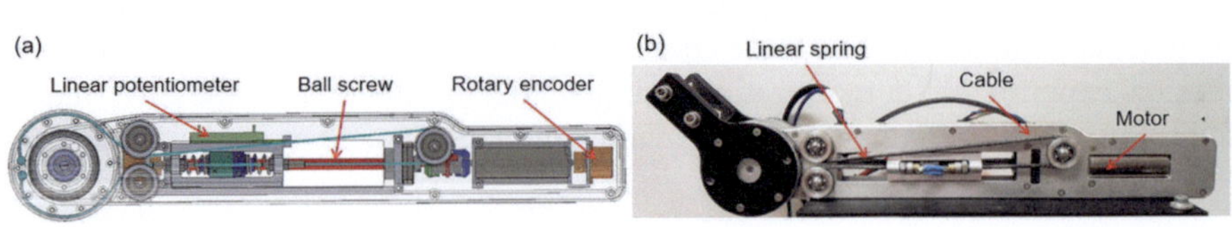

Fig. 2.1 Mechanical structure of the compliant actuator: **a** Principle of the actuator design (CAD model); **b** Prototype

$$\boldsymbol{\tau}_o = \boldsymbol{K}(\boldsymbol{\theta} - \boldsymbol{q}), \tag{2.3}$$

which is proportional to the offset between the rotor shaft position and the robot joint angle $\boldsymbol{\theta} - \boldsymbol{q}$.

Some important properties of the dynamic model described by (2.1) and (2.2) are listed as follows.

(1) The matrices $\boldsymbol{M}(\boldsymbol{q})$ and $\boldsymbol{B}$ are positive definite;
(2) The matrix $\dot{\boldsymbol{M}}(\boldsymbol{q}) - 2\boldsymbol{C}(\dot{\boldsymbol{q}}, \boldsymbol{q})$ is skew-symmetric;
(3) The matrices $\boldsymbol{D}_q$, $\boldsymbol{D}_\theta$ are diagonal and positive definite;
(4) The first four terms of robot dynamics (2.1) are linear in a set of physical parameters $\boldsymbol{\psi}_q = [\psi_{q1}, \ldots, \psi_{qn_q}]^T \in \Re^{n_q}$ as [2, 3]:

$$\boldsymbol{M}(\boldsymbol{q})\ddot{\boldsymbol{q}} + \boldsymbol{C}(\dot{\boldsymbol{q}}, \boldsymbol{q})\dot{\boldsymbol{q}} + \boldsymbol{D}_q\dot{\boldsymbol{q}} + \boldsymbol{g}(\boldsymbol{q}) = \boldsymbol{Y}_q(\ddot{\boldsymbol{q}}, \dot{\boldsymbol{q}}, \dot{\boldsymbol{q}}, \boldsymbol{q})\boldsymbol{\psi}_q, \tag{2.4}$$

where $\boldsymbol{Y}_q(\ddot{\boldsymbol{q}}, \dot{\boldsymbol{q}}, \dot{\boldsymbol{q}}, \boldsymbol{q}) \in \Re^{n \times n_q}$ is a known dynamic regressor matrix, and also [3]:

$$\frac{\lambda}{\lambda + p}(\boldsymbol{M}(\boldsymbol{q})\ddot{\boldsymbol{q}} + \boldsymbol{C}(\dot{\boldsymbol{q}}, \boldsymbol{q})\dot{\boldsymbol{q}} + \boldsymbol{D}_q\dot{\boldsymbol{q}} + \boldsymbol{g}(\boldsymbol{q})) = \boldsymbol{W}_q(\dot{\boldsymbol{q}}, \boldsymbol{q})\boldsymbol{\psi}_q, \tag{2.5}$$

where λ is a positive constant, p is the derivative operator, $\frac{\lambda}{\lambda+p}$ is a first-order low-pass operator, and $\boldsymbol{W}_q(\dot{\boldsymbol{q}}, \boldsymbol{q}) \in \Re^{n \times n_q}$ is another known regressor matrix.
(5) The actuator friction matrix in (2.2) is also linear in a set of physical parameters $\boldsymbol{\psi}_\theta = [d_{\theta 1}, \ldots, d_{\theta n}]^T \in \Re^n$ as:

$$\boldsymbol{D}_\theta\dot{\boldsymbol{\theta}} = \boldsymbol{Y}_\theta(\dot{\boldsymbol{\theta}})\boldsymbol{\psi}_\theta, \tag{2.6}$$

where $\boldsymbol{Y}_\theta(\dot{\boldsymbol{\theta}}) = \mathrm{diag}(\dot{\theta}_1, \ldots, \dot{\theta}_n) \in \Re^{n \times n}$ is a known and diagonal regressor matrix.

The overall dynamic model described by (2.1) and (2.2) is a high-order system, which is dealt with by using the backstepping approach in this chapter. Backstepping is a control design technique particularly suited for nonlinear systems, where traditional control strategies may struggle due to the system's inherent complexity [4]. This method is both systematic and modular, enabling controllers to be developed progressively by "stepping back" from the final desired state through intermediate stabilization steps [5, 6].

The backstepping approach begins by isolating the simplest subsystem or "substate" and designing a control input to stabilize it independently. Then, additional control laws are incrementally constructed for each successive subsystem, ultimately leading to a comprehensive controller for the entire system. In the context of SEA control, the objective is to regulate the robot's joint positions, denoted as $\boldsymbol{q}$, by manipulating the motor position $\boldsymbol{\theta}$ in the relationship $\dot{\boldsymbol{q}} = f(\boldsymbol{q}, \dot{\boldsymbol{q}}, \boldsymbol{\theta})$, where $\boldsymbol{\theta}$ is the substate governed by the control input $\boldsymbol{u}$. The backstepping method proceeds by first defining a virtual control $\boldsymbol{\theta}_d$ to stabilize $\boldsymbol{q}$ and then designing the final control input to stabilize the entire system, taking into account the motor dynamics $\dot{\boldsymbol{\theta}} = g(\boldsymbol{\theta}, \boldsymbol{u})$. This sequential stabilization process is illustrated in Fig. 2.2.

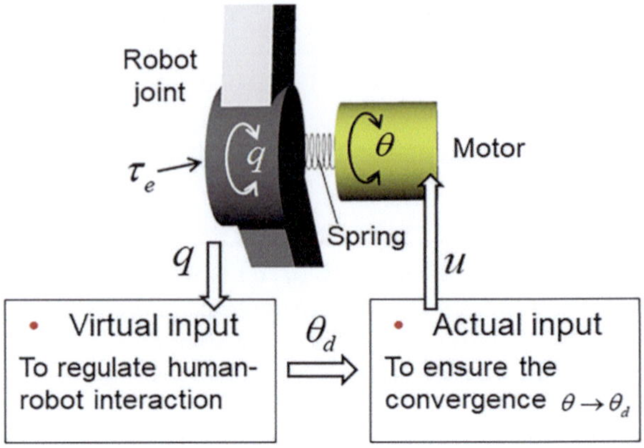

Fig. 2.2 Illustration of backstepping approach

Under the formulation of backstepping, (2.1) can be written as

$$\boldsymbol{M}(\boldsymbol{q})\ddot{\boldsymbol{q}}+\boldsymbol{C}(\dot{\boldsymbol{q}},\boldsymbol{q})\dot{\boldsymbol{q}}+\boldsymbol{D}_q\dot{\boldsymbol{q}}+\boldsymbol{g}(\boldsymbol{q})+\boldsymbol{K}\boldsymbol{q} = \boldsymbol{K}\boldsymbol{\theta}_d + \boldsymbol{K}\Delta\boldsymbol{\theta} + \boldsymbol{\tau}_e, \tag{2.7}$$

where $\Delta\boldsymbol{\theta} = \boldsymbol{\theta} - \boldsymbol{\theta}_d$ and $\boldsymbol{\theta}_d$ denote the desired position input, which is to be designed according to interaction tasks. Then, the actual control input $\boldsymbol{u}$ is designed such that $\Delta\boldsymbol{\theta} \to \boldsymbol{0}$ and hence the control objective is achieved. Hence, (2.7) can be viewed as being controlled by the input $\boldsymbol{K}\boldsymbol{\theta}_d$ with the perturbation of $\boldsymbol{K}\Delta\boldsymbol{\theta}$.

This chapter considers the dynamic uncertainties in both the robot and the actuator, in the sense that the dynamic models $\boldsymbol{M}(\boldsymbol{q})$, $\boldsymbol{C}(\dot{\boldsymbol{q}},\boldsymbol{q})$, $\boldsymbol{D}_q$, $\boldsymbol{g}(\boldsymbol{q})$, and $\boldsymbol{D}_\theta$ are unknown. The stiffness $\boldsymbol{K}$ and the inertia matrix $\boldsymbol{B}$ of actuator are assumed to be well defined. By referring to the above properties, adaptation laws will be developed to estimate unknown dynamic parameters of the robot and the actuator respectively.

2.2 Adaptive Interaction Control

We are now introducing an adaptive human-robot interaction controller proposed for SEA-driven robots, which smoothly integrates both modes of co-manipulation, i.e., *robot-in-charge* and *human-in-charge*, into a single controller.

2.2.1 Dual-Mode Interaction

In this control scheme, the two modes of human-robot interaction are defined as:

(1) *Robot-in-charge*: the robot takes the lead to follow a desired trajectory, while the human does not intervene or manage to follow the moment of the robot such that the interaction force is not large;

Table 2.1 Robot's role in dual-mode interaction

MODE	Robot-in-charge	Human-in-charge
OBJECTIVE	Predefined task	Human guidance
CONTROL	Adaptive tracking control	Damping control + gravity compensation

(2) *Human-in-charge*: the human exerts interaction forces on the robot to take control actions, while the robot becomes passive to complement the human's intention and movement.

In the proposed system, the transition between the two modes is automatically achieved by monitoring the variation of interaction forces. The concepts of *robot-in-charge* and *human-in-charge* are general and can be specified to suit different human-robot interaction scenarios. The change of the robot's role in dual-mode interaction is shown in Table 2.1.

For example, in the mode of *robot-in-charge*, the environment is well known, and the robot is able to track the desired trajectory precisely without the supervision of humans. In the mode of *human-in-charge*, a moving obstacle suddenly appears within the robot workspace, and the human is able to react intelligently to the unforeseen change and guide the robot for obstacle avoidance. The dual-mode co-manipulation can also be applied in sensory-feedback robotic manipulation [7] to improve the robot's performance, where the robot works in the *robot-in-charge* mode when it performs the manipulation task with sensory feedback, and the human takes control of the robot in the *human-in-charge* mode if the robot moves outside the limited sensing zone.

To regulate two different modes in a unified scheme, a force region function is formulated to monitor the variation of the interaction force $\boldsymbol{f}_e$ as:

$$h(\boldsymbol{f}_e) = ||\boldsymbol{f}_e||^2 - R^2, \tag{2.8}$$

where R is a positive constant which denotes the radius of the force region. When $\boldsymbol{f}_e$ is inside the force region, $h(\boldsymbol{f}_e) \leq 0$, and vice versa. Based on the force region, a weight factor is defined as:

$$w(\boldsymbol{f}_e) = 1 - \frac{(\min(0, (\min(0, h(\boldsymbol{f}_e)))^4 - ((\kappa \cdot R)^2 - R^2)^4))^4}{((\kappa \cdot R)^2 - R^2)^{16}}, \tag{2.9}$$

where $0 < \kappa < 1$ is a positive constant. An illustration of the weight factor $w(\boldsymbol{f}_e)$ is shown in Fig. 2.3. From Eq. (2.9) and Fig. 2.3, when $\boldsymbol{f}_e$ is inside the force region such that $h(\boldsymbol{f}_e) \leq 0$, $w(\boldsymbol{f}_e) = 1$. When $\boldsymbol{f}_e$ exceeds the force region such that $h(\boldsymbol{f}_e) > 0$, it equals to 0. Since $w(\boldsymbol{f}_e)$ is second-order continuous, the transition between $w(\boldsymbol{f}_e) = 1$ and $w(\boldsymbol{f}_e) = 0$ is smooth. Different values of weight factor will be utilized in the modes of *robot-in-charge* and the *human-in-charge* respectively.

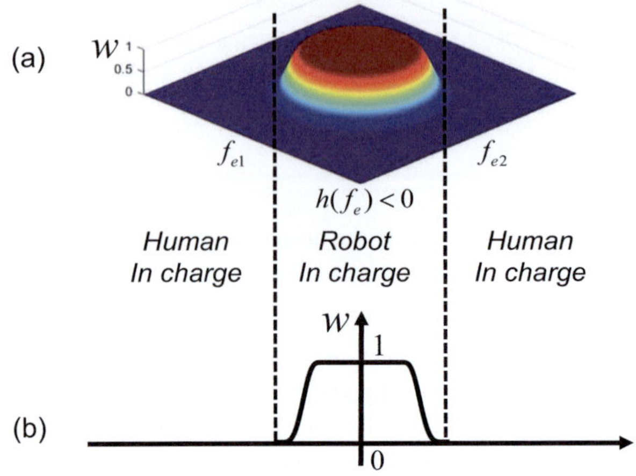

Fig. 2.3 An illustration of weight factor $w(\boldsymbol{f}_e)$, where $\boldsymbol{f}_e=[f_{e1}, f_{e2}]^T \in \Re^2$: **a** $w(\boldsymbol{f}_e)$ transits smoothly between 0 and 1, and $w(\boldsymbol{f}_e)$=1 and $w(\boldsymbol{f}_e)$=0 correspond to the modes of *robot-in-charge* and the *human-in-charge* respectively; **b** side view

The force feedback is only required in the weight factor, which is specified as a measure for the transition between both modes and embedded inside the control scheme. Then, the usage of force feedback follows a qualitative manner, i.e., if the interaction force is not large, the *robot-in-charge* mode is in effect, and if the force is large, the *human-in-charge* mode is activated. In addition, the proposed method provides a general solution, in the sense that other sensory feedback, which is able to describe the change of human motion intention qualitatively, can also be specified as the measure of transition instead of interaction force, such as EMG/EEG sensors for the measurement of electric activity from human muscles or brains, or inertial measurement units or 3D sensors capable of detecting human gestures.

2.2.2 Control Scheme

By using the backstepping approach, a fictitious desired input $\boldsymbol{\theta}_d$ is proposed to achieve the *robot-in-charge* and *human-in-charge* modes. Next, the control input $\boldsymbol{\tau}$ is derived to realize the fictitious signal, such that the actual shaft position of actuator $\boldsymbol{\theta}$ tracks the desired input $\boldsymbol{\theta}_d$.

First, a sliding vector $\boldsymbol{s}_q$ for the robot dynamics is given as:

$$\begin{aligned}\boldsymbol{s}_q =&\dot{\boldsymbol{q}}-\dot{\boldsymbol{q}}_r\\ =&\dot{\boldsymbol{q}}-w(\boldsymbol{f}_e)\boldsymbol{J}^+(\boldsymbol{q})\dot{\boldsymbol{x}}_d+\alpha_q w(\boldsymbol{f}_e)\boldsymbol{J}^+(\boldsymbol{q})\Delta\boldsymbol{x},\end{aligned} \tag{2.10}$$

where $\boldsymbol{J}^+(\boldsymbol{q})$ is the pseudo-inverse of $\boldsymbol{J}(\boldsymbol{q})$, $\Delta\boldsymbol{x}=\boldsymbol{x}-\boldsymbol{x}_d$, $\boldsymbol{x}_d$ is the desired position, $\dot{\boldsymbol{x}}_d$ is the time-derivative of $\boldsymbol{x}_d$, and $\dot{\boldsymbol{q}}_r$ is a reference vector defined as:

$$\dot{\boldsymbol{q}}_r = w(\boldsymbol{f}_e)\boldsymbol{J}^{+}(\boldsymbol{q})\dot{\boldsymbol{x}}_d - \alpha_q w(\boldsymbol{f}_e)\boldsymbol{J}^{+}(\boldsymbol{q})\Delta\boldsymbol{x}, \tag{2.11}$$

where α_q is a positive constant.

From (2.10), we have $\dot{\boldsymbol{q}} = \boldsymbol{s}_q + \dot{\boldsymbol{q}}_r$ and $\ddot{\boldsymbol{q}} = \dot{\boldsymbol{s}}_q + \ddot{\boldsymbol{q}}_r$. Therefore, the robot dynamics (2.1) can be rewritten as:

$$\begin{aligned}
&\boldsymbol{M}(\boldsymbol{q})\dot{\boldsymbol{s}}_q + (\boldsymbol{C}(\dot{\boldsymbol{q}},\boldsymbol{q}) + \boldsymbol{D}_q)\boldsymbol{s}_q + \boldsymbol{M}(\boldsymbol{q})\ddot{\boldsymbol{q}}_r \\
&+ (\boldsymbol{C}(\dot{\boldsymbol{q}},\boldsymbol{q}) + \boldsymbol{D}_q)\dot{\boldsymbol{q}}_r + \boldsymbol{g}(\boldsymbol{q}) + \boldsymbol{K}\boldsymbol{q} \\
&= \boldsymbol{K}\boldsymbol{\theta}_d + \boldsymbol{K}\Delta\boldsymbol{\theta} + \boldsymbol{\tau}_e,
\end{aligned} \tag{2.12}$$

where $\Delta\boldsymbol{\theta} = \boldsymbol{\theta} - \boldsymbol{\theta}_d$. The desired input $\boldsymbol{\theta}_d$ is considered as a fictitious input for the robot, and $\Delta\boldsymbol{\theta}$ represents an input perturbation to the robot dynamics.

From (2.4), the terms $\boldsymbol{M}(\boldsymbol{q})\ddot{\boldsymbol{q}}_r + (\boldsymbol{C}(\dot{\boldsymbol{q}},\boldsymbol{q}) + \boldsymbol{D}_q)\dot{\boldsymbol{q}}_r + \boldsymbol{g}(\boldsymbol{q})$ in (2.12) can be expressed in terms of dynamic parameters as: $\boldsymbol{M}(\boldsymbol{q})\ddot{\boldsymbol{q}}_r + (\boldsymbol{C}(\dot{\boldsymbol{q}},\boldsymbol{q}) + \boldsymbol{D}_q)\dot{\boldsymbol{q}}_r + \boldsymbol{g}(\boldsymbol{q}) = \boldsymbol{Y}_q(\ddot{\boldsymbol{q}}_r, \dot{\boldsymbol{q}}_r, \dot{\boldsymbol{q}}, \boldsymbol{q})\boldsymbol{\psi}_q$. Hence, equation (2.12) leads to:

$$\begin{aligned}
&\boldsymbol{M}(\boldsymbol{q})\dot{\boldsymbol{s}}_q + (\boldsymbol{C}(\boldsymbol{q},\dot{\boldsymbol{q}}) + \boldsymbol{D}_q)\boldsymbol{s}_q + \boldsymbol{Y}_q(\ddot{\boldsymbol{q}}_r, \dot{\boldsymbol{q}}_r, \dot{\boldsymbol{q}}, \boldsymbol{q})\boldsymbol{\psi}_q + \boldsymbol{K}\boldsymbol{q} \\
&= \boldsymbol{K}\boldsymbol{\theta}_d + \boldsymbol{K}\Delta\boldsymbol{\theta} + \boldsymbol{\tau}_e.
\end{aligned} \tag{2.13}$$

We can now propose the desired input $\boldsymbol{\theta}_d$ as:

$$\boldsymbol{\theta}_d = \boldsymbol{q} - \boldsymbol{K}^{-1}(w(\boldsymbol{f}_e)\boldsymbol{K}_q\boldsymbol{s}_q - \boldsymbol{Y}_q(\ddot{\boldsymbol{q}}_r, \dot{\boldsymbol{q}}_r, \dot{\boldsymbol{q}}, \boldsymbol{q})\hat{\boldsymbol{\psi}}_q), \tag{2.14}$$

where $\boldsymbol{K}_q$ is a diagonal and positive definite matrix, and $\hat{\boldsymbol{\psi}}_q$ represents the estimate of $\boldsymbol{\psi}_q$ which is updated by:

$$\dot{\hat{\boldsymbol{\psi}}}_q = -\boldsymbol{L}_q(\boldsymbol{Y}_q^T(\ddot{\boldsymbol{q}}_r, \dot{\boldsymbol{q}}_r, \dot{\boldsymbol{q}}, \boldsymbol{q})\boldsymbol{s}_q + \boldsymbol{W}_q^T(\dot{\boldsymbol{q}}, \boldsymbol{q})\boldsymbol{e}_p), \tag{2.15}$$

where $\boldsymbol{L}_q \in \Re^{n_q \times n_q}$ is a positive definite matrix, $\boldsymbol{e}_p = \boldsymbol{W}_q(\dot{\boldsymbol{q}}, \boldsymbol{q})\Delta\boldsymbol{\psi}_q$ is the prediction error, and $\Delta\boldsymbol{\psi}_q = \boldsymbol{\psi}_q - \hat{\boldsymbol{\psi}}_q$. The update law in Eq. (2.15) is designed as a composite updating law [8, 9], where the convergence of estimated parameters is driven by both the sliding vector $\boldsymbol{s}_q$ and the prediction error $\boldsymbol{e}_p$. That is, the parameter updating does not stop until both $\boldsymbol{s}_q$ and $\boldsymbol{e}_p$ reduce to zero. Note that the knowledge of $\boldsymbol{W}_q(\dot{\boldsymbol{q}}, \boldsymbol{q})\boldsymbol{\psi}_q$ is computed from (2.1) and (2.5) as: $\boldsymbol{W}_q(\dot{\boldsymbol{q}}, \boldsymbol{q})\boldsymbol{\psi}_q = \frac{\lambda}{\lambda + p}(\boldsymbol{\tau}_e + \boldsymbol{K}(\boldsymbol{\theta} - \boldsymbol{q}))$.

The desired input $\boldsymbol{\theta}_d$ integrates both modes of human-robot interaction as:

(1) *Robot-in-charge*: When the interaction force $\boldsymbol{f}_e$ is inside the force region where $h(\boldsymbol{f}_e) \leq 0$, $w(\boldsymbol{f}_e) = 1$, the sliding vector $\boldsymbol{s}_q = \dot{\boldsymbol{q}} - \boldsymbol{J}^{+}(\boldsymbol{q})\dot{\boldsymbol{x}}_d + \alpha_q \boldsymbol{J}^{+}(\boldsymbol{q})\Delta\boldsymbol{x}$, and the desired input becomes:

$$\boldsymbol{\theta}_d = \boldsymbol{q} - \boldsymbol{K}^{-1}(\boldsymbol{K}_q\boldsymbol{s}_q - \boldsymbol{Y}_q(\ddot{\boldsymbol{q}}_r, \dot{\boldsymbol{q}}_r, \dot{\boldsymbol{q}}, \boldsymbol{q})\hat{\boldsymbol{\psi}}_q), \tag{2.16}$$

where the position control term $\alpha_q \boldsymbol{J}^{+}(\boldsymbol{q})\Delta\boldsymbol{x}$ and the velocity control term $\dot{\boldsymbol{q}} - \boldsymbol{J}^{+}(\boldsymbol{q})\dot{\boldsymbol{x}}_d$ in the sliding vector drive the robot to track the desired position $\boldsymbol{x}_d$.

(2) *Human-in-charge*: When the interaction force $\boldsymbol{f}_e$ is outside the force region where $h(\boldsymbol{f}_e)>0$, $w(\boldsymbol{f}_e)=0$, and hence $\dot{\boldsymbol{q}}_r$ reduces to $\boldsymbol{0}$ from (2.11) and also $\ddot{\boldsymbol{q}}_r=\boldsymbol{0}$. Hence, $\boldsymbol{Y}_q(\ddot{\boldsymbol{q}}_r, \dot{\boldsymbol{q}}_r, \dot{\boldsymbol{q}}, \boldsymbol{q})\hat{\boldsymbol{\psi}}_q=\hat{\boldsymbol{M}}(\boldsymbol{q})\ddot{\boldsymbol{q}}_r + (\hat{\boldsymbol{C}}(\dot{\boldsymbol{q}}, \boldsymbol{q}) + \hat{\boldsymbol{D}}_q)\dot{\boldsymbol{q}}_r + \hat{\boldsymbol{g}}(\boldsymbol{q}) \rightarrow \hat{\boldsymbol{g}}(\boldsymbol{q})$, where $\hat{\boldsymbol{M}}(\boldsymbol{q})$, $\hat{\boldsymbol{C}}(\boldsymbol{q}, \dot{\boldsymbol{q}})$, $\hat{\boldsymbol{D}}_q$, and $\hat{\boldsymbol{g}}(\boldsymbol{q})$ represent the approximate models for $\boldsymbol{M}(\boldsymbol{q})$, $\boldsymbol{C}(\dot{\boldsymbol{q}}, \boldsymbol{q})$, $\boldsymbol{D}_q$, and $\boldsymbol{g}(\boldsymbol{q})$ respectively. Then, the desired input is simplified as:

$$\boldsymbol{\theta}_d = \boldsymbol{q} + \boldsymbol{K}^{-1}\hat{\boldsymbol{g}}(\boldsymbol{q}). \tag{2.17}$$

The desired input in Eq. (2.17) leads to the dynamic equation of the robot as: $\boldsymbol{M}(\boldsymbol{q})\ddot{\boldsymbol{q}} + (\boldsymbol{C}(\dot{\boldsymbol{q}}, \boldsymbol{q}) + \boldsymbol{D}_q)\dot{\boldsymbol{q}} + \boldsymbol{g}(\boldsymbol{q}) - \hat{\boldsymbol{g}}(\boldsymbol{q}) = \boldsymbol{K}\Delta\boldsymbol{\theta} + \boldsymbol{\tau}_e$. The use of the composite adaptation in Eq. (2.15) ensures the convergence of the prediction error [8, 9], thus $\hat{\boldsymbol{g}}(\boldsymbol{q})\rightarrow\boldsymbol{g}(\boldsymbol{q})$. In addition, if the actual position of the rotor shaft $\boldsymbol{\theta}$ tracks the desired input $\boldsymbol{\theta}_d$ such that $\Delta\boldsymbol{\theta}=\boldsymbol{\theta}-\boldsymbol{\theta}_d\rightarrow\boldsymbol{0}$, we have:

$$\boldsymbol{M}(\boldsymbol{q})\ddot{\boldsymbol{q}} + (\boldsymbol{C}(\dot{\boldsymbol{q}}, \boldsymbol{q}) + \boldsymbol{D}_q)\dot{\boldsymbol{q}} = \boldsymbol{\tau}_e, \tag{2.18}$$

which is a damping system in the sense that the human is able to take control of the robot by providing the directional information for the robot.

The impedance of the robot varies from a high value to a low value, when the desired input transits from the mode of *robot-in-charge* to the mode of *human-in-charge*, and then the human can easily adjust or guide the movement of the robot while the robot provides assistive forces. In addition, the transition stage where $0<w(\boldsymbol{f}_e)<1$ is useful for shared-control scenarios to combine human control efforts and robot controllers, and the proportion of the transition stage during interaction can be increased by setting the parameter κ small. Since the weight factor $w(\boldsymbol{f}_e)$ is continuous, the proposed desired input $\boldsymbol{\theta}_d$ is also continuous throughout human-robot interaction.

The radius of the force region is adjustable and should be specified to suit specific scenarios. It can be set large where the robot plays a more active role during human-robot interaction, and set small where the human is more dominant. For example, in robotic rehabilitation [10], when patients are fully paralyzed, and the robot is expected to take the lead, a large force region is desired such that most signals and interactions from humans are treated as disturbances. When motion-related functions are partially restored, the force region can be scaled down such that the human motion intention is encouraged [11].

Moreover, when the force region is set arbitrarily large, the proposed control method is specified as a fully *robot-in-charge* controller, which is useful for applications where the desired position is well-defined and the robot's behavior is highly reliable. When the force region reduces to a point, the proposed controller becomes a fully *human-in-charge* controller, and the robotic system is then treated as a human direct teaching system where the directional information is always provided by the human.

Substituting the desired input $\boldsymbol{\theta}_d$ (2.14) into the dynamic equation of a robot (2.13) yields:

$$\begin{aligned}&\boldsymbol{M}(\boldsymbol{q})\dot{\boldsymbol{s}}_q+(\boldsymbol{C}(\dot{\boldsymbol{q}},\boldsymbol{q})+\boldsymbol{D}_q+w(\boldsymbol{f}_e)\boldsymbol{K}_q)\boldsymbol{s}_q+\boldsymbol{Y}_q(\ddot{\boldsymbol{q}}_r,\dot{\boldsymbol{q}}_r,\dot{\boldsymbol{q}},\boldsymbol{q})\Delta\boldsymbol{\psi}_q\\&=\boldsymbol{K}\Delta\boldsymbol{\theta}+\boldsymbol{\tau}_e.\end{aligned} \tag{2.19}$$

After designing the controller at the joint side, the control input for the actuator $\boldsymbol{\tau}$ is developed such that the actual position of the actuator $\boldsymbol{\theta}$ tracks the desired input $\boldsymbol{\theta}_d$ and hence $\Delta\boldsymbol{\theta}\rightarrow\boldsymbol{0}$. Similarly, another sliding vector is introduced for the actuator as

$$\begin{aligned}\boldsymbol{s}_\theta&=\dot{\boldsymbol{\theta}}-\dot{\boldsymbol{\theta}}_r\\&=\dot{\boldsymbol{\theta}}-\dot{\boldsymbol{\theta}}_d+\alpha_\theta(\boldsymbol{\theta}-\boldsymbol{\theta}_d)\\&=\Delta\dot{\boldsymbol{\theta}}+\alpha_\theta\Delta\boldsymbol{\theta}.\end{aligned} \tag{2.20}$$

where α_θ is a positive constant, and $\dot{\boldsymbol{\theta}}_r$ is a reference vector which is defined as:

$$\dot{\boldsymbol{\theta}}_r=\dot{\boldsymbol{\theta}}_d-\alpha_\theta\Delta\boldsymbol{\theta}. \tag{2.21}$$

The control input of the actuator is now proposed as:

$$\boldsymbol{\tau}=\boldsymbol{K}(\boldsymbol{\theta}-\boldsymbol{q})-\boldsymbol{K}_\theta\boldsymbol{s}_\theta+\boldsymbol{B}\ddot{\boldsymbol{\theta}}_r+\boldsymbol{Y}_\theta(\dot{\boldsymbol{\theta}}_r)\hat{\boldsymbol{\psi}}_\theta, \tag{2.22}$$

where $\boldsymbol{K}_\theta$ is a positive definite matrix, and the uncertain dynamic parameters $\hat{\boldsymbol{\psi}}_\theta$ are updated by:

$$\dot{\hat{\boldsymbol{\psi}}}_\theta=-\boldsymbol{L}_\theta\boldsymbol{Y}_\theta^T(\dot{\boldsymbol{\theta}}_r)\boldsymbol{s}_\theta, \tag{2.23}$$

where $\boldsymbol{L}_\theta\in\Re^{n\times n}$ is a positive definite matrix. The update law in Eq. (2.23) is specified as a direct adaption law [3]. Hence, the convergence of $\Delta\boldsymbol{\theta}\rightarrow\boldsymbol{0}$ does not need the convergence of estimated parameters.

Using the sliding vector $\boldsymbol{s}_\theta$, equation (2.2) is expressed as:

$$\boldsymbol{B}\dot{\boldsymbol{s}}_\theta+\boldsymbol{D}_\theta\boldsymbol{s}_\theta+\boldsymbol{K}(\boldsymbol{\theta}-\boldsymbol{q})+\boldsymbol{B}\ddot{\boldsymbol{\theta}}_r+\boldsymbol{Y}_\theta(\dot{\boldsymbol{\theta}}_r)\boldsymbol{\psi}_\theta=\boldsymbol{\tau}. \tag{2.24}$$

Substituting (2.22) into (2.24), the closed-loop equation for the human-robot interaction system is obtained as:

$$\boldsymbol{B}\dot{\boldsymbol{s}}_\theta+(\boldsymbol{D}_\theta+\boldsymbol{K}_\theta)\boldsymbol{s}_\theta+\boldsymbol{Y}_\theta(\dot{\boldsymbol{\theta}}_r)\Delta\boldsymbol{\psi}_\theta=\boldsymbol{0}, \tag{2.25}$$

where $\Delta\boldsymbol{\psi}_\theta=\boldsymbol{\psi}_\theta-\hat{\boldsymbol{\psi}}_\theta$. The block diagram of the closed-loop system is shown in Fig. 2.4.

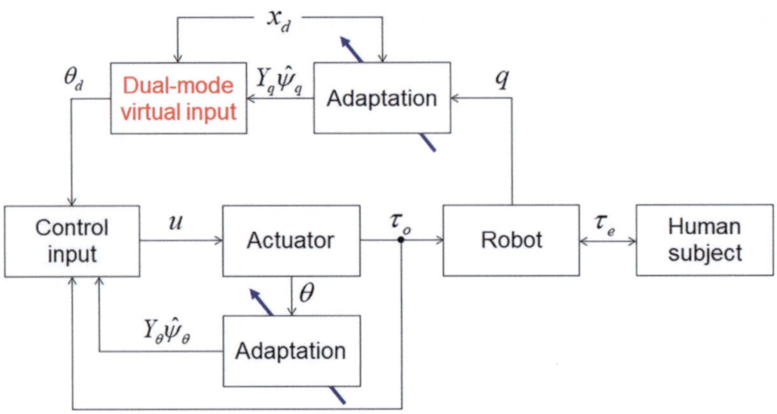

Fig. 2.4 Block diagram of the closed-loop system, where the desired position input consists of both phases of *human-in-charge* and *robot-in-charge*

2.2.3 Stability Analysis

A Lyapunov-like candidate is developed to analyze the stability of the closed-loop system as:

$$V = V_q + \frac{1}{2}\boldsymbol{s}_\theta^T \boldsymbol{B}\boldsymbol{s}_\theta + \frac{1}{2}\Delta\boldsymbol{\psi}_\theta^T \boldsymbol{L}_\theta^{-1}\Delta\boldsymbol{\psi}_\theta + \alpha_\theta \Delta\boldsymbol{\theta}^T \boldsymbol{K}_\theta \Delta\boldsymbol{\theta}, \tag{2.26}$$

where

$$V_q = \frac{1}{2}\boldsymbol{s}_q^T \boldsymbol{M}(\boldsymbol{q})\boldsymbol{s}_q + \frac{1}{2}\Delta\boldsymbol{\psi}_q^T \boldsymbol{L}_q^{-1}\Delta\boldsymbol{\psi}_q. \tag{2.27}$$

Differentiating equation (2.26) with respect to time, and substituting Eqs. (2.23) and (2.25) into it yield:

$$\begin{aligned} \dot{V} &= \dot{V}_q + \boldsymbol{s}_\theta^T \boldsymbol{B}\dot{\boldsymbol{s}}_\theta - \Delta\boldsymbol{\psi}_\theta^T \boldsymbol{L}_\theta^{-1}\dot{\hat{\boldsymbol{\psi}}}_\theta + 2\alpha_\theta \Delta\boldsymbol{\theta}^T \boldsymbol{K}_\theta \Delta\dot{\boldsymbol{\theta}} \\ &= -\boldsymbol{s}_q^T(\boldsymbol{D}_q + w(\boldsymbol{f}_e)\boldsymbol{K}_q)\boldsymbol{s}_q + \boldsymbol{s}_q^T(\boldsymbol{K}\Delta\boldsymbol{\theta} + \boldsymbol{\tau}_e) \\ &\quad - \Delta\boldsymbol{\psi}_q^T \boldsymbol{W}_q^T(\dot{\boldsymbol{q}}, \boldsymbol{q})\boldsymbol{W}_q(\dot{\boldsymbol{q}}, \boldsymbol{q})\Delta\boldsymbol{\psi}_q \\ &\quad + 2\alpha_\theta \Delta\boldsymbol{\theta}^T \boldsymbol{K}_\theta \Delta\dot{\boldsymbol{\theta}} - \boldsymbol{s}_\theta^T(\boldsymbol{D}_\theta + \boldsymbol{K}_\theta)\boldsymbol{s}_\theta. \end{aligned} \tag{2.28}$$

Note that $\boldsymbol{s}_\theta = \Delta\dot{\boldsymbol{\theta}} + \alpha_\theta \Delta\boldsymbol{\theta}$, (2.28) can be written as:

$$\begin{aligned} \dot{V} &= -\boldsymbol{s}_q^T(\boldsymbol{D}_q + w(\boldsymbol{f}_e)\boldsymbol{K}_q)\boldsymbol{s}_q + \boldsymbol{s}_q^T \boldsymbol{K}\Delta\boldsymbol{\theta} \\ &\quad + \boldsymbol{s}_q^T\boldsymbol{\tau}_e - \boldsymbol{s}_\theta^T \boldsymbol{D}_\theta \boldsymbol{s}_\theta - \Delta\dot{\boldsymbol{\theta}}^T \boldsymbol{K}_\theta \Delta\dot{\boldsymbol{\theta}} - \alpha_\theta^2 \Delta\boldsymbol{\theta}^T \boldsymbol{K}_\theta \Delta\boldsymbol{\theta} \\ &\quad - \Delta\boldsymbol{\psi}_q^T \boldsymbol{W}_q^T(\dot{\boldsymbol{q}}, \boldsymbol{q})\boldsymbol{W}_q(\dot{\boldsymbol{q}}, \boldsymbol{q})\Delta\boldsymbol{\psi}_q \\ &= \boldsymbol{s}_q^T\boldsymbol{\tau}_e - \boldsymbol{s}_\theta^T \boldsymbol{D}_\theta \boldsymbol{s}_\theta - \Delta\dot{\boldsymbol{\theta}}^T \boldsymbol{K}_\theta \Delta\dot{\boldsymbol{\theta}} \\ &\quad - \Delta\boldsymbol{\psi}_q^T \boldsymbol{W}_q^T(\dot{\boldsymbol{q}}, \boldsymbol{q})\boldsymbol{W}_q(\dot{\boldsymbol{q}}, \boldsymbol{q})\Delta\boldsymbol{\psi}_q - [s_q; \Delta\boldsymbol{\theta}]^T \boldsymbol{P}[s_q; \Delta\boldsymbol{\theta}], \end{aligned} \tag{2.29}$$

where $\boldsymbol{P} \in \Re^{4\times4}$ is:

$$\boldsymbol{P}=\begin{bmatrix} \boldsymbol{D}_q+w(\boldsymbol{f}_e)\boldsymbol{K}_q & -\frac{1}{2}\boldsymbol{K} \\ -\frac{1}{2}\boldsymbol{K} & \alpha_\theta^2\boldsymbol{K}_\theta \end{bmatrix}. \tag{2.30}$$

If the controller parameters α_θ and $\boldsymbol{K}_\theta$ are chosen so that

$$\alpha_\theta^2\lambda_{\min}\{\boldsymbol{K}_\theta\boldsymbol{D}_q\} > \frac{1}{4}\lambda_{\max}\{\boldsymbol{K}^2\}, \tag{2.31}$$

where $\lambda_{\min}$ and $\lambda_{\max}$ represent the minimum and the maximum eigenvalues, respectively, then $\boldsymbol{P}$ is positive definite. Next, (2.29) is written as:

$$\dot{V}+W=\boldsymbol{s}_q^T\boldsymbol{\tau}_e, \tag{2.32}$$

where

$$\begin{aligned} W =&\boldsymbol{s}_\theta^T\boldsymbol{D}_\theta\boldsymbol{s}_\theta+\Delta\dot{\boldsymbol{\theta}}^T\boldsymbol{K}_\theta\Delta\dot{\boldsymbol{\theta}}+[\boldsymbol{s}_q;\,\Delta\boldsymbol{\theta}]^T\boldsymbol{P}[\boldsymbol{s}_q;\,\Delta\boldsymbol{\theta}] \\ &+\Delta\boldsymbol{\psi}_q^T\boldsymbol{W}_q^T(\dot{\boldsymbol{q}},\boldsymbol{q})\boldsymbol{W}_q(\dot{\boldsymbol{q}},\boldsymbol{q})\Delta\boldsymbol{\psi}_q, \end{aligned} \tag{2.33}$$

and W is positive if condition (2.31) is satisfied. Integrating equation (2.32) over $[0, t]$ yields:

$$\int_0^t \boldsymbol{s}_q^T(\varsigma)\boldsymbol{\tau}_e(\varsigma)\mathrm{d}\varsigma=V(t)-V(0)+\int_0^t W(\varsigma)\mathrm{d}\varsigma. \tag{2.34}$$

Since W is positive, we have:

$$\int_0^t \boldsymbol{s}_q^T(\varsigma)\boldsymbol{\tau}_e(\varsigma)\mathrm{d}\varsigma\geq V(t)-V(0). \tag{2.35}$$

The above inequality demonstrates the passivity of the dynamics between the input $\boldsymbol{\tau}_e$ and output $\boldsymbol{s}_q$ [1]. The passivity shows that at any time the energy stored in the system $V(t)-V(0)$ is less than or equal to the energy introduced from humans. Therefore, the proposed control scheme guarantees the safe human-robot co-manipulation.

Now we have the following theorem to state the boundedness of the sliding vector $\boldsymbol{s}_q$:

Theorem 2.2.1 *The closed-loop system (2.19) and (2.25) gives rise to the boundedness of the sliding vector $\boldsymbol{s}_q$, if the interaction force $\boldsymbol{f}_e$ is bounded, and the control parameters α_θ and $\boldsymbol{K}_\theta$ are chosen such that*

$$2\lambda_{\min}\{\boldsymbol{P}\}-1>0. \tag{2.36}$$

Proof If $\boldsymbol{f}_e$ is bounded, the torque $\boldsymbol{\tau}_e = \boldsymbol{J}^T(\boldsymbol{q})\boldsymbol{f}_e$ is also bounded since $\boldsymbol{J}(\boldsymbol{q})$ is a trigonometric function of $\boldsymbol{q}$. Note that

$$\int_0^t \boldsymbol{s}_q^T(\varsigma)\boldsymbol{\tau}_e(\varsigma)\mathrm{d}\varsigma \le \frac{1}{2}\int_0^t ||\boldsymbol{s}_q(\varsigma)||^2\mathrm{d}\varsigma + \frac{1}{2}\int_0^t ||\boldsymbol{\tau}_e(\varsigma)||^2\mathrm{d}\varsigma. \tag{2.37}$$

From (2.34) and (2.37), we have:

$$-V(0) + \int_0^t W(\varsigma)\mathrm{d}\varsigma \le \frac{1}{2}\int_0^t ||\boldsymbol{s}_q(\varsigma)||^2\mathrm{d}\varsigma + \frac{1}{2}\int_0^t ||\boldsymbol{\tau}_e(\varsigma)||^2\mathrm{d}\varsigma. \tag{2.38}$$

If condition (2.36) is satisfied, $\boldsymbol{P}$ is positive definite, thus W is positive. From (2.33), $W \ge \lambda_{\min}\{\boldsymbol{P}\}||\boldsymbol{s}_q||^2$. Therefore, we have:

$$\begin{aligned}&\int_0^t \lambda_{\min}[\boldsymbol{P}]||\boldsymbol{s}_q(\varsigma)||^2\mathrm{d}\varsigma \\ &\le \frac{1}{2}\int_0^t ||\boldsymbol{s}_q(\varsigma)||^2\mathrm{d}\varsigma + \frac{1}{2}\int_0^t ||\boldsymbol{\tau}_e(\varsigma)||^2\mathrm{d}\varsigma + V(0),\end{aligned} \tag{2.39}$$

which also implies that:

$$\int_0^t (2\lambda_{\min}\{\boldsymbol{P}\}-1)||\boldsymbol{s}_q(\varsigma)||^2\mathrm{d}\varsigma \le \int_0^t ||\boldsymbol{\tau}_e(\varsigma)||^2\mathrm{d}\varsigma + 2V(0). \tag{2.40}$$

Since $2\lambda_{\min}\{\boldsymbol{P}\} - 1 > 0$, we can conclude that the boundedness of $\boldsymbol{\tau}_e$ ensures the boundedness of $\boldsymbol{s}_q$. □

Specifically, in the *robot-in-charge* mode ($||\boldsymbol{f}_e|| < R$, $w(\boldsymbol{f}_e) = 1$), $\boldsymbol{s}_q = \dot{\boldsymbol{q}} - \boldsymbol{J}^+(\boldsymbol{q})\dot{\boldsymbol{x}}_d + \alpha_q \boldsymbol{J}^+(\boldsymbol{q})\Delta\boldsymbol{x}$, and it can be then derived from (2.40) that:

$$\int_0^t ||\Delta\boldsymbol{x}(\varsigma)||^2\mathrm{d}\varsigma \le \int_0^t \frac{\eta_1}{\alpha_q^2}||\boldsymbol{f}_e(\varsigma)||^2\mathrm{d}\varsigma + \frac{\eta_2 + \alpha_q||\Delta\boldsymbol{x}(0)||^2}{\alpha_q^2}, \tag{2.41}$$

where $\eta_1 \triangleq \frac{||\boldsymbol{J}(\boldsymbol{q})\boldsymbol{J}^T(\boldsymbol{q})||^2}{2\lambda_{\min}\{\boldsymbol{P}\}-1}$ and $\eta_2 \triangleq \frac{2||\boldsymbol{J}(\boldsymbol{q})||^2 V(0)}{2\lambda_{\min}\{\boldsymbol{P}\}-1}$ which are positive. Hence, the boundedness of $\boldsymbol{f}_e$ ensures the boundedness of tracking errors $\Delta\boldsymbol{x}$.

In the *human-in-charge* mode ($R \le ||\boldsymbol{f}_e|| < B_f$, $w(\boldsymbol{f}_e) = 0$, where B_f is the upper bound of $\boldsymbol{f}_e$), $\boldsymbol{s}_q = \dot{\boldsymbol{q}}$, and thus inequality (2.40) leads to:

$$\int_0^t ||\dot{\boldsymbol{q}}(\varsigma)||^2\mathrm{d}\varsigma \le \int_0^t \eta_3||\boldsymbol{f}_e(\varsigma)||^2\mathrm{d}\varsigma + \eta_4, \tag{2.42}$$

where $\eta_3 \overset{\triangle}{=} \frac{\|\boldsymbol{J}^T(\boldsymbol{q})\|^2}{2\lambda_{\min}\{\boldsymbol{P}\}-1}$ and $\eta_4 \overset{\triangle}{=} \frac{2V(0)}{2\lambda_{\min}\{\boldsymbol{P}\}-1}$, which are also positive. The inequality (2.42) implies that the boundedness of $\boldsymbol{f}_e$ ensures the boundedness of $\dot{\boldsymbol{q}}$. That is, the joint-space velocity of the robot is limited when the human guides the movement of the robot.

Then, the following theorem states the stability of the overall SEA-driven robot system consisting of both the robot dynamics and the actuator dynamics:

Theorem 2.2.2 *The proposed control scheme described by (2.14), (2.15), (2.22), and (2.23) ensures the stability of the SEA-driven robotic system described by (2.1) and (2.2) and the convergence $\Delta\boldsymbol{\theta}\to\mathbf{0}$, if the interaction force $\boldsymbol{f}_e$ is bounded, and the control parameters are chosen such that condition (2.36) is satisfied.*

Proof The boundedness of $\boldsymbol{f}_e$ ensures the boundedness of $\boldsymbol{\tau}_e$, and the boundedness of $\boldsymbol{\tau}_e$ ensures the boundedness of $\boldsymbol{s}_q$ if condition (2.36) is satisfied, as shown in Theorem 1. From inequality (2.35), we have:

$$V(t) \leq \int_0^t \boldsymbol{s}_q^T(\varsigma)\boldsymbol{\tau}_e(\varsigma)\mathrm{d}\varsigma + V(0). \tag{2.43}$$

Since both $\boldsymbol{s}_q$ and $\boldsymbol{\tau}_e$ are bounded, V is also bounded. From Eq. (2.26), the boundedness of V ensures the boundedness of $\boldsymbol{s}_q$, $\Delta\boldsymbol{\psi}_q$, $\boldsymbol{s}_\theta$, $\Delta\boldsymbol{\theta}$, and $\Delta\boldsymbol{\psi}_\theta$. Since all state variables are bounded, the closed-loop system is stable.

In addition, the boundedness of $\boldsymbol{s}_\theta$ and $\Delta\boldsymbol{\theta}$ ensures the boundedness of $\Delta\dot{\boldsymbol{\theta}}$. Hence, $\Delta\boldsymbol{\theta}$ is uniformly continuous. Then, from (2.34), we have:

$$\begin{aligned}\int_0^{+\infty} (\lambda_{\min}\{\boldsymbol{P}\}\Delta\boldsymbol{\theta}^T(t)\Delta\boldsymbol{\theta}(t))\mathrm{d}t &\leq \int_0^{+\infty} W(t)\mathrm{d}t \\ &= V(0) - V(+\infty) + \int_0^{+\infty} \boldsymbol{s}_q^T(t)\boldsymbol{\tau}_e(t)\mathrm{d}t,\end{aligned} \tag{2.44}$$

where the right side is bounded. Hence, $\Delta\boldsymbol{\theta}\in L_2(0, +\infty)$. It then follows [2, 3] that $\Delta\boldsymbol{\theta} \to \mathbf{0}$. That is, the convergence of $\boldsymbol{\theta} \to \boldsymbol{\theta}_d$ is guaranteed. □

Note that the stability analysis is not dependent on specific values of the weight factor $w(\boldsymbol{f}_e)$, and thus applicable to the *robot-in-charge* mode, the *human-in-charge* mode, and also the transition stage. If the interaction force is further negligible, the proposed controller works in the *robot-in-charge* mode. Then, the following theorem states the convergence of tracking errors:

Theorem 2.2.3 *The closed-loop system (2.19) and (2.25) gives rise to the convergence of tracking errors in the robot-in-charge mode, if the interaction force $\boldsymbol{f}_e$ is negligible, and the control parameters are chosen such that condition (2.31) is satisfied.*

Proof Since the interaction force is negligible, we have $\boldsymbol{\tau}_e \rightarrow \mathbf{0}$. If condition (2.31) is satisfied, we have $\dot{V} = -W \leq 0$. Since $V > 0$, and $\dot{V} \leq 0$, V is bounded. The boundedness of V ensures the boundedness of $\boldsymbol{s}_q$, $\Delta\boldsymbol{\psi}_q$, $\boldsymbol{s}_\theta$, $\Delta\boldsymbol{\psi}_\theta$, and $\Delta\boldsymbol{\theta}$. The boundedness of $\boldsymbol{s}_\theta$ and $\Delta\boldsymbol{\theta}$ ensures the boundedness of $\Delta\dot{\boldsymbol{\theta}}$, and thus $\Delta\boldsymbol{\theta}$ is uniformly continuous. Since $\boldsymbol{s}_q = \dot{\boldsymbol{q}} - \boldsymbol{J}^+(\boldsymbol{q})\dot{\boldsymbol{x}}_d + \alpha_q \boldsymbol{J}^+(\boldsymbol{q})\Delta\boldsymbol{x}$ in the *robot-in-charge* mode, we have $\boldsymbol{J}(\boldsymbol{q})\boldsymbol{s}_q = \Delta\dot{\boldsymbol{x}} + \alpha_q \Delta\boldsymbol{x}$. The boundedness of $\boldsymbol{s}_q$ ensures the boundedness of $\Delta\boldsymbol{x}$ and $\Delta\dot{\boldsymbol{x}}$, and thus the boundedness of $\dot{\boldsymbol{x}}$. The boundedness of $\Delta\boldsymbol{x}$ and $\boldsymbol{s}_q$ ensures the boundedness of $\dot{\boldsymbol{q}}$ from (2.10). Then, both $\dot{\boldsymbol{q}}_r$ and $\ddot{\boldsymbol{q}}_r$ are bounded, as seen from equation (2.11). From the robot dynamic equation (2.19), it is obtained that $\dot{\boldsymbol{s}}_q$ is bounded, since $\boldsymbol{\tau}_e \rightarrow \mathbf{0}$, and $\Delta\boldsymbol{\theta}, \boldsymbol{s}_q, \Delta\boldsymbol{\psi}_q, \dot{\boldsymbol{q}}_r$, and $\ddot{\boldsymbol{q}}_r$ are all bounded. Therefore, $\boldsymbol{s}_q$ is also uniformly continuous.

From Eq. (2.34), it is easy to verify that:

$$\int_0^{+\infty} (\lambda_{\min}\{\boldsymbol{P}\}(\Delta\boldsymbol{\theta}^T(t)\Delta\boldsymbol{\theta}(t) + \boldsymbol{s}_q^T(t)\boldsymbol{s}_q(t)))\mathrm{d}t$$
$$\leq \int_0^{+\infty} W(t)\mathrm{d}t = V(0) - V(+\infty), \tag{2.45}$$

where $V(0) - V(+\infty)$ is bounded. Hence, $\boldsymbol{s}_q, \Delta\boldsymbol{\theta} \in L_2(0, +\infty)$. Then it follows [2, 3] that $\boldsymbol{s}_q, \Delta\boldsymbol{\theta} \rightarrow \mathbf{0}$. The convergence of $\boldsymbol{s}_q \rightarrow \mathbf{0}$ ensures that $\boldsymbol{x} \rightarrow \boldsymbol{x}_d$ and $\dot{\boldsymbol{x}} \rightarrow \dot{\boldsymbol{x}}_d$ as $t \rightarrow \infty$. That is, the convergence of tracking errors to zero is achieved. □

2.2.4 Adaptive Observer

Note that the control input (2.22) for the SEA is dependent on the derivatives of the desired input $\dot{\boldsymbol{\theta}}_d$ and $\ddot{\boldsymbol{\theta}}_d$, and hence the acceleration information $\ddot{\boldsymbol{q}}$ and its derivatives are required. An observer is constructed to estimate the desired input as $\hat{\boldsymbol{\theta}}_d$, and thus eliminate the requirement of the acceleration and its derivatives.

Observer techniques for LTI systems, e.g., classic Luenberger observers [12, 13], have been well explored. In this chapter, we consider the problem of uncertain dynamics and propose an adaptive observer to estimate the desired input $\boldsymbol{\theta}_d$, which follows a similar development in [14] as:

$$\begin{cases} \dot{\hat{\boldsymbol{\theta}}}_d = \boldsymbol{\eta} + \beta \boldsymbol{e}_o \\ \dot{\boldsymbol{\eta}} = \boldsymbol{B}^{-1}(\boldsymbol{u} - \boldsymbol{K}(\boldsymbol{\theta} - \boldsymbol{q}) - \hat{\boldsymbol{D}}_{\theta o}\boldsymbol{\eta} + \boldsymbol{K}_e \boldsymbol{e}_o), \end{cases} \tag{2.46}$$

where $\hat{\boldsymbol{\theta}}_d$ denotes the estimated signal of $\boldsymbol{\theta}_d$, $\boldsymbol{e}_o = \boldsymbol{\theta}_d - \hat{\boldsymbol{\theta}}_d$ represents the observation error, $\boldsymbol{\eta}$ is an auxiliary variable, β is a positive constant, and $\boldsymbol{K}_e$ is a positive-definite matrix. The matrix $\hat{\boldsymbol{D}}_{\theta o}$ denotes the approximate model for $\boldsymbol{D}_\theta$, which is updated through another adaptation law for the observer as:

$$\dot{\hat{\boldsymbol{\psi}}}_o = -\boldsymbol{L}_o \boldsymbol{Y}_\theta(\boldsymbol{\eta})\boldsymbol{z}, \tag{2.47}$$

where $z = \dot{\boldsymbol{\theta}} - \boldsymbol{\eta}$, $\hat{\boldsymbol{\psi}}_o = [\hat{\psi}_{o1}, \ldots, \hat{\psi}_{on}] \in \Re^n$ is the vector of estimated parameters for the observer, and $\boldsymbol{L}_o \in \Re^{n \times n}$ is a positive definite matrix. Note that $\hat{\boldsymbol{\theta}}_d$ is obtained by integrating $\dot{\hat{\boldsymbol{\theta}}}_d$, while $\ddot{\hat{\boldsymbol{\theta}}}_d$ is obtained by differentiating $\dot{\hat{\boldsymbol{\theta}}}_d$ with respect to time. By using the estimated signals, a sliding vector is then defined for the SEA as:

$$
\begin{aligned}
\hat{\boldsymbol{s}}_\theta &= \dot{\boldsymbol{\theta}} - \dot{\hat{\boldsymbol{\theta}}}_r \\
&= \dot{\boldsymbol{\theta}} - \dot{\hat{\boldsymbol{\theta}}}_d + \alpha_\theta (\boldsymbol{\theta} - \hat{\boldsymbol{\theta}}_d) \\
&= \Delta \dot{\hat{\boldsymbol{\theta}}} + \alpha_\theta \Delta \hat{\boldsymbol{\theta}},
\end{aligned}
\tag{2.48}
$$

where the reference vector $\dot{\hat{\boldsymbol{\theta}}}_r$ is defined as:

$$
\dot{\hat{\boldsymbol{\theta}}}_r = \dot{\hat{\boldsymbol{\theta}}}_d - \alpha_\theta \Delta \hat{\boldsymbol{\theta}},
\tag{2.49}
$$

and $\Delta \hat{\boldsymbol{\theta}} = \boldsymbol{\theta} - \hat{\boldsymbol{\theta}}_d$.

The control input for the SEA is now proposed as:

$$
\boldsymbol{u} = \boldsymbol{K}(\boldsymbol{\theta} - \boldsymbol{q}) - \boldsymbol{K}_\theta \hat{\boldsymbol{s}}_\theta + \boldsymbol{B} \ddot{\hat{\boldsymbol{\theta}}}_r + \boldsymbol{Y}_\theta (\dot{\hat{\boldsymbol{\theta}}}_r) \hat{\boldsymbol{\psi}}_\theta,
\tag{2.50}
$$

where the uncertain parameters $\hat{\boldsymbol{\psi}}_\theta$ are updated by:

$$
\dot{\hat{\boldsymbol{\psi}}}_\theta = -\boldsymbol{L}_\theta \boldsymbol{Y}_\theta^T (\dot{\hat{\boldsymbol{\theta}}}_r) \hat{\boldsymbol{s}}_\theta.
\tag{2.51}
$$

By using the estimated signals $\hat{\boldsymbol{\theta}}_d$, $\dot{\hat{\boldsymbol{\theta}}}_d$, and $\ddot{\hat{\boldsymbol{\theta}}}_d$, the control input $\boldsymbol{u}$ in Eq. (2.50) does not require the acceleration information and its derivatives.

Equation (2.2) can be expressed in terms of $\hat{\boldsymbol{s}}_\theta$ as:

$$
\boldsymbol{B} \dot{\hat{\boldsymbol{s}}}_\theta + \boldsymbol{D}_\theta \hat{\boldsymbol{s}}_\theta + \boldsymbol{K}(\boldsymbol{\theta} - \boldsymbol{q}) + \boldsymbol{B} \ddot{\hat{\boldsymbol{\theta}}}_r + \boldsymbol{Y}_\theta (\dot{\hat{\boldsymbol{\theta}}}_r) \boldsymbol{\psi}_\theta = \boldsymbol{u}.
\tag{2.52}
$$

Substituting Eq. (2.50) into equation (2.52), the closed-loop equation is obtained as:

$$
\boldsymbol{B} \dot{\hat{\boldsymbol{s}}}_\theta + (\boldsymbol{D}_\theta + \boldsymbol{K}_\theta) \hat{\boldsymbol{s}}_\theta + \boldsymbol{Y}_\theta (\dot{\hat{\boldsymbol{\theta}}}_r) \Delta \boldsymbol{\psi}_\theta = \boldsymbol{0}.
\tag{2.53}
$$

Next, multiplying both sides of Eq. (2.46) with $\boldsymbol{B}$ and using (2.2), we can get the closed-loop observer dynamics for $\dot{\hat{\boldsymbol{\theta}}}_d$ as:

$$
\begin{aligned}
& \boldsymbol{B} \dot{z} + \boldsymbol{D}_\theta z + \boldsymbol{K}_e \boldsymbol{e}_o \\
&= (\hat{\boldsymbol{D}}_{\theta o} - \boldsymbol{D}_\theta) \boldsymbol{\eta} \\
&= - \boldsymbol{Y}_\theta (\boldsymbol{\eta}) \Delta \boldsymbol{\psi}_o,
\end{aligned}
\tag{2.54}
$$

where $\Delta \boldsymbol{\psi}_o = \boldsymbol{\psi}_o - \hat{\boldsymbol{\psi}}_o$, and $\boldsymbol{\psi}_o$ is the dynamic model for $\boldsymbol{D}_\theta$.

In actual implementations, the control parameters are tuned according to the robot's performance. In general, better velocity tracking can be achieved by tuning $\boldsymbol{K}_q$ and $\boldsymbol{K}_\theta$ up, better position tracking and smaller steady-state error can be achieved with higher α_q and α_θ, and faster convergence of observation errors can be obtained by tuning β and $\boldsymbol{K}_e$ up. Finally, the adaptation gains $\boldsymbol{L}_q$, $\boldsymbol{L}_\theta$, and $\boldsymbol{L}_o$ should start with very small values and be tuned up when faster adaptation is desired. The control gains should be fine-tuned together to achieve the desired transient and steady-state performance.

To prove the stability of the overall system, a Lyapunov-like candidate is proposed as:

$$\begin{aligned}\hat{V} =& V_q + \frac{1}{2}\hat{\boldsymbol{s}}_\theta^T \boldsymbol{B}\hat{\boldsymbol{s}}_\theta + \frac{1}{2}\Delta\boldsymbol{\psi}_\theta^T \boldsymbol{L}_\theta^{-1}\Delta\boldsymbol{\psi}_\theta + \alpha_\theta \Delta\hat{\boldsymbol{\theta}}^T \boldsymbol{K}_\theta \Delta\hat{\boldsymbol{\theta}}\\ &+ \frac{1}{2}\boldsymbol{z}^T \boldsymbol{B}\boldsymbol{z} + \frac{1}{2}\Delta\boldsymbol{\psi}_o^T \boldsymbol{L}_o^{-1}\Delta\boldsymbol{\psi}_o.\end{aligned} \tag{2.55}$$

Differentiating equation (2.55) with respect to time and substituting (2.47), (2.51), (2.53), and (2.54) into it, one obtains:

$$\begin{aligned}\dot{\hat{V}} =& -\boldsymbol{s}_q^T(\boldsymbol{D}_q + w(\boldsymbol{f}_e)\boldsymbol{K}_q)\boldsymbol{s}_q + \boldsymbol{s}_q^T(\boldsymbol{K}\Delta\boldsymbol{\theta} + \boldsymbol{\tau}_e)\\ &- \Delta\boldsymbol{\psi}_q^T \boldsymbol{W}_q^T(\dot{\boldsymbol{q}}, \boldsymbol{q})\boldsymbol{W}_q(\dot{\boldsymbol{q}}, \boldsymbol{q})\Delta\boldsymbol{\psi}_q\\ &- \hat{\boldsymbol{s}}_\theta^T(\boldsymbol{D}_\theta + \boldsymbol{K}_\theta)\hat{\boldsymbol{s}}_\theta + 2\alpha_\theta\Delta\hat{\boldsymbol{\theta}}^T \boldsymbol{K}_\theta \Delta\dot{\hat{\boldsymbol{\theta}}}\\ &- \boldsymbol{z}^T \boldsymbol{D}_\theta \boldsymbol{z} - \boldsymbol{z}^T \boldsymbol{K}_e \boldsymbol{e}_o.\end{aligned} \tag{2.56}$$

Note that $\Delta\boldsymbol{\theta} = \Delta\hat{\boldsymbol{\theta}} - \boldsymbol{e}_o$, $\hat{\boldsymbol{s}}_\theta = \Delta\dot{\hat{\boldsymbol{\theta}}} + \alpha_\theta\Delta\hat{\boldsymbol{\theta}}$, and also $\boldsymbol{z} = \Delta\dot{\hat{\boldsymbol{\theta}}} + \beta\boldsymbol{e}_o$. Therefore, Eq. (2.56) can be written as:

$$\begin{aligned}\dot{\hat{V}} =& \boldsymbol{s}_q^T\boldsymbol{\tau}_e - \hat{\boldsymbol{s}}_\theta^T \boldsymbol{D}_\theta \hat{\boldsymbol{s}}_\theta - \boldsymbol{z}^T \boldsymbol{D}_\theta \boldsymbol{z}\\ &- \Delta\boldsymbol{\psi}_q^T \boldsymbol{W}_q^T(\dot{\boldsymbol{q}}, \boldsymbol{q})\boldsymbol{W}_q(\dot{\boldsymbol{q}}, \boldsymbol{q})\Delta\boldsymbol{\psi}_q\\ &- [\boldsymbol{s}_q; \Delta\hat{\boldsymbol{\theta}}; \Delta\dot{\hat{\boldsymbol{\theta}}}; \boldsymbol{e}_o]^T \hat{\boldsymbol{P}}[\boldsymbol{s}_q; \Delta\hat{\boldsymbol{\theta}}; \Delta\dot{\hat{\boldsymbol{\theta}}}; \boldsymbol{e}_o],\end{aligned} \tag{2.57}$$

where $\hat{\boldsymbol{P}} \in \Re^{8\times 8}$ is given by:

$$\hat{\boldsymbol{P}} = \begin{bmatrix} \boldsymbol{D}_q + w(\boldsymbol{f}_e)\boldsymbol{K}_q & -\frac{1}{2}\boldsymbol{K} & \boldsymbol{0} & \frac{1}{2}\boldsymbol{K}\\ -\frac{1}{2}\boldsymbol{K} & \alpha_\theta^2\boldsymbol{K}_\theta & \boldsymbol{0} & \boldsymbol{0}\\ \boldsymbol{0} & \boldsymbol{0} & \boldsymbol{K}_\theta & \frac{1}{2}\boldsymbol{K}_e\\ \frac{1}{2}\boldsymbol{K} & \boldsymbol{0} & \frac{1}{2}\boldsymbol{K}_e & \beta\boldsymbol{K}_e \end{bmatrix}. \tag{2.58}$$

If the parameters α_θ, $\boldsymbol{K}_\theta$, β, and $\boldsymbol{K}_e$ are chosen so that

$$\alpha_\theta^2\lambda_{\min}\{\boldsymbol{K}_\theta\boldsymbol{D}_q\} > \frac{1}{4}\lambda_{\max}\{\boldsymbol{K}^2\},$$

$$
\begin{aligned}
\beta\lambda_{\min}\{\boldsymbol{K}_\theta\} >& \frac{1}{4}\lambda_{\max}\{\boldsymbol{K}_e\}, \\
\lambda_{\min}\{\boldsymbol{K}_e\}(\beta\lambda_{\min}\{\boldsymbol{K}_\theta\} - \frac{1}{4}\lambda_{\max}\{\boldsymbol{K}_e\})(\alpha_\theta^2\lambda_{\min}\{\boldsymbol{K}_\theta\boldsymbol{D}_q\} &- \frac{1}{4}\lambda_{\max}\{\boldsymbol{K}^2\}) \\
>& \frac{1}{4}\alpha_\theta^2\lambda_{\max}\{\boldsymbol{K}_\theta^2\boldsymbol{K}^2\},
\end{aligned}
\tag{2.59}
$$

then $\hat{\boldsymbol{P}}$ is positive definite.

The proposed adaptive tracking control scheme in Eq. (2.50) also guarantees the passivity of the human-robot interaction system. Similarly, it can be shown that the boundedness of the interaction force $\boldsymbol{f}_e$ ensures the boundedness of the sliding vector $\hat{\boldsymbol{s}}_q$, when the control parameters are chosen such that $2\lambda_{\min}\{\hat{\boldsymbol{P}}\} - 1 > 0$. In addition, the stability of the overall system and the convergence of $\Delta\boldsymbol{\theta} \to \boldsymbol{0}$ are guaranteed. When the interaction force is further negligible such that $\boldsymbol{\tau}_e \to \boldsymbol{0}$, the controller works in the mode of *robot-in-charge*, and the convergence of tracking errors is also guaranteed, that is, $\boldsymbol{x} \to \boldsymbol{x}_d$ and $\dot{\boldsymbol{x}} \to \dot{\boldsymbol{x}}_d$ as $t \to \infty$.

2.2.5 Results

Numerical simulations and experimental results are presented to demonstrate the effectiveness of the proposed control method. The simulation studies and experiments were carried out in different SEA-driven robotic systems.

First, the proposed control method is implemented in the two-link robot as illustrated in Fig. 2.5. The kinematic and dynamic parameters of this two-link system were specified as shown in Table 2.2. Both links have identical lengths, center of mass distances, moments of inertia, and masses. The dynamic parameters of the SEAs include the inertia matrix $\boldsymbol{B}$, friction matrix $\boldsymbol{D}_\theta$, and stiffness matrix $\boldsymbol{K}$, which are also detailed in the table. The human interacts with the robot by exerting forces

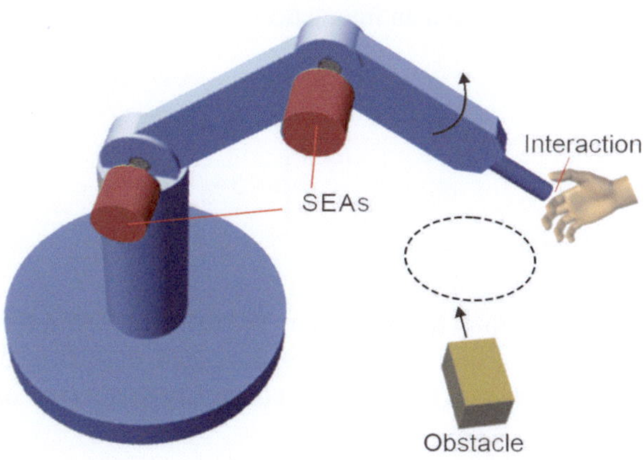

Fig. 2.5 An illustration of two-link compliantly-actuated robot

Table 2.2 Dynamic and kinematic parameters

Parameter	Value
Link lengths (l_1, l_2)	0.42 m
Center of mass distances (l_{c1}, l_{c2})	0.21 m
Moments of inertia (I_1, I_2)	$0.5\ \mathrm{kg \cdot m^2}$
Masses (m_1, m_2)	2 kg
Inertia matrix ($\boldsymbol{B}$)	$0.2\boldsymbol{I}_2\ \mathrm{kg \cdot m^2}$
Friction matrix ($\boldsymbol{D}_\theta$)	$\boldsymbol{I}_2\ \mathrm{kg \cdot m^2/s}$
Stiffness matrix ($\boldsymbol{K}$)	$50\boldsymbol{I}_2\ \mathrm{N \cdot m/rad}$

Table 2.3 Controller parameters

Parameter	κ	α_q	$\boldsymbol{K}_q$	$\boldsymbol{L}_q$	α_θ	$\boldsymbol{K}_\theta$	$\boldsymbol{L}_\theta$
Value	0.7	4	$20\boldsymbol{I}_2$	$5 \times 10^{-5}\boldsymbol{I}_5$	4	$20\boldsymbol{I}_2$	$5 \times 10^{-4}\boldsymbol{I}_2$

on the end effector of the robot, and the position of the end effector is specified in Cartesian space.

The robot end-effector started from (0.19, 0.7) m, and was controlled to track the following trajectory as:

$$x_{d1} = 0.2 + 0.2\sin(0.5t)\ \mathrm{m} \tag{2.60}$$

$$x_{d2} = 0.4 + 0.2\cos(0.5t)\ \mathrm{m} \tag{2.61}$$

The force region in Eq. (2.8) was specified as: $h(\boldsymbol{f}_e) = f_{e1}^2 + f_{e2}^2 - 2^2 \leq 0$, where f_{e1} and f_{e2} represent the forces in two coordinates respectively, the radius R=2 N. The parameters for the weight factor and controller were configured as summarized in Table 2.3.

The path of the robot end effector is shown in Fig. 2.6a, and the tracking error is shown in Fig. 2.6b, which reduces to zero at around 2 s. During the manipulation, the human exerted an interaction force on the robot end effector as:

$$f_{ei} = \begin{cases} 0, t \leq t_1 \\ 5\frac{t-t_1}{\delta t}, t_1 < t < t_1 + \delta t \\ 5, t_1 + \delta t \leq t \leq t_2 - \delta t \\ 5 - 5\frac{t_2 - t}{\delta t}, t_2 - \delta t < t < t_2 \\ 0, t \geq t_2 \end{cases} \tag{2.62}$$

where $i = 1, 2$, $t_1 = 8.5$ s, $t_2 = 11.6$ s, and $\delta t = 0.004$ s. Equation (2.62) describes a very steep ramp input reaching the value of 5 N in 0.004 s.

When the interaction force exceeded the force region such that $h(\boldsymbol{f}_e) > 0$, the weight factor $w(\boldsymbol{f}_e)$ decreased from 1 to 0 as shown in Fig. 2.6c. Then, the proposed

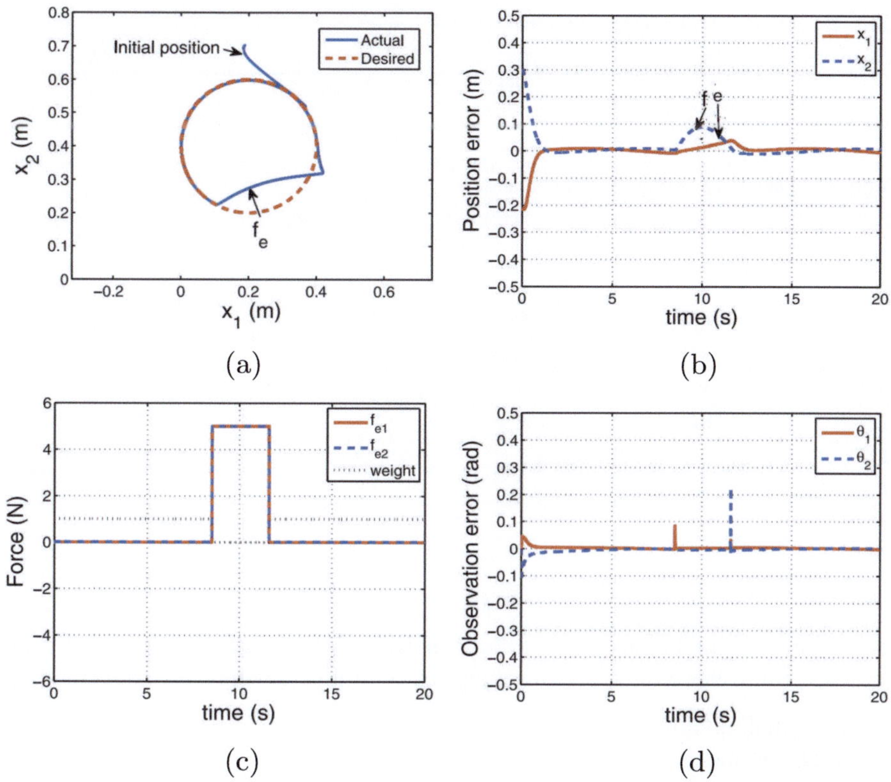

Fig. 2.6 The proposed control method was implemented in a 2 DOF robotic manipulator driven by SEAs, and the robot end effector was controlled to track the desired trajectory as: $x_{d1} = 0.2 + 0.2\sin(0.5t)$ m, and $x_{d2} = 0.4 + 0.2\cos(0.5t)$ m. During the manipulation, interaction forces were exerted on the robot end effector to guide the movement of robot. **a** Path of robot end effector; **b** Tracking error $\Delta \boldsymbol{x} = \boldsymbol{x} - \boldsymbol{x}_d$; **c** Interaction forces $\boldsymbol{f}_e$ and weight factor $w(\boldsymbol{f}_e)$; **d** Observation error $\boldsymbol{e}_o = \boldsymbol{\theta}_d - \hat{\boldsymbol{\theta}}_d$

controller transited from the mode of *robot-in-charge* to the mode of *human-in-charge*, and the robot became passive. The human pushed the end effector away from the desired trajectory $\boldsymbol{x}_d$, to guide the robot to avoid suddenly appearing obstacles as illustrated in Fig. 2.5. When the human force $\boldsymbol{f}_e$ disappeared such that $h(\boldsymbol{f}_e) \leq 0$, the controller transited to the mode of *robot-in-charge* again, which drove the robot end effector to track the desired trajectory.

To eliminate the requirement of acceleration information and its derivatives, the observer (2.46) was constructed to estimate the desired input $\boldsymbol{\theta}_d$. The observer parameters were configured as summarized in Table 2.4. The estimated signals $\hat{\boldsymbol{\theta}}_d$, $\dot{\hat{\boldsymbol{\theta}}}_d$, and $\ddot{\hat{\boldsymbol{\theta}}}_d$ were used instead of $\boldsymbol{\theta}_d$, $\dot{\boldsymbol{\theta}}_d$, and $\ddot{\boldsymbol{\theta}}_d$ in the control input $\boldsymbol{u}$, and the observation error is shown in Fig. 2.6d, which is less than 0.2 rad throughout the course of manipulation.

Table 2.4 Observer parameters and initial settings

Parameter	β	$\boldsymbol{K}_e$	$\boldsymbol{L}_o$	Initial $\hat{\boldsymbol{D}}_{\theta o}$
Value	100	$80\boldsymbol{I}_2$	$20\boldsymbol{I}_2$	$\mathbf{0}$

Table 2.5 Parameter settings in different cases

Parameter	Setting I	Setting II	Setting III	Setting IV
α_q	1	1	–	10
α_θ	1	4	–	4
$\boldsymbol{K}_q$	$15\boldsymbol{I}_2$	$15\boldsymbol{I}_2$	–	$20\boldsymbol{I}_2$
$\boldsymbol{K}_\theta$	$1\boldsymbol{I}_2$	$20\boldsymbol{I}_2$	–	$20\boldsymbol{I}_2$
β	1	100	–	100
$\boldsymbol{K}_e$	$8\boldsymbol{I}_2$	$80\boldsymbol{I}_2$	–	$80\boldsymbol{I}_2$
$\boldsymbol{L}_q$	–	–	–	–
$\boldsymbol{L}_\theta$	–	–	–	–
$\boldsymbol{L}_o$	–	–	–	–

A series of simulations were further carried out to test the performance of the proposed controller with different settings of control parameters, which include:

(1) very small parameters;
(2) small parameters;
(3) well-tuned parameters (all remain the same);
(4) too large parameters.

The parameter settings for each case are summarized in Table 2.5, where – represents the parameters that remain the same.

Random noises were added in simulations and treated as measurement noises, and the results are shown in Fig. 2.7. When the control parameters were set very small (setting I) such that conditions in (2.59) were not satisfied, the closed-loop system was unstable, as shown in Fig. 2.7a. When conditions in (2.59) were satisfied but the parameters were still small (setting II), the tracking errors in the steady state were very large, as shown in Fig. 2.7b. Tracking errors were reduced by using well-tuned parameters (setting III) as shown in Fig. 2.7c. However, if the parameters were set too large (setting IV), noises were amplified, resulting in an unstable system as in Fig. 2.7d.

To illustrate the transition performance of the proposed controller, a comparison was also performed with an online adjustment of desired impedance in adaptive impedance control [15]. In the comparison, the interaction force was specified as a sinusoid as: $\boldsymbol{f}_e$=[2.1sin(0.4t), 0]^T. The desired impedance was switched from a large value in the *robot-in-charge* mode ($w(\boldsymbol{f}_e)$=1), to a low value in the *human-in-charge* mode ($w(\boldsymbol{f}_e)$=0). The comparison results are shown in Figs. 2.8 and 2.9, where the control output for the proposed controller is always bounded, which ensures

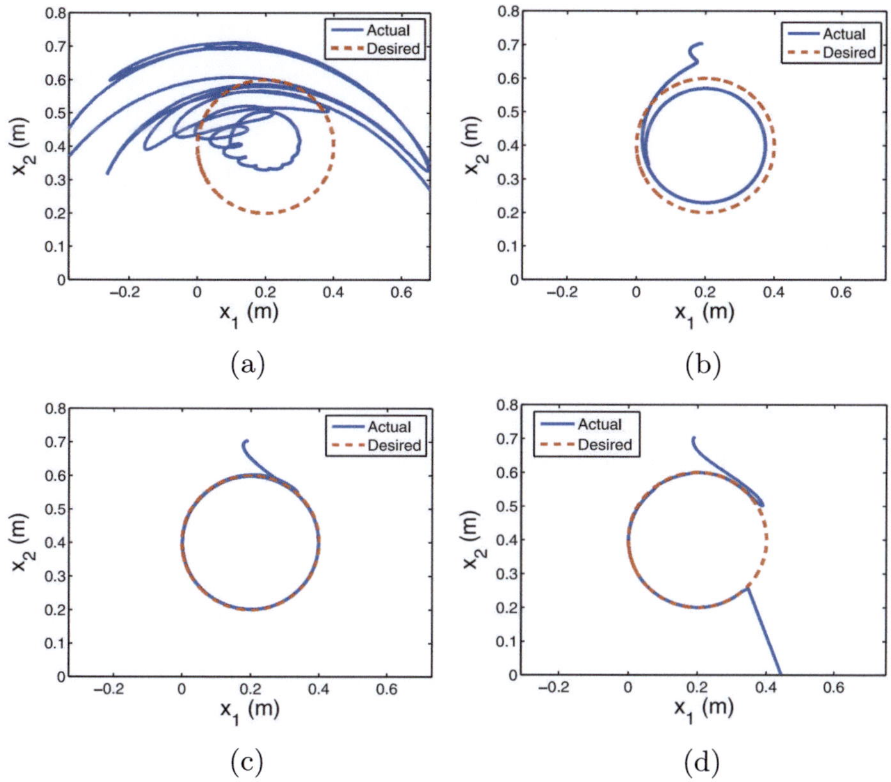

Fig. 2.7 The performance of the proposed controller with different settings of control parameters: **a** very small; **b** small; **c** well-tuned; **d** too large

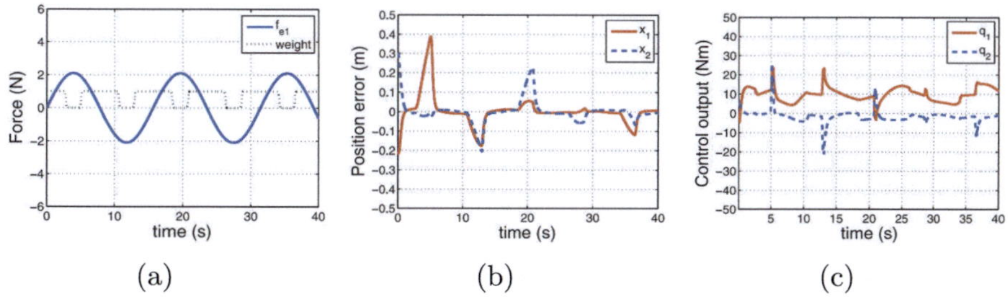

Fig. 2.8 Transition of the proposed controller between *robot-in-charge* ($w(\boldsymbol{f}_e)$=1) and *human-in-charge* ($w(\boldsymbol{f}_e)$=0)

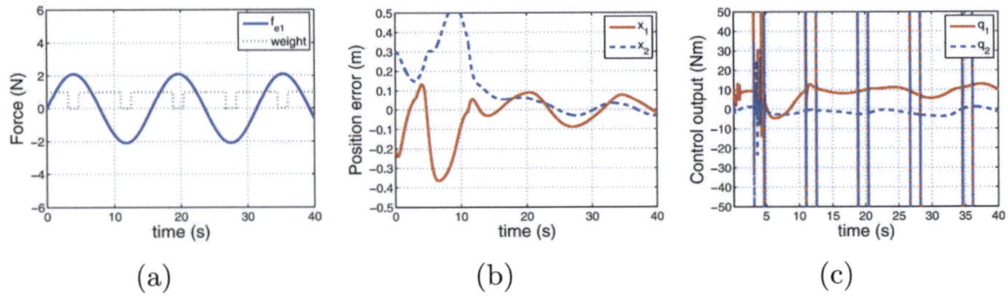

Fig. 2.9 Online adjustment of desired impedance: a high impedance is specified for *robot-in-charge* ($w(\boldsymbol{f}_e)$=1) while a low impedance is specified for *human-in-charge* ($w(\boldsymbol{f}_e)$=0)

a stable and smooth transition between both modes, but the control output of the online adjustment is discontinuous, resulting in a large position errors around t=7 s (see Fig. 2.9b).

Then, the proposed control method was implemented in a one-DOF robot arm as shown in Fig. 2.10. The single-link robot is driven by our SEA introduced in Fig. 2.1, with a calibrated stiffness K=0.0024 N · m/rad, and the human interacts with the robot by exerting forces on the end point of the robot link. Due to hardware limitations, it is unable to measure the interaction force directly. Based on the general formulation, the output torque of the actuator is measured to reflect the change of human intent as:

(1) In the *robot-in-charge* mode, the human does not intervene such that $\boldsymbol{\tau}_e$ is very small, and $\boldsymbol{\tau}_o = \boldsymbol{K}(\boldsymbol{\theta} - \boldsymbol{q}) \approx \boldsymbol{M}(\boldsymbol{q}_d)\ddot{\boldsymbol{q}}_d + \boldsymbol{C}(\dot{\boldsymbol{q}}_d, \boldsymbol{q}_d)\dot{\boldsymbol{q}}_d + \boldsymbol{D}_q\dot{\boldsymbol{q}}_d + \boldsymbol{q}(\boldsymbol{q}_d)$ ($\boldsymbol{q}_d$ denotes the desired joint angle) in steady state from Eq. (2.1), which is also small because of the light robot link and the low desired velocity;
(2) In the *robot-in-charge* mode, the human guides the robot's movement, resulting in a large relative position $\boldsymbol{\theta} - \boldsymbol{q}$ and thus a large output torque of actuator.

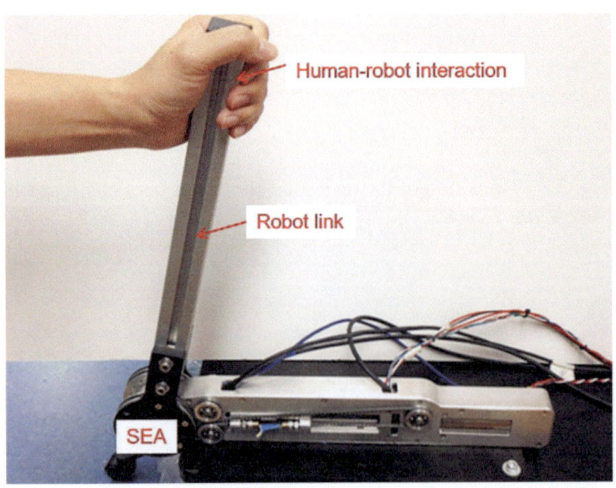

Fig. 2.10 A single-link robot driven by SEA, where the human interacts with the robot by exerting forces on the endpoint of the robot link

Table 2.6 Parameters setting of SEA

Parameter	R	α_q	K_q	L_q	K_e	β	L_o	α_θ	K_θ	L_θ
Value	$14.3\,\mathrm{N}\cdot\mathrm{m}$	10	100	10^{-5}	100	60	2	18	120	10^{-5}

First, a torque region function was formulated similarly as:

$$h(\tau_o) = \tau_o^2 - R^2, \tag{2.63}$$

where R denotes the radius of the torque region. When the torque τ_o is inside the torque region such that $h(\tau_o) \leq 0$, the controller works in the mode of *robot-in-charge*. When the torque is very large such that $h(\tau_o) > 0$, the controller works in the mode of *human-in-charge*. The desired angle for the robot q_d was specified as a sine wave as:

$$q_d = 0.28 \cdot \sin(\pi \cdot t) \text{ rad}. \tag{2.64}$$

In the first experiment, the radius of the torque region R was set sufficiently large, ensuring the weight factor $w(\tau_o)$ remained 1. The proposed controller was specified as a *fully robot-in-charge* controller, where the robot always played a dominant role. To eliminate the requirement for acceleration information and its derivatives, an observer was developed in Eq. (2.46) to estimate the desired input $\hat{\theta}_d$. The parameters used in both the observer and control input settings are summarized in Table 2.6. The angular position of robot is shown in Fig. 2.11a, and the tracking error is shown in Fig. 2.11b which is less than 0.1 rad. The output torque is shown in Fig. 2.11c. The observation error is shown in Fig. 2.11d, which is around 0.02 rad. The results show that the robot successfully tracked the desired trajectory in the mode of *robot-in-charge*.

In the second experiment, the radius R was set arbitrarily small such that $R \to 0$, the weight factor $w(\tau_o)$ remained 0, and the proposed controller was specified as a *fully human-in-charge* controller. The robot became passive, and the human guided the movement of robot by providing the directional information for the robot. The control parameters remained the same. The angular position of robot is shown in Fig. 2.12a, which implies that the robot was controlled by the human, regardless of the desired time-varying trajectory. The error between the robot joint angle and the desired angle is shown in Fig. 2.12b. The output torque of SEA is shown in Fig. 2.12c. Since the human kept exerting force on the robot end effector, the relative angular displacement between the robot link and the rotor shaft of the actuator $\theta - q$ was very large, and the output torque of SEA was also very large. The observation error is shown in Fig. 2.12d, which is around 0.02 rad.

In the third experiment, the torque region $h(\tau_o)$ was specified with the radius of R=1.17 Nm, and the parameter of the weight factor was set as κ=0.67. Therefore, when the output torque τ_o exceeded the torque region due to the interaction with humans, $h(\tau_o)>0$, $w(\tau_o)=0$, the controller transited from the mode of *robot-in-charge* to the mode of *human-in-charge*, and vice versa. The control parameters

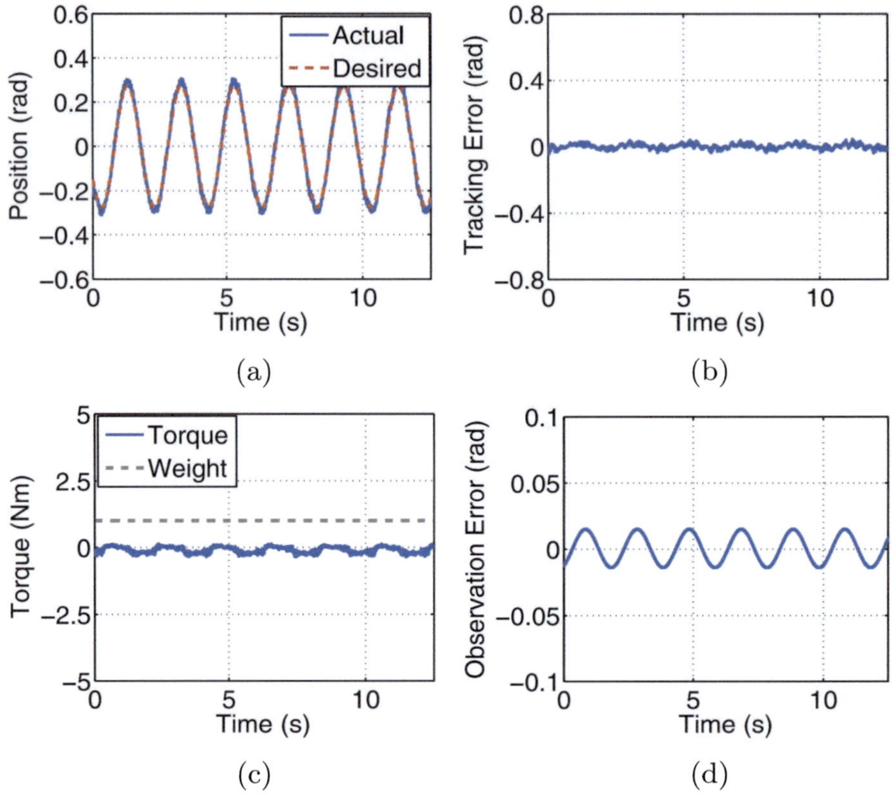

Fig. 2.11 Experiment 1: The torque region was specified sufficiently large such that the weight factor remained 1, and the proposed controller was treated as a fully *robot-in-charge* controller. (**a** Angular position of robot; **b** Tracking error $\Delta q = q - q_d$; **c** Output torque of actuator $\tau_o = K(\theta - q)$; (**d** Observation error $e_o = \theta_d - \hat{\theta}_d$

also remained the same. The angular position of the robot and the tracking error are shown in Fig. 2.13a, b respectively. In the beginning, the controller was working in the mode of *robot-in-charge*, and the robot was able to track the desired trajectory where the tracking error was less than 0.1 rad. During the manipulation (t=2~3 s, 6~7 s, etc.), the human exerted force on the robot, which led to a large output torque τ_o, the weight factor $w(\tau_o)$ reduced to zero, and the robot became passive. Therefore, the human was able to push the robot away from the desired trajectory. When the human control effort was removed, $w(\tau_o)$ smoothly increased to 1, and the controller transited to the mode of *robot-in-charge* again, which drove the robot to track the desired trajectory. The output torque of SEA is shown in Fig. 2.13c. The observation error is shown in Fig. 2.13d, which is less than 0.08 rad.

The continuous transition between both modes is illustrated in the following snapshots (see Fig. 2.14). In the experiment, a comparison was also performed with the

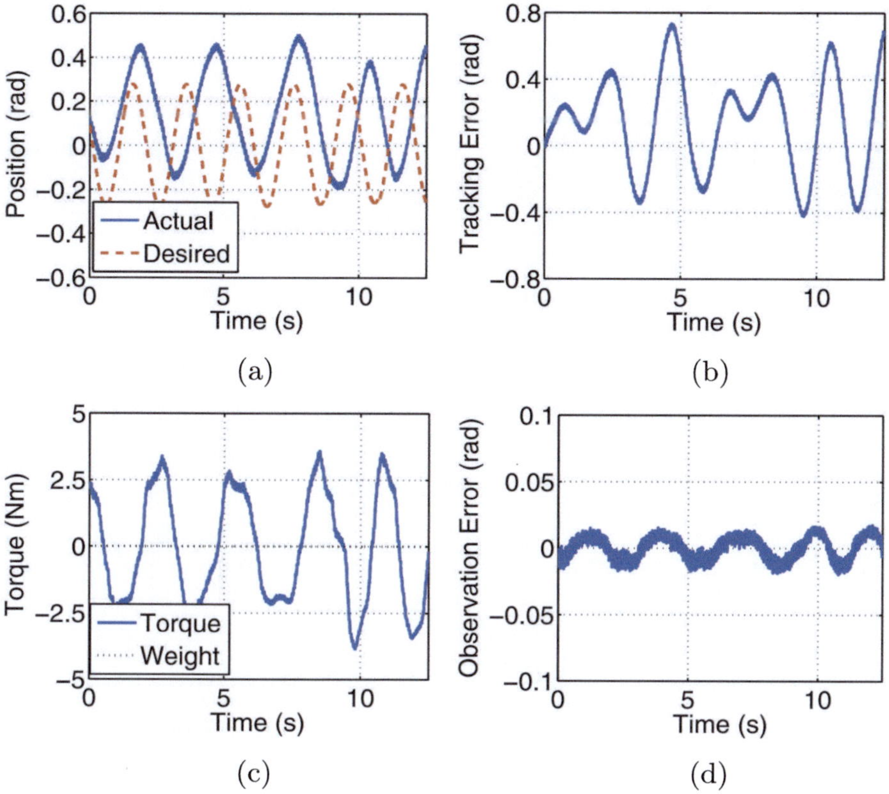

Fig. 2.12 Experiment 2: The torque region was specified arbitrarily small such that the weight factor remained 0, and the proposed controller was treated as a fully *human-in-charge* controller. (**a** Angular position of robot; **b** Tracking error $\Delta q = q - q_d$; **c** Output torque of actuator $\tau_o = K(\theta - q)$; (**d** Observation error $e_o = \theta_d - \hat{\theta}_d$

hard-switching of a proxy-based controller and a damping controller [16]. The proxy-based controller works in the *robot-in-charge* mode ($w(\tau_o)=1$), and the damping controller is activated in the *human-in-charge* mode ($w(\tau_o)=0$). The results are shown in Fig. 2.15. As seen from Fig. 2.15a, the proxy-based controller was able to deal with the interaction torque when it was not large (at $t=0.7$ s and $t=1.8$ s). However, when the interaction torque became larger such that the control scheme was switched to the damping controller, the overall control output was discontinuous during the transition stage, which then resulted in a series of chattering at $t=4$ s, $t=6$ s, and $t=12$ s (see Fig. 2.15b).

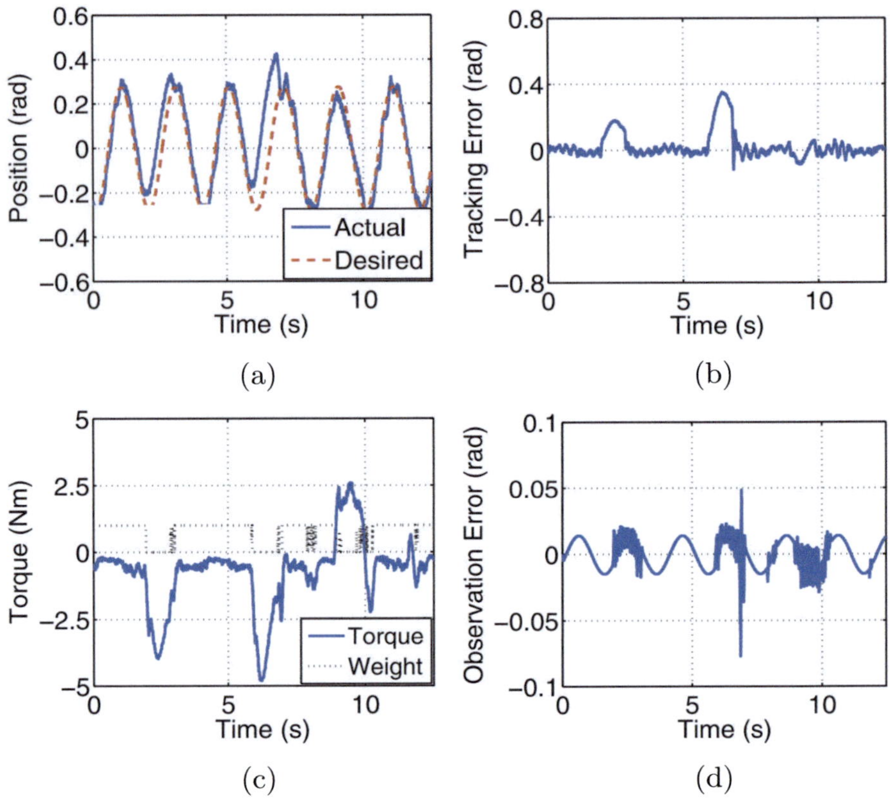

Fig. 2.13 Experiment 3: The torque region was specified as $h(\tau_o) = \tau_o^2 - 1.17^2 \leq 0$, and the proposed controller transited from the mode of *robot-in-charge* to the mode of *human-in-charge* when the weight factor reduced to 0, and vice versa. **a** Angular position of robot; **b** Tracking error $\Delta q = q - q_d$; **c** Output torque of actuator $\tau_o = K(\theta - q)$; (**d** Observation error $e_o = \theta_d - \hat{\theta}_d$

2.3 Iterative Learning Control

This section follows the backstepping approach and presents another iterative learning impedance controller for rehabilitation robots driven by series elastic actuators (SEAs), where the control objective is specified as a desired impedance model such that more parameters can be specified (e.g., inertia, damping, stiffness) to suit the human subject. The desired impedance model is achieved in an iterative manner, which suits the repeating nature of several applications (e.g., patients' task through therapeutic process) and also guarantees the transient performance of robot. The stability of the overall system is rigorously proved with Lyapunov methods by taking into account both the robot and actuator dynamics. Experimental results are presented to illustrate the performance of the proposed iterative control scheme.

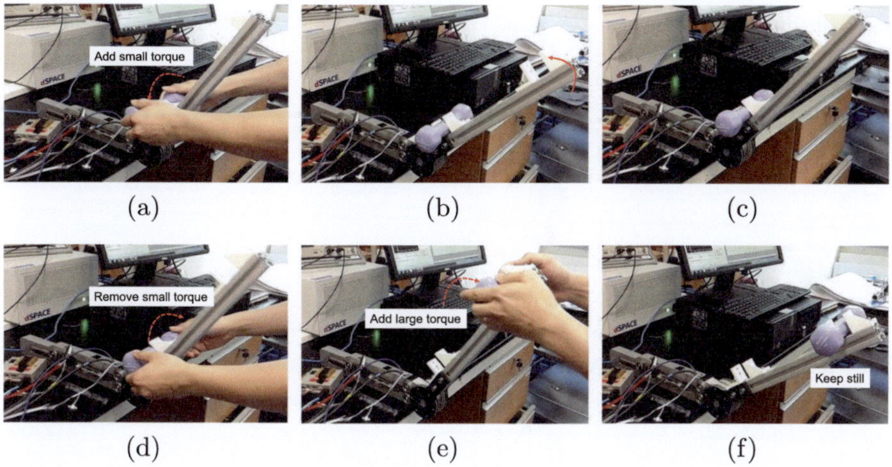

Fig. 2.14 Snapshots of transition between robot-in-charge and human-in-charge: **a** A weight is placed close to the rotation center; **b** The resulting external torque is small and within the region; **c** The robot is working under the *robot-in-charge* and tracking the predefined trajectory; **d** The weight is moved to the end of the link; **e** The resulting external torque is large and within the region; **f** The robot smoothly transits to the *human-in-charge* and stops tracking

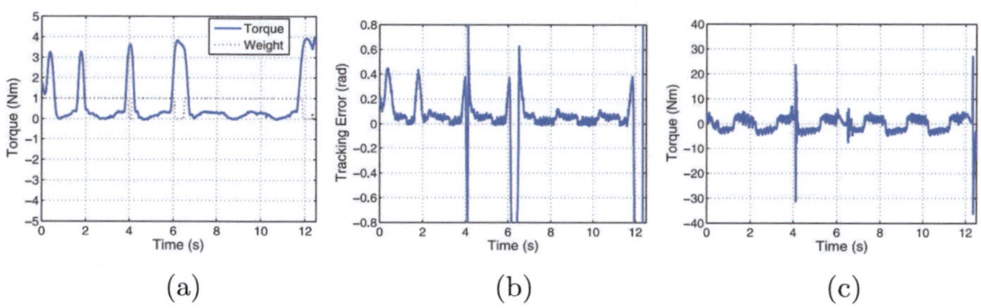

Fig. 2.15 Hard-switching of a proxy-based controller for *robot-in-charge* ($w(\tau_o)$=1) and a damping controller for *human-in-charge* ($w(\tau_o)$=0)

2.3.1 Desired Impedance Model

The impedance control problem is formulated in terms of a reference trajectory and a desired dynamic relationship between the position error and the interaction force [17], shown in Fig. 2.16. This section considers the desired impedance model defined in joint space as

$$\boldsymbol{M}_d(\ddot{\boldsymbol{q}}-\ddot{\boldsymbol{q}}_d)+\boldsymbol{C}_d(\dot{\boldsymbol{q}}-\dot{\boldsymbol{q}}_d)+\boldsymbol{K}_d(\boldsymbol{q}-\boldsymbol{q}_d) = \boldsymbol{\tau}_e, \tag{2.65}$$

where $\boldsymbol{q}_d \in \Re^{\bar{n}}$ is the vector of the desired joint angles which are time-varying, $\boldsymbol{M}_d, \boldsymbol{C}_d, \boldsymbol{K}_d \in \Re^{\bar{n}\times\bar{n}}$ denote the desired inertia, the desired damping, and the desired stiffness matrices respectively, which are diagonal and positive definite.

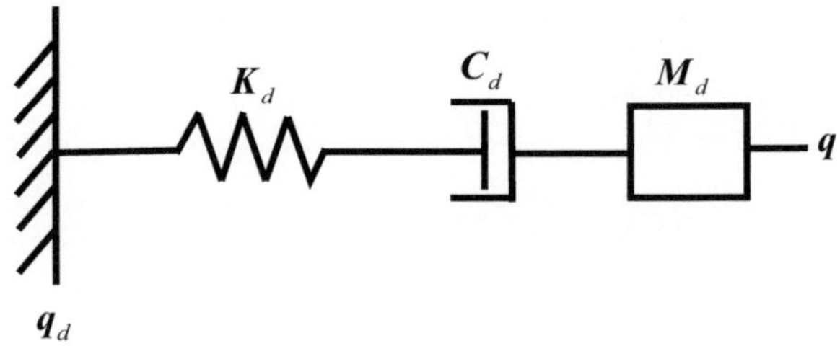

Fig. 2.16 Illustration of desired impedance model

Let $\Delta \boldsymbol{q} = \boldsymbol{q} - \boldsymbol{q}_d$ represent the joint-space position error. Then, two matrices $\boldsymbol{\Gamma} \in \Re^{\bar{n}\times\bar{n}}$ and $\boldsymbol{\Lambda} \in \Re^{\bar{n}\times\bar{n}}$ and a vector $\boldsymbol{\tau}_l \in \Re^{\bar{n}}$ are defined as

$$\begin{aligned}\boldsymbol{\Lambda} + \boldsymbol{\Gamma} &= \boldsymbol{M}_d^{-1}\boldsymbol{C}_d, &(2.66)\\ \boldsymbol{\Gamma}\boldsymbol{\Lambda} &= \boldsymbol{M}_d^{-1}\boldsymbol{K}_d, &(2.67)\\ \dot{\boldsymbol{\tau}}_l + \boldsymbol{\Gamma}\boldsymbol{\tau}_l &= \boldsymbol{M}_d^{-1}\boldsymbol{\tau}_e, &(2.68)\end{aligned}$$

where both $\boldsymbol{\Lambda}$ and $\boldsymbol{\Gamma}$ are diagonal and positive definite.

The parameters of the desired impedance model should be specified such that both $\boldsymbol{\Lambda}$ and $\boldsymbol{\Gamma}$ exist. Given that

$$\begin{aligned}\boldsymbol{\Lambda} &= \mathrm{diag}\{\lambda_1, \ldots, \lambda_{\bar{n}}\},\\ \boldsymbol{\Gamma} &= \mathrm{diag}\{\gamma_1, \ldots, \gamma_{\bar{n}}\},\\ \boldsymbol{M}_d &= \mathrm{diag}\{m_{d1}, \ldots, m_{d\bar{n}}\},\\ \boldsymbol{C}_d &= \mathrm{diag}\{c_{d1}, \ldots, c_{d\bar{n}}\},\\ \boldsymbol{K}_d &= \mathrm{diag}\{k_{d1}, \ldots, k_{d\bar{n}}\} \qquad (2.69)\end{aligned}$$

where λ_i, γ_i, m_{di}, c_{di}, and k_{di} (i=1, ..., $\bar{n}$) are positive constants, the parameters should be specified such that $c_{di}^2 \geq 4m_{di}k_{di}$. Then, the elements of $\boldsymbol{\Lambda}$ and $\boldsymbol{\Gamma}$ can specified as:

$$\begin{aligned}\lambda_i &= \frac{m_{di}^{-1}c_{di} \pm \sqrt{m_{di}^{-2}c_{di}^2 - 4m_{di}^{-1}k_{di}}}{2}, &(2.70)\\ \gamma_i &= \frac{m_{di}^{-1}c_{di} \mp \sqrt{m_{di}^{-2}c_{di}^2 - 4m_{di}^{-1}k_{di}}}{2}. &(2.71)\end{aligned}$$

Using (2.66)-(2.68), an augmented impedance error can be rewritten as

$$\begin{aligned}\bar{\boldsymbol{w}} &= \Delta\ddot{\boldsymbol{q}} + (\boldsymbol{\Lambda} + \boldsymbol{\Gamma})\Delta\dot{\boldsymbol{q}} + \boldsymbol{\Gamma}\boldsymbol{\Lambda}\Delta\boldsymbol{q} - \dot{\boldsymbol{\tau}}_l - \boldsymbol{\Gamma}\boldsymbol{\tau}_l\\ &= \dot{\boldsymbol{z}} + \boldsymbol{\Gamma}\boldsymbol{z}. \qquad (2.72)\end{aligned}$$

where

$$z = \Delta\dot{q} + \Lambda\Delta q - \tau_l, \quad (2.73)$$

is the impedance vector. It can be derived that the convergence of $\bar{w} \to 0$ leads to the realization of the desired impedance model. From (2.72), the impedance vector z can be treated as the low-pass-filtered signal of $\bar{w}$.

In this section, the control objective is formulated as $z \to 0$, indicating the realization of the desired impedance model in the low-frequency range (this is reasonable in the scenario of rehabilitation, as the frequency of patients' movement is usually low, e.g., < 2 Hz). The formulation of $z \to 0$ has been extensively used in [18, 19], but those results are limited to rigid-joint robots. The overall SEA-driven robot, consisting of both the rigid-link subsystem and the actuator subsystem, is a high-order system, and finding a solution to stabilize both subsystems is not trivial, and neglecting the coupling dynamics is exposed to stability issues [1].

In actual implementations, the parameters of the desired impedance model should also be initialized according to the assessment of human subjects [11]. For example, in the scenario of rehabilitation, if the patient is at early stage of the therapy and has minimal motion control, the desired impedance should be increased by setting larger M_d, C_d, K_d, to allow the robot to take a lead and thus maximize its assistance; As the patient's motion control ability is being restored, the desired impedance could be decreased by setting smaller M_d, C_d, K_d, to encourage patient's active motions and also minimize the robot's assistance.

2.3.2 Control Scheme

In this section, a new iterative learning impedance controller is proposed for SEA-driven rehabilitation robots, to guarantee the convergence of $z \to 0$ as actions are repeated.

The development of the proposed controller follows a backstepping approach. First, (2.1) is rewritten as

$$\begin{aligned} &M(q)\ddot{q} + C(\dot{q}, q)\dot{q} + g(q) + Kq \\ &= K\theta_d + K\Delta\theta + d_r + \tau_e, \end{aligned} \quad (2.74)$$

where d_r denotes the unmodeled disturbance, which is added into the dynamic model, $\Delta\theta = \theta - \theta_d \in \Re^{\bar{n}}$, and $\theta_d \in \Re^{\bar{n}}$ represents a fictitious desired input. Then, the desired input θ_d is specified to stabilize the rigid-link subsystem (2.74) and achieve the impedance control task, while the control input u is developed to stabilize the actuator subsystem (2.2) such that the actual position of the motor tracks the desired input, that is, $\theta \to \theta_d$.

The impedance vector is rewritten as

$$z = \dot{q} - \dot{q}_r, \tag{2.75}$$

where

$$\dot{q}_r = \dot{q}_d - \Lambda \Delta q + \tau_l \tag{2.76}$$

is a reference vector. By using (2.75), (2.74) can be rewritten as

$$\begin{aligned} &M(q)\dot{z} + C(\dot{q}, q)z + M(q)\ddot{q}_r + C(\dot{q}, q)\dot{q}_r + g(q) + Kq \\ &= K\theta_d + K\Delta\theta + d_r + \tau_e, \end{aligned} \tag{2.77}$$

where $\ddot{q}_r$ is the time derivative of $\dot{q}_r$.

Then, we can propose the desired input as

$$\begin{aligned} \theta_d = &q + K^{-1}(-K_z z - k_r sgn(z) - \tau_e \\ &+ M(q)\ddot{q}_r + C(\dot{q}, q)\dot{q}_r + g(q) + m), \end{aligned} \tag{2.78}$$

where $K_z \in \Re^{\bar{n}\times\bar{n}}$ is positive definite, k_r is a positive constant, and $sgn(\cdot)$ is a sign function [20], as illustrated in Fig. 2.17. The sign function enhances the robustness of the controller against model uncertainties and disturbances by introducing a corrective term to counteract them.

The term $-K_z z$ is to realize the impedance control task, the terms $-\tau_e + M(q)\ddot{q}_r + C(\dot{q}, q)\dot{q}_r + g(q)$ are to compensate for the dynamics. The vector $m \in \mathbb{R}^{\bar{n}}$ represents a feedforward learning control input, which is constructed by following the iterative learning control (ILC) principle shown in Fig. 2.18. The feedforward control input is updated according to the following law:

$$m_{k+1} = m_k - \beta_m z_k, \tag{2.79}$$

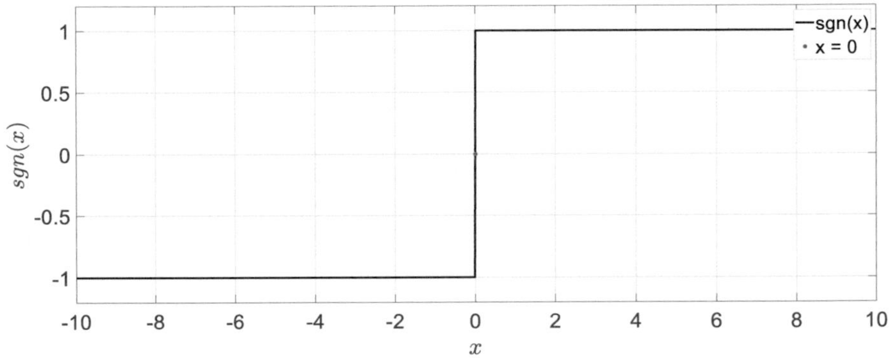

Fig. 2.17 Illustration of sign function

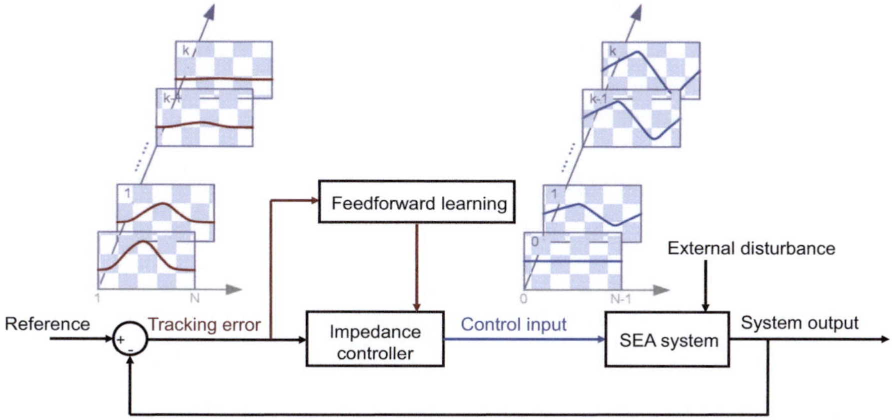

Fig. 2.18 Principle of iterative learning controller, where the feedforward compensation term is computed and updated before the running of the controller and then added into the control input at the corresponding time instants

where β_m is a positive constant, and the subscript of k denotes the kth iteration.

Substituting (2.78) into (2.77), we obtain the dynamic equation of the rigid-link subsystem as

$$\boldsymbol{M}(\boldsymbol{q})\dot{z}+(\boldsymbol{C}(\dot{\boldsymbol{q}},\boldsymbol{q})+\boldsymbol{K}_z)z+k_r\boldsymbol{sgn}(z)=\boldsymbol{K}\Delta\boldsymbol{\theta}+\boldsymbol{d}_r+\boldsymbol{m}. \tag{2.80}$$

Next, a sliding vector is proposed for the actuator dynamics as

$$\begin{aligned}\boldsymbol{s}=&\dot{\boldsymbol{\theta}}-\dot{\boldsymbol{\theta}}_r\\=&\dot{\boldsymbol{\theta}}-\dot{\boldsymbol{\theta}}_d+\alpha_\theta\Delta\boldsymbol{\theta},\end{aligned} \tag{2.81}$$

where

$$\dot{\boldsymbol{\theta}}_r=\dot{\boldsymbol{\theta}}_d-\alpha_\theta\Delta\boldsymbol{\theta} \tag{2.82}$$

is another reference vector, $\dot{\boldsymbol{\theta}}_d$ is the time derivative of $\boldsymbol{\theta}_d$ in (2.78), and α_θ is a positive constant. By using the sliding vector (2.81), the actuator dynamics can be rewritten as

$$\boldsymbol{B}\dot{\boldsymbol{s}}+\boldsymbol{B}\ddot{\boldsymbol{\theta}}_r+\boldsymbol{K}(\boldsymbol{\theta}-\boldsymbol{q})=\boldsymbol{d}_a+\boldsymbol{u}, \tag{2.83}$$

where $\ddot{\boldsymbol{\theta}}_r$ is the time derivative of $\dot{\boldsymbol{\theta}}_r$ as

$$\ddot{\boldsymbol{\theta}}_r=\ddot{\boldsymbol{\theta}}_d-\alpha_\theta\Delta\dot{\boldsymbol{\theta}}, \tag{2.84}$$

where $\ddot{\boldsymbol{\theta}}_d$ is the second-order time derivative of $\boldsymbol{\theta}_d$. Note that the vector $\ddot{\boldsymbol{\theta}}_r$ is a function of $\ddot{\boldsymbol{\theta}}_d$. From (2.78), the desired input $\boldsymbol{\theta}_d$ requires the information of $\ddot{\boldsymbol{q}}_r$ which is a function of $\dot{\boldsymbol{q}}$, and hence $\ddot{\boldsymbol{\theta}}_d$ is dependent on the third-order time derivative, i.e.,

$\dddot{q}$. That is, the high-order derivatives are required in the control input (2.85). To eliminate the requirement of high-order derivatives, a model-based observer can be developed to estimate the desired input by following ([21]).

Then, the control input $\boldsymbol{u}$ is proposed as

$$\begin{aligned}\boldsymbol{u} =& \boldsymbol{K}(\boldsymbol{\theta} - \boldsymbol{q}) - k_a \boldsymbol{sgn}(\boldsymbol{s}) + \boldsymbol{B}\ddot{\boldsymbol{\theta}}_r \\ & - \boldsymbol{K}_s \boldsymbol{s} - \boldsymbol{K}_\theta \Delta\boldsymbol{\theta} + \boldsymbol{n},\end{aligned} \tag{2.85}$$

where $\boldsymbol{K}_s \in \Re^{\bar{n}\times\bar{n}}$ and $\boldsymbol{K}_\theta \in \Re^{\bar{n}\times\bar{n}}$ are positive definite matrices, and k_a is a positive constant. In (2.85), the terms of $\boldsymbol{K}(\boldsymbol{\theta} - \boldsymbol{q})$ and $\boldsymbol{B}\ddot{\boldsymbol{\theta}}_r$ represent the dynamic compensation, and the terms of $-\boldsymbol{K}_s\boldsymbol{s} - \boldsymbol{K}_\theta\Delta\boldsymbol{\theta}$ include the velocity control and the position control.

Then, the $\boldsymbol{n}$ is the learning control input for the actuator subsystem, which is updated by

$$\boldsymbol{n}_{k+1} = \boldsymbol{n}_k - \beta_n \boldsymbol{s}_k, \tag{2.86}$$

where β_n is a positive constant.

Substituting (2.85) into (2.83), we have the dynamic equation of the actuator subsystem as

$$\boldsymbol{B}\dot{\boldsymbol{s}} + \boldsymbol{K}_s\boldsymbol{s} + k_a\boldsymbol{sgn}(\boldsymbol{s}) + \boldsymbol{K}_\theta\Delta\boldsymbol{\theta} = \boldsymbol{d}_a + \boldsymbol{n}. \tag{2.87}$$

The key novelty of the proposed controller is that two learning strategies (2.79) and (2.86) are developed to update feedforward control terms for the rigid-link subsystem (2.80) and the actuator subsystem (2.87) respectively. Note that $\boldsymbol{m}_k$ and $\boldsymbol{n}_k$ are updated in the next iteration by incorporating error information (i.e., z_k and $\boldsymbol{s}_k$ respectively). As the iterations go to infinity, both the impedance error and the tracking error reduce to zero, and the iterative control terms converge to cancel out unmodelled disturbances. Then, the ratio of $\boldsymbol{m}_k$ over $\boldsymbol{\theta}_d$ and the ratio of $\boldsymbol{n}_k$ over $\boldsymbol{u}$ start from zero and progressively increase, as the feedback control terms converge to zero, while the feedforward control terms are playing more dominant roles. Different from adaptive/robust control approaches, the proposed iterative learning controller explores and utilizes the information contained in repetitive actions, and taking into account the repetitive information can achieve better transient tracking performance and attenuate repetitive disturbance [22, 23].

2.3.3 *Stability Analysis*

We can now state the following theorem.

Theorem 2.3.1 *Consider the learning control system given by (2.80) and (2.87). Let the control parameters be chosen such that the following conditions are satisfied:*

$$\lambda_{\min}\{\boldsymbol{K}_s\} \geq \frac{\beta_n}{2}, \tag{2.88}$$

$$\lambda_{\min}\{\boldsymbol{K}_z\} \geq \frac{\beta_m}{2}, \tag{2.89}$$

$$\frac{2\alpha_\theta}{\beta_n}\lambda_{\min}\{(\frac{2}{\beta_m}\boldsymbol{K}_z - \boldsymbol{I}_n)\boldsymbol{K}_\theta\} \geq \frac{1}{\beta_m^2}\lambda_{\max}\{\boldsymbol{K}^2\}, \tag{2.90}$$

$$k_r \geq B_{\max}\{\boldsymbol{d}_r\}, \tag{2.91}$$

$$k_a \geq B_{\max}\{\boldsymbol{d}_a\}, \tag{2.92}$$

where $\lambda_{\min}\{\cdot\}$ and $\lambda_{\max}\{\cdot\}$ represent the minimum and maximum eigenvalues respectively, $B_{\max}\{\cdot\}$ denote the upper bounds. Then, a sequence of control inputs will be generated such that the impedance control task is achieved. That is

$$z_k(t) \rightarrow \boldsymbol{0}, \tag{2.93}$$

for all $t \in [0, t_f]$ as $k \rightarrow \infty$, where t_f is the iteration period, and $\boldsymbol{I}_n$ is an identity matrix.

Proof First, an index function $V_k(t)$ is defined as:

$$V_k(t) = V_k(0) + \int_0^t (\frac{1}{\beta_m^2}\boldsymbol{m}_k^T(\varsigma)\boldsymbol{m}_k(\varsigma) + \frac{1}{\beta_n^2}\boldsymbol{n}_k^T(\varsigma)\boldsymbol{n}_k(\varsigma))\mathrm{d}\varsigma, \tag{2.94}$$

for all $t \in [0, t_f]$, where $V_k(0)$ represents the initial value at $t = 0$.

Define $\Delta V_k = V_{k+1} - V_k$. From (2.79), (2.86), and (2.94), it is obtained that:

$$\begin{aligned}
\Delta V_k =& V_{k+1}(0) - V_k(0) + \\
& \int_0^t (\frac{1}{\beta_m^2}\boldsymbol{m}_{k+1}^T(\varsigma)\boldsymbol{m}_{k+1}(\varsigma) + \frac{1}{\beta_n^2}\boldsymbol{n}_{k+1}^T(\varsigma)\boldsymbol{n}_{k+1}(\varsigma))\mathrm{d}\varsigma \\
& - \int_0^t (\frac{1}{\beta_m^2}\boldsymbol{m}_k^T(\varsigma)\boldsymbol{m}_k(\varsigma) + \frac{1}{\beta_n^2}\boldsymbol{n}_k^T(\varsigma)\boldsymbol{n}_k(\varsigma))\mathrm{d}\varsigma \\
=& V_{k+1}(0) - V_k(0) + \int_0^t z_k^T(\varsigma)z_k(\varsigma)\mathrm{d}\varsigma + \int_0^t \boldsymbol{s}_k^T(\varsigma)\boldsymbol{s}_k(\varsigma)\mathrm{d}\varsigma \\
& - \frac{2}{\beta_m}\int_0^t z_k^T(\varsigma)\boldsymbol{m}_k(\varsigma)\mathrm{d}\varsigma - \frac{2}{\beta_n}\int_0^t \boldsymbol{s}_k^T(\varsigma)\boldsymbol{n}_k(\varsigma)\mathrm{d}\varsigma.
\end{aligned} \tag{2.95}$$

From (2.80) and (2.87), the two learning inputs, i.e., $\boldsymbol{m}$ and $\boldsymbol{n}$, can be specified as:

$$\begin{aligned}
\boldsymbol{m}_k(t) =& \boldsymbol{M}(\boldsymbol{q}_k(t))\dot{z}_k(t) + (\boldsymbol{C}(\dot{\boldsymbol{q}}_k(t), \boldsymbol{q}_k(t)) + \boldsymbol{K}_z)z_k(t) \\
& - \boldsymbol{d}_r(t) + k_r\boldsymbol{sgn}(z_k(t)) - \boldsymbol{K}\Delta\boldsymbol{\theta}_k(t),
\end{aligned} \tag{2.96}$$

$$\begin{aligned}
\boldsymbol{n}_k(t) =& \boldsymbol{B}\dot{\boldsymbol{s}}_k(t) + \boldsymbol{K}_s\boldsymbol{s}_k(t) + \boldsymbol{K}_\theta\Delta\boldsymbol{\theta}_k(t) \\
& - \boldsymbol{d}_a(t) + k_a\boldsymbol{sgn}(\boldsymbol{s}_k(t)).
\end{aligned} \tag{2.97}$$

Substituting (2.96) and (2.97) into (2.95) yields:

$$\begin{aligned}\Delta V_k =& V_{k+1}(0) - V_k(0) - \frac{2}{\beta_m}\int_0^t z_k^T(\varsigma)(\boldsymbol{M}(\boldsymbol{q}_k(\varsigma))\dot{z}_k(\varsigma)\\ &+ (\boldsymbol{C}(\dot{\boldsymbol{q}}_k(\varsigma), \boldsymbol{q}_k(\varsigma)) + \boldsymbol{K}_z)z_k(\varsigma) - \boldsymbol{K}\Delta\boldsymbol{\theta}_k(\varsigma)\\ &- \boldsymbol{d}_r(\varsigma) + k_r\boldsymbol{sgn}(z_k(\varsigma)))\mathrm{d}\varsigma\\ &+ \int_0^t z_k^T(\varsigma)z_k(\varsigma)\mathrm{d}\varsigma + \int_0^t \boldsymbol{s}_k^T(\varsigma)\boldsymbol{s}_k(\varsigma)\mathrm{d}\varsigma\\ &- \frac{2}{\beta_n}\int_0^t \boldsymbol{s}_k^T(\varsigma)(\boldsymbol{B}\dot{\boldsymbol{s}}_k(\varsigma) + \boldsymbol{K}_s\boldsymbol{s}_k(\varsigma) + \boldsymbol{K}_\theta\Delta\boldsymbol{\theta}_k(\varsigma)\\ &- \boldsymbol{d}_a(\varsigma) + k_a\boldsymbol{sgn}(\boldsymbol{s}_k(\varsigma)))\mathrm{d}\varsigma. \end{aligned} \tag{2.98}$$

Since

$$\begin{aligned} z^T(k_r\boldsymbol{sgn}(z) - \boldsymbol{d}_r) &\geq (k_r - B_{\max}\{\boldsymbol{d}_r\})||z||,\\ \boldsymbol{s}^T(k_a\boldsymbol{sgn}(\boldsymbol{s}) - \boldsymbol{d}_a) &\geq (k_a - B_{\max}\{\boldsymbol{d}_a\})||\boldsymbol{s}||, \end{aligned} \tag{2.99}$$

and

$$\begin{aligned} &\int_0^t z_k^T(\varsigma)\boldsymbol{M}(\boldsymbol{q}_k(\varsigma))\dot{z}_k(\varsigma)\mathrm{d}\varsigma\\ =& \frac{1}{2}z_k^T(t)\boldsymbol{M}(\boldsymbol{q}_k(t))z_k(t) - \frac{1}{2}z_k^T(0)\boldsymbol{M}(\boldsymbol{q}_k(0))z_k(0)\\ &- \frac{1}{2}\int_0^t z_k^T(\varsigma)\dot{\boldsymbol{M}}(\boldsymbol{q}_k(\varsigma))z_k(\varsigma)\mathrm{d}\varsigma, \end{aligned} \tag{2.100}$$

by setting k_r and k_a such that (2.91) and (2.92) are satisfied and using initial conditions [22, 23], we have:

$$\begin{aligned} \Delta V_k \leq& \int_0^t z_k^T(\varsigma)z_k(\varsigma)\mathrm{d}\varsigma - \frac{1}{\beta_m}z_k^T(t)\boldsymbol{M}(\boldsymbol{q}_k(t))z_k(t)\\ &- \frac{2}{\beta_m}\int_0^t (z_k^T(\varsigma)\boldsymbol{K}_z z_k(\varsigma) - z_k^T(\varsigma)\boldsymbol{K}\Delta\boldsymbol{\theta}_k(\varsigma))\mathrm{d}\varsigma +\\ &\int_0^t \boldsymbol{s}_k^T(\varsigma)\boldsymbol{s}_k(\varsigma)\mathrm{d}\varsigma - \frac{1}{\beta_n}\boldsymbol{s}_k^T(t)\boldsymbol{B}\boldsymbol{s}_k(t) - \frac{1}{\beta_n}\Delta\boldsymbol{\theta}_k^T(t)\boldsymbol{K}_\theta\Delta\boldsymbol{\theta}_k(t)\\ &- \frac{2}{\beta_n}\int_0^t (\boldsymbol{s}_k^T(\varsigma)\boldsymbol{K}_s\boldsymbol{s}_k(\varsigma) + \alpha_\theta\Delta\boldsymbol{\theta}_k^T(\varsigma)\boldsymbol{K}_\theta\Delta\boldsymbol{\theta}_k(\varsigma))\mathrm{d}\varsigma, \end{aligned} \tag{2.101}$$

which can be rewritten as:

$$
\begin{aligned}
\Delta V_k \leq & -\frac{1}{\beta_m} z_k^T(t) \boldsymbol{M}(\boldsymbol{q}(t)) z_k(t) - \frac{1}{\beta_n} \boldsymbol{s}_k^T(t) \boldsymbol{B} \boldsymbol{s}_k(t) \\
& - \frac{1}{\beta_n} \Delta\boldsymbol{\theta}_k^T(t) \boldsymbol{K}_\theta \Delta\boldsymbol{\theta}_k(t) - \int_0^t (\boldsymbol{s}_k^T(\varsigma)(\frac{2}{\beta_n}\boldsymbol{K}_s - \boldsymbol{I}_n)\boldsymbol{s}_k(\varsigma))\mathrm{d}\varsigma \\
& - \int_0^t \{[z_k(\varsigma);\ \Delta\boldsymbol{\theta}_k(\varsigma)]^T \boldsymbol{P}[z_k(\varsigma);\ \Delta\boldsymbol{\theta}_k(\varsigma)]\}\mathrm{d}\varsigma,
\end{aligned}
\tag{2.102}
$$

where

$$
\boldsymbol{P} = \begin{bmatrix} \frac{2}{\beta_m}\boldsymbol{K}_z - \boldsymbol{I}_n & -\frac{1}{\beta_m}\boldsymbol{K} \\ -\frac{1}{\beta_m}\boldsymbol{K} & \frac{2\alpha_\theta}{\beta_n}\boldsymbol{K}_\theta \end{bmatrix}. \tag{2.103}
$$

If the control parameters are chosen such that conditions (2.88)–(2.90) are satisfied, $\boldsymbol{P}$ is non-negative definite, and $\Delta V_k \leq 0$. This implies that $V_k(t)$ converges to a nonnegative constant because $V_0(t)$ is bounded. Therefore, $\Delta V_k(t) \to 0$ as $k \to \infty$. From (2.102), we have:

$$
z_k(t) \to \mathbf{0}, \tag{2.104}
$$

for all $t \in [0, t_f]$ as $k \to \infty$. That is, the impedance control objective is achieved. □

The control parameters should be set to guarantee both the performance and the stability of the closed-loop system. In particular, better tracking performance can be achieved by tuning $\boldsymbol{K}_s$ and $\boldsymbol{K}_z$ up, and smaller steady-state error can be achieved with higher α_θ. Faster convergence can be obtained by increasing β_n and β_m, but it should not be set too large such that (2.88) and (2.89) are satisfied. Then, (2.90) is satisfied by setting $\boldsymbol{K}_\theta$ large, and (2.91) and (2.92) can be satisfied by setting k_r and k_a large. The control parameters should be fine-tuned together. Note that parameters that are too large should be avoided. Otherwise, the control input will be saturated, and the measurement noises may be amplified.

2.3.4 Results

The proposed control method was implemented in the experimental setup of an upper-limb rehabilitation robot, as shown in Fig. 2.19. The upper-limb rehabilitation robot mainly consists of a robot link driven by a SEA and a dSPACE control system. The subject is asked to put their forearm in the upright vertical position, and to make a cyclical flexion/extension movement around this position. A force sensor (ATI) is mounted on the end of the robot link and used to measure the interaction force. The SEA introduced in Fig. 2.1is employed. Specifically, a DC brushless motor (EC 4-pole, Maxon) is used in the SEA, due to its lightweight (0.175 kg) and low moment of inertia. The ball screw (Eichenberger Gewinde AG) has a pitch of 2 mm/rev and

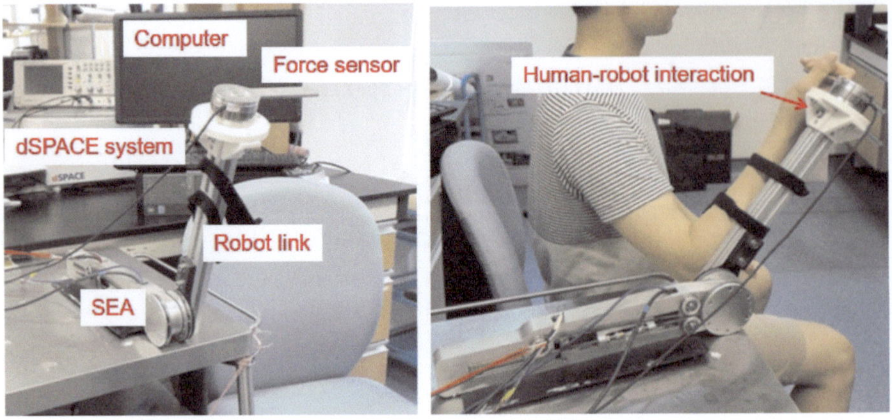

(a) (b)

Fig. 2.19 An upper-limb compliantly-driven exoskeleton. **a** The system consists of the SEA, the robot link, the force sensor, the dSPACE control system, and the computer; **b** The subject is asked to put his forearm in the upright vertical position

Table 2.7 Desired impedance parameters

Impedance					Backstepping						ILC	
M_d	C_d	K_d	Γ	Λ	K_z	α_θ	K_s	K_θ	k_r	k_a	β_m	β_n
1	15.4	6	15	0.4	20	2	2	0.004	0.02	0.04	0.2	0.04

can output over 1100 N force. The stiffness of the spring is K=0.0024 N · m/rad. The resolution of the rotary encoder is 1024 lines/rev, and the total mass of the actuator is less than 0.85 kg.

The desired trajectory was specified as a sinusoid as: $q_d = 0.26\sin(\pi t)$ rad, where q_d denotes the desired joint angle. The parameters used in the proposed impedance controller were configured as summarized in Table 2.7.

In the experiments, the iteration period was set as $t_f = 50$ s. The results of the iterative control terms m, n are shown in Fig. 2.20, which are bounded and converge to cancel out disturbances as the iterations continue. The position errors and the impedance errors after different iterations ($k = 3$, $k = 15$, $k = 30$) are shown in Figs. 2.21 and 2.22 respectively. When $k = 3$, the position error is around 0.003 rad, and the impedance error is around 0.02 Nm. When $k = 30$, the position error reduces to 0.001 rad (no interaction forces exerted) and the impedance error reduces to 0.01 Nm. That is, both the position error and the impedance error converge to zero as the iteration goes on. During the experiments, the subject exerted forces on the robot by holding the robot link at a certain position (e.g., $t \approx 14 \sim 20$ s in Fig. 2.22a and $t \approx 5 \sim 14$ s in Fig. 2.22c). While the robot joint angle is drifted away from the sinusoid trajectory because of the interaction force, the convergence of the impedance

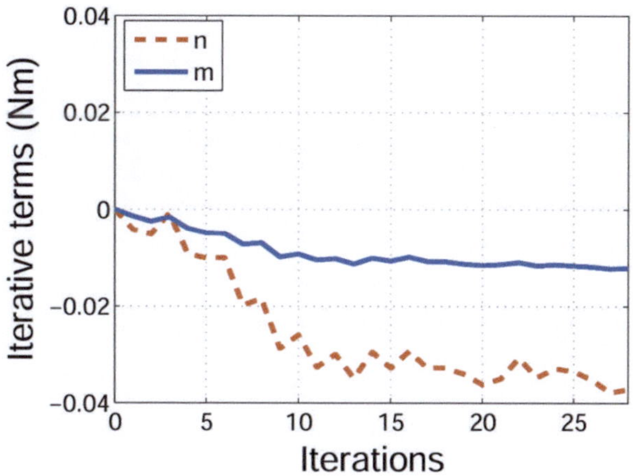

Fig. 2.20 Iterative control terms m and n

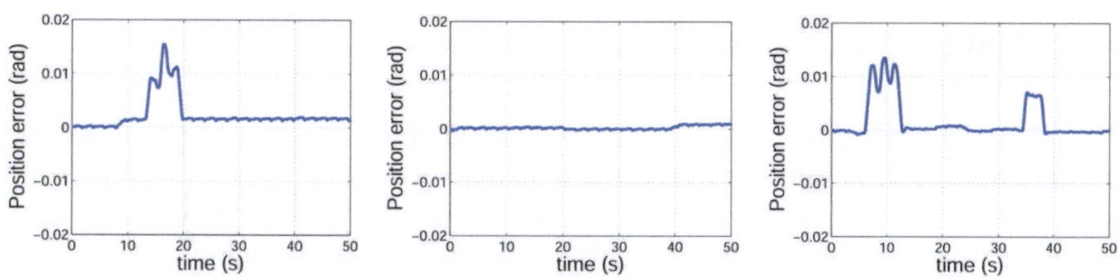

(a) (b) (c)

Fig. 2.21 Position error Δq: **a** 3 iterations; **b** 15 iterations; **c** 30 iterations

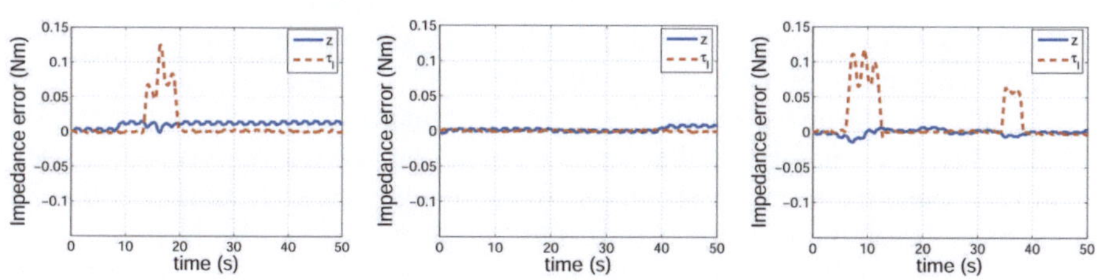

(a) (b) (c)

Fig. 2.22 Impedance vector z: **a** 3 iterations; **b** 15 iterations; **c** 30 iterations

error to zero is guaranteed, which shows the realization of the impedance control task. The control inputs after different iterations are shown in Fig. 2.23. Therefore, the experimental results validate the performance of the proposed controller.

From Figs. 2.20 and 2.23, it is seen that the ratio of $\boldsymbol{m}$ over $\boldsymbol{\theta}_d$ after different iterations (i.e., 3, 15, 30) is 0.0006, 0.0043, 0.008 respectively, while the ratio of $\boldsymbol{n}$ over $\boldsymbol{\tau}$ after different iterations is 0.0064, 0.1616, 0.2803 respectively. That is, both

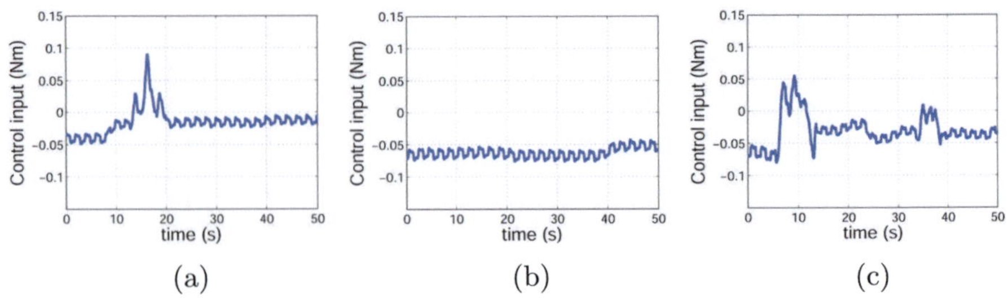

Fig. 2.23 Control input τ: **a** 3 iterations; **b** 15 iterations; **c** 30 iterations

ratios increase as the iteration continues, indicating the iterative feedforward control is playing a more dominant role in the proposed controller.

2.4 Summary and Notes

Compliant actuators have been extensively adopted in robotic systems that engage in physical human-robot interaction. These actuators connect robot joints to the actuation system through elastic elements, resulting in a high-order dynamic system. While the compliant design enhances interaction safety and adaptability, it also significantly complicates control design due to the system's inherent nonlinearities and high-order dynamics.

Several control methodologies have been proposed to address challenges in human-robot interaction. Proxy-based approaches [24] are commonly employed to allow robots to follow desired trajectories with smooth and damped responses to unexpected external impacts. Considering one representative application of compliantly-driven robots is to deal with physical interaction in rehabilitation training, an important strategy in that application is the "Assist-As-Needed" (AAN) paradigm, where the robot adapts its role dynamically—either leading or following the human—based on real-time performance assessments [25, 26]. However, these methods often simplify the overall dynamics and fail to account for the high-order characteristics of compliantly-driven robots. Additionally, some approaches rely on hard switching between multiple controllers, potentially leading to robot chattering that compromises both safety and performance [27].

In this chapter, we present a novel control framework for compliantly driven robots designed using the backstepping method. Backstepping is a recursive control design strategy that constructs a stabilizing control law by systematically addressing the system's higher-order dynamics through virtual control variables. This approach not only facilitates the management of complex nonlinearities but also inherently ensures system stability during the design process.

Building on the backstepping principle, this chapter introduces two innovative control strategies. The first is an adaptive control method that incorporates two distinct operational modes: a robot-in-charge mode and a human-in-charge mode. A key feature of this controller is its seamless and stable transition between these modes, ensuring a continuous and safe interaction experience. The second strategy leverages iterative learning control, which employs feedforward learning to compensate for unknown disturbances within the system. This approach enhances the precision of the desired impedance model by iteratively refining the control input based on repeated tasks. Together, these contributions demonstrate the effectiveness of backstepping-based control for compliantly-driven robots in addressing the challenges of high-order dynamics, safety, and adaptability in human-robot interaction.

References

1. F. Petit, A. Dietrich, A. Albu-Schaffer, Generalizing torque control concepts: Using well-established torque control methods on variable stiffness robots. IEEE Robot. Autom. Mag. **22**(4), 37–51 (2015)
2. S. Arimoto, *Control Theory of Non-Linear Mechanical Systems: A Passivity-Based and Circuit-Theoretic Approach* (Oxford University Press, London, U.K., 1996)
3. J.-J.E. Slotine, W. Li et al., *Applied Nonlinear Control*, vol. 199 (Prentice hall Englewood Cliffs, NJ, 1991)
4. I. Kanellakopoulos, P.V. Kokotovic, A.S. Morse, Systematic design of adaptive controllers for feedback linearizable systems, in *1991 American Control Conference* (IEEE, 1991), pp. 649–654
5. A. Astolfi, D. Karagiannis, R. Ortega, *Nonlinear and Adaptive Control with Applications*, vol. 187 (Springer, 2008)
6. R. Sepulchre, M. Jankovic, P.V. Kokotovic, *Constructive Nonlinear Control* (Springer Science & Business Media, 2012)
7. C.C. Cheah, X. Li, *Task-Space Sensory Feedback Control of Robot Manipulators*, vol. 73 (Springer, 2015)
8. Y. Pan, Y. Zhou, T. Sun, M.J. Er, Composite adaptive fuzzy h^{∞} tracking control of uncertain nonlinear systems. Neurocomputing **99**, 15–24 (2013)
9. Y. Pan, H. Yu, Composite learning from adaptive dynamic surface control. IEEE Trans. Autom. Control **61**(9), 2603–2609 (2015)
10. H. Yu, S. Huang, G. Chen, Y. Pan, Z. Guo, Human-robot interaction control of rehabilitation robots with series elastic actuators. IEEE Trans. Robot. **31**(5), 1089–1100 (2015)
11. J. Zhang, C.C. Cheah, Passivity and stability of human-robot interaction control for upper-limb rehabilitation robots. IEEE Trans. Robot. **31**(2), 233–245 (2015)
12. D. Luenberger, Observers for multivariable systems. IEEE Trans. Autom. Control **11**(2), 190–197 (1966)
13. D. Luenberger, An introduction to observers. IEEE Trans. Autom. Control **16**(6), 596–602 (1971)
14. C. Liu, C.C. Cheah, J.J.E. Slotine, Adaptive task-space regulation of rigid-link flexible-joint robots with uncertain kinematics. Automatica **44**(7), 1806–1814 (2008)
15. W.-S. Lu, Q.-H. Meng, Impedance control with adaptation for robotic manipulations. IEEE Trans. Robot. Autom. **7**(3), 408–415 (1991)
16. Y. Maeda, T. Hara, T. Arai, Human-robot cooperative manipulation with motion estimation, in *Proceedings 2001 IEEE/RSJ International Conference on Intelligent Robots and Systems.*

Expanding the Societal Role of Robotics in the the Next Millennium (Cat. No. 01CH37180), vol. 4 (IEEE, 2001), pp. 2240–2245
17. N. Hogan, Impedance control - an approach to manipulation. i - theory. ii - implementation. iii - applications. J. Dyn. Syst. Meas. Control **107**(1), 1–24 (1985)
18. C.C. Cheah, D.W. Wang, Learning impedance control for robotic manipulators (1998)
19. D.W. Wang, C.C. Cheah, An iterative learning- control scheme for impedance control of robotic manipulators. Int. J. Robot. Res. **17**, 1091–1104 (1998)
20. X. Li, C.C. Cheah, Adaptive neural network control of robot based on a unified objective bound. IEEE Trans. Control Syst. Technol. **22**(3), 1032–1043 (2013)
21. X. Li, Y. Pan, G. Chen, H. Yu, Adaptive human–robot interaction control for robots driven by series elastic actuators. IEEE Trans. Robot. **33**, 169–182 (2017)
22. S. Arimoto, S. Kawamura, F. Miyazaki, Better operation of robots by learning. J. Robot. Syst. **1**, 123–140 (1984)
23. D.A. Bristow, M. Tharayil, A.G. Alleyne, A survey of iterative learning control. IEEE Control Syst. Mag. **26**(3), 96–114 (2006)
24. R. Kikuuwe, S. Yasukouchi, H. Fujimoto, M. Yamamoto, Proxy-based sliding mode control: a safer extension of pid position control. IEEE Trans. Robot. **26**(4), 670–683 (2010)
25. J.F. Veneman, R. Kruidhof, E.E.G. Hekman, R. Ekkelenkamp, E.H.F. van Asseldonk, H. van der Kooij, Design and evaluation of the lopes exoskeleton robot for interactive gait rehabilitation. IEEE Trans. Neural Syst. Rehabil. Eng. **15**(3), 379–386 (2007)
26. A. Duschau-Wicke, J. von Zitzewitz, A. Caprez, L. Lunenburger, R. Riener, Path control: A method for patient-cooperative robot-aided gait rehabilitation. IEEE Trans. Neural Syst. Rehabil. Eng. **18**(1), 38–48 (2010)
27. P. Beyl, K. Knaepen, S. Duerinck, M. Van Damme, B. Vanderborght, R. Meeusen, D. Lefeber, Safe and compliant guidance by a powered knee exoskeleton for robot-assisted rehabilitation of gait. Adv. Robot. **25**(5), 513–535 (2011)

Chapter 3
Exoskeleton Robots

An exoskeleton robot is a wearable mechanical device designed to augment, assist, or enhance human movement and physical capabilities. These robotic systems typically feature external frameworks that support and replicate the structure and function of the human body, often integrating sensors, actuators, and control systems to enable dynamic interaction with the wearer. That is, the robot assesses the wear's motion intention and provides the appropriate assistance via physical interaction to suit that intention.

As the exoskeleton robot closely and continuously interacts with the wearer, safety is always the main concern. The safety of the wearer is usually realized in two ways. In hardware, many exoskeleton robots adopt compliant actuators to drive the joint [1, 2], and then its dynamic model has a similar structure as the previous chapter. In software, the interaction force is explicitly considered in the controller development (e.g., a dynamic relationship between force and position is formulated in impedance control) [3, 4]. After safety is achieved, another important issue is to promptly and accurately estimate the human motion intention, which is the directional guidance for the robot and also the prior to its effective assistance.

This chapter considers the compliantly-driven exoskeleton robot and introduces the method of intention estimation and interaction control for it. First, a dual-channel structure is constructed for the estimation of wearer motion intention; Channel 1 is based on the interaction model and then outputs the estimated desired position using adaptive NNs; Channel 2 is based on the GMM and outputs the predicted position of the human limb; Hence, such a structure achieves both the fast response and the high accuracy. Second, a multi-model interaction control scheme is proposed for exoskeleton robots; three control modes are smoothly integrated into the controller, where the *robot-assisted* mode allows the human to exert voluntary efforts within a desired region, the *robot-dominant* mode is activated to correct the movement of human when the robot is outside the region, and the *safety-stop* mode is to stop the

X. Li and H. Yu, *Compliantly-Driven Robots with Application to Exoskeletons*,
https://doi.org/10.1007/978-981-96-9544-7_3

robot whenever the interaction force is too large which may result in injuries; By using the regional position and force feedback, the proposed controller achieves the paradigm of "AAN" and also guarantees the safety of human.

3.1 Existing Works

Exoskeletons can be broadly categorized into upper-limb and lower-limb exoskeletons based on the region of the body they are designed to assist. Lower-limb exoskeletons primarily support body weight or assist leg movement during walking tasks to achieve effective assistance. These exoskeletons are further classified according to the key joints they target, including ankle [5], knee [6], hip [7], and multi-joint designs [8], as illustrated in Fig. 3.1. Ankle exoskeletons are lightweight and provide primary propulsion during walking, making them suitable for a wide range of daily applications. Knee exoskeletons often require additional mechanisms to connect with the foot to compensate for inadequate support when worn directly. However, this design may hinder the natural force generation of the lower leg, potentially affecting gait naturalness. Hip exoskeletons are relatively easy to wear but offer limited assistance. Multi-joint lower-limb exoskeletons, incorporating at least the hip and

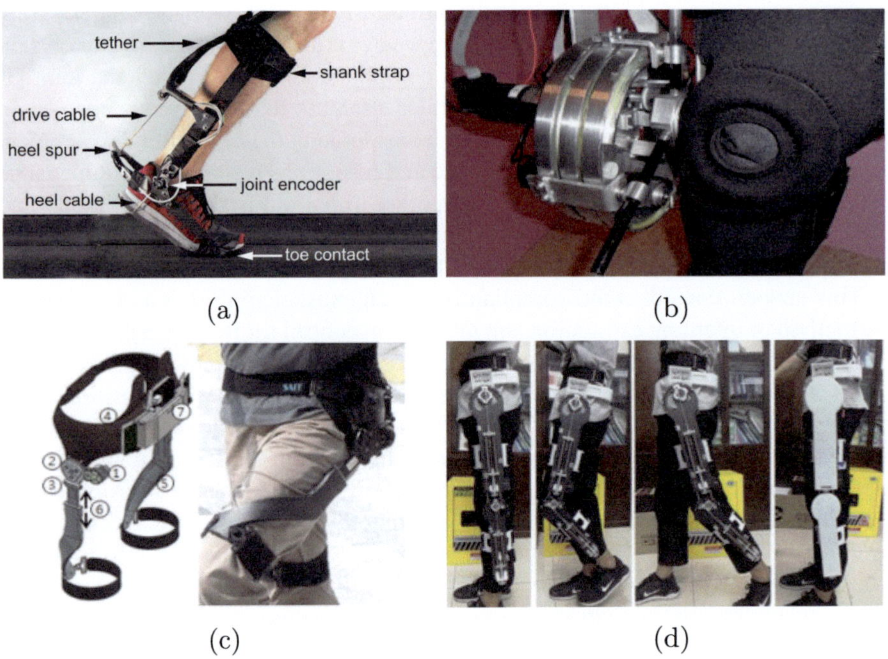

Fig. 3.1 Different types of lower-limb exoskeleton: **a** Ankle Exoskeleton [5] **b** Knee Exoskeleton [6] **c** Hip Exoskeleton [7] **d** Whole-leg Exoskeleton [8]

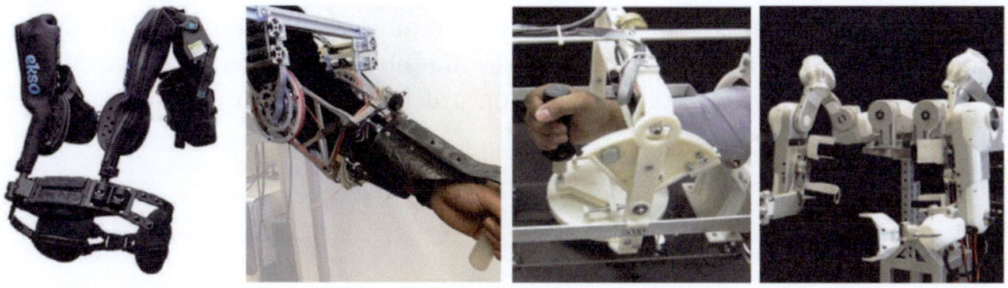

(a) (b) (c) (d)

Fig. 3.2 Different types of upper-limb exoskeleton: **a** Shoulder Exoskeleton [9] **b** Elbow Exoskeleton [10] **c** Wrist Exoskeleton [11] **d** Whole-limb Exoskeleton [12]

Fig. 3.3 Other types of exoskeleton: **a** Upper-limb Exosuit [13] **b** Low-limb Exosuit [14] **c** Thumb Exoskeleton [15]

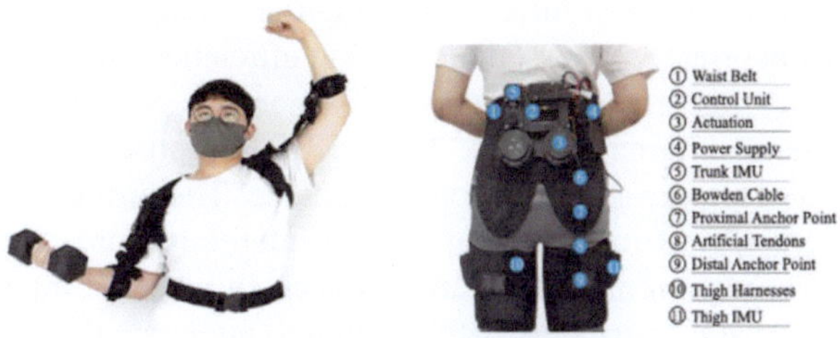

(a) (b)

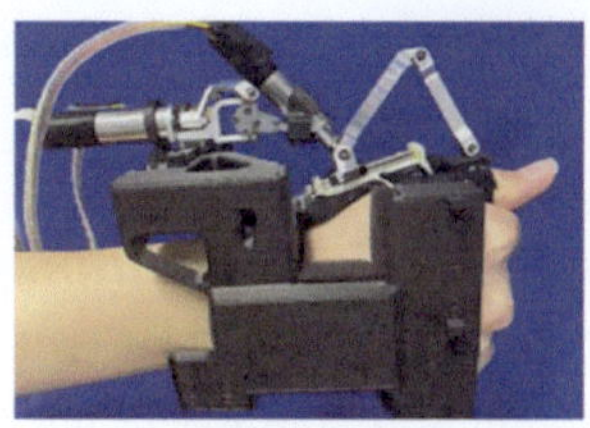

(c)

knee joints, provide a complete walking trajectory, enhancing interaction between the device and the wearer.

Unlike lower-limb exoskeletons, which focus on walking assistance, upper-limb rehabilitation demands a broader range of motion. Thus, upper-limb exoskeleton robots must possess sufficient degrees of freedom to accommodate complex and varied movements. Multi-degree-of-freedom designs introduce redundancy that significantly enhances patient comfort [16]. As shown in Fig. 3.2, most current upper-limb exoskeletons are designed for single-joint rehabilitation, targeting joints such as the shoulder [9], elbow [10], or wrist [11]. In contrast, multi-joint upper-limb exoskeletons can cater to patients with injuries in multiple joints. By assisting the entire arm's movement, these exoskeletons help rebuild neural connections in the joints and improve motor coordination. In addition, there are exoskeleton designs tailored to specific needs. As shown in Fig. 3.3, an exosuit, as a lightweight wearable robotic design, can provide assistance while minimizing the impact of the device's weight on movement [13, 14]. Similarly, a thumb exoskeleton [15] is specifically designed to align with the biological motion structure of the thumb, offering effective assistance.

3.2 Intention Estimation

The exoskeleton robot works in an intention-driven way. However, issues regarding modeling and predicting human motion intention limit the performance of the robot and present challenges. Due to its uncertain and stochastic nature [17–19], in most physical interaction situations, human motion intention may be inaccurate or incorrectly assessed, degrading efficiency and even compromising safety. At the same time, the wearable robot is unable to provide effective assistance to match the intended human motion. This section proposes a new intention estimation method and the corresponding control scheme. The proposed structure for intention estimation will estimate human motion intention more accurately. It explores the immediate intentional information hidden in physical interaction and also considers the stochastic distribution of intention to attenuate the influence of random noise. With such estimated human motion intention, the proposed controller allows the robot to provide fast, effective, and safe assistance.

3.2.1 *Human Motion Intention During pHRI*

Because the wearable robot closely interacts with the human subject, the dynamic model of the human limb must be taken into consideration to estimate human motion intention during physical interaction. By referring to the general physiological system [20] and noting that the damper and spring terms are dominant [21, 22], the dynamic model of the human limb can be given as

$$\boldsymbol{D}_h(\dot{\boldsymbol{q}}_h - \dot{\boldsymbol{q}}_d) + \boldsymbol{K}_h(\boldsymbol{q}_h - \boldsymbol{q}_d) = \boldsymbol{\tau}_e, \tag{3.1}$$

where $\boldsymbol{q}_h \in \Re^n$ is the vector of the joint angles of the human limb; $\boldsymbol{D}_h, \boldsymbol{K}_h \in \Re^{n\times n}$ denote the parameters of damping and stiffness, respectively, which varies between human subjects and is thus hard to obtain exactly, and $\boldsymbol{q}_d \in \Re^n$ is the desired position of the human limb, which reflects human motion intention, and $\boldsymbol{\tau}_e \in \Re^n$ denotes the interaction torque.

From (3.1), it is obtained that

$$\boldsymbol{q}_d = \boldsymbol{q}_h - \boldsymbol{K}_h^{-1}(\boldsymbol{\tau}_e - \boldsymbol{D}_h(\dot{\boldsymbol{q}}_h - \dot{\boldsymbol{q}}_d)), \tag{3.2}$$

which specifies the *immediate intentional information* during HRI (i.e., $\boldsymbol{q}_d$), including the information of both the interaction torque $\boldsymbol{\tau}_e$ and the joint angles of the human limb $\boldsymbol{q}_h$.

Note that the joint angles of the human limb are subject to random variations because of uncertainty in human motion intention and sensory noise. When a human wears a robot to carry out repeated movements, variations also gradually arise due to fatigue or adaptation. Therefore, modeling the limb positions as random variables is more effective [23]. Accordingly, the corresponding probability density function (PDF) can be modeled using the Gaussian mixture model (GMM) as

$$p(\boldsymbol{q}_h) = \sum_{m=1}^{M} \pi_m \mathcal{N}(\boldsymbol{q}_h | \boldsymbol{\mu}_m, \boldsymbol{\sigma}_m), \tag{3.3}$$

where $p(\cdot)$ represents the PDF, M is the number of the mixture, $\mathcal{N}(\cdot)$ represents the normal distribution with the mean vector $\boldsymbol{\mu}_m$ and the covariance matrix $\boldsymbol{\sigma}_m$, and π_m $(0 \leq \pi_m \leq 1)$ is the mixing coefficient, which is set such that $\sum_{m=1}^{M} \pi_m = 1$.

Then, the future position of the human limb can be predicted based on (3.3) and specified as the conditional probability of joint angles with respect to past positions, thereby representing the *long-term effect of human motion intention*. As such, this chapter models human motion intention in a wearable robotic system from two aspects: the immediate desired position of the human limb and its long-term prediction.

As mentioned, this chapter considers an exoskeleton robot driven by compliant actuators, and its dynamic model can be reproduced as [3, 24]:

$$\boldsymbol{M}(\boldsymbol{q})\ddot{\boldsymbol{q}} + \boldsymbol{C}(\dot{\boldsymbol{q}}, \boldsymbol{q})\dot{\boldsymbol{q}} + \boldsymbol{g}(\boldsymbol{q}) = \boldsymbol{K}(\boldsymbol{\theta} - \boldsymbol{q}) + \boldsymbol{\tau}_e + \boldsymbol{\tau}_s, \tag{3.4}$$

$$\boldsymbol{B}\ddot{\boldsymbol{\theta}} + \boldsymbol{K}(\boldsymbol{\theta} - \boldsymbol{q}) = \boldsymbol{u}, \tag{3.5}$$

where $\boldsymbol{q} \in \Re^n$ denotes the vector of the robot joint angles; $\boldsymbol{\theta} \in \Re^n$ is the vector of the motor rotor shaft positions; $\boldsymbol{M}(\boldsymbol{q}) \in \Re^{n\times n}$, $\boldsymbol{C}(\dot{\boldsymbol{q}}, \boldsymbol{q})\dot{\boldsymbol{q}} \in \Re^n$, and $\boldsymbol{g}(\boldsymbol{q}) \in \Re^n$ denote the inertia matrix, centripetal and Coriolis torque, and gravitational torque of the robot, respectively, $\boldsymbol{\tau}_s \in \Re^n$ represents the disturbance, $\boldsymbol{K} \in \Re^{n\times n}$ denotes the stiffness matrix, $\boldsymbol{B} \in \Re^{n\times n}$ is the inertia matrix of the actuator, and $\boldsymbol{u} \in \Re^n$ is the control input exerted on the actuator. An illustration of the compliantly-driven robot joint is shown in Fig. 3.4. Equation (3.4) represents the rigid-joint system, and Eq. (3.5) represents the compliant actuator system.

Some important properties of the dynamic model described by (3.4) and (3.5) are listed as follows [25, 26]:

(1) the matrix $\boldsymbol{M}(\boldsymbol{q})$ is bounded, symmetric, and positive definite;
(2) $\dot{\boldsymbol{M}}(\boldsymbol{q}) - 2\boldsymbol{C}(\dot{\boldsymbol{q}}, \boldsymbol{q})$ is skew-symmetric, in the sense that $\boldsymbol{a}^T(\dot{\boldsymbol{M}}(\boldsymbol{q}) - 2\boldsymbol{C}(\dot{\boldsymbol{q}}, \boldsymbol{q}))\boldsymbol{a} = 0$, where $\boldsymbol{a} \in \Re^n$ is a non-zero vector;
(3) the matrices $\boldsymbol{B}$ and $\boldsymbol{K}$ are constant, diagonal, and positive definite; and
(4) the left side of (3.4) can be parameterized as

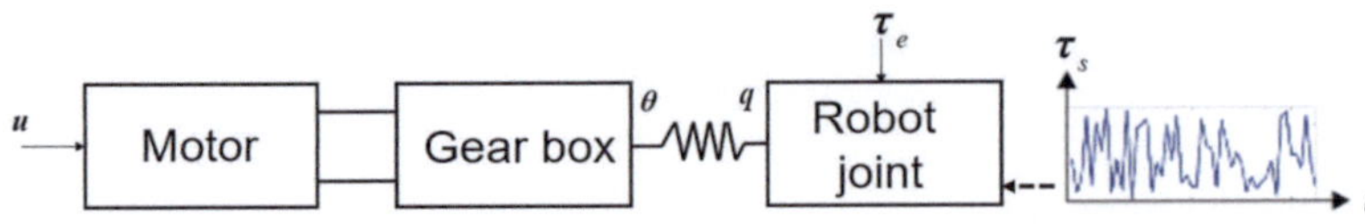

Fig. 3.4 An illustration of the compliantly-driven robot joint, where $\boldsymbol{q}$ is the vector of the joint angles, $\boldsymbol{\theta}$ is the vector of the robot shaft positions, $\boldsymbol{\tau}_e$ denotes the interaction torque, and $\boldsymbol{\tau}_s$ is the vector of external disturbances

$$\boldsymbol{M}(\boldsymbol{q})\ddot{\boldsymbol{q}} + \boldsymbol{C}(\dot{\boldsymbol{q}}, \boldsymbol{q})\dot{\boldsymbol{q}} + \boldsymbol{g}(\boldsymbol{q}) = \boldsymbol{Y}_q(\ddot{\boldsymbol{q}}, \dot{\boldsymbol{q}}, \dot{\boldsymbol{q}}, \boldsymbol{q})\boldsymbol{\psi}_q, \tag{3.6}$$

where $\boldsymbol{Y}_q(\cdot) \in \Re^{n \times n_q}$ represents the regressor matrix, n_q is related to the regressor structure, and $\boldsymbol{\psi}_q = [\psi_{q1}, \ldots, \psi_{qn_q}]^T \in \Re^{n_q}$ is the vector of the dynamic parameters.

The stiffness $\boldsymbol{K}$ and the inertia matrix $\boldsymbol{B}$ of the actuator can be obtained with sufficient accuracy (e.g., from the datasheet). However, the dynamic parameters ($\boldsymbol{M}(\boldsymbol{q})$, $\boldsymbol{C}(\dot{\boldsymbol{q}}, \boldsymbol{q})$, and $\boldsymbol{g}(\boldsymbol{q})$) are usually unknown, hence the exact knowledge of $\boldsymbol{\psi}_q$ is also unknown. The unknown dynamic parameters of the robot will be estimated online by referring to the above property.

In this chapter, the disturbance $\boldsymbol{\tau}_s$ specifies unintended interaction between the robot and other parts of the human body. It is then described using a stochastic process as

$$\boldsymbol{\tau}_s = [\dot{d}_1, \dot{d}_2, \ldots, \dot{d}_n]^T, \tag{3.7}$$

where $d_i(t)$ has the following properties [27, 28]:

(1) $d_i(t)$ is continuous in the parameter t, and $d_i(0) = 0$;
(2) for every $0 \le t_1 < t_2 < t_3 < \cdots < t_n$, the increments $d_i(t_2) - d_i(t_1), d_i(t_3) - d_i(t_2)$, ..., $d_i(t_n) - d_i(t_{n-1})$ are independent random variables, and
(3) for $0 \le s < t < +\infty$, $d_i(t) - d_i(s)$ is a Gaussian random variable with mean 0 and variance $t - s$.

3.2.2 Dual-Channel Structure

This section proposes a new type of variable impedance controller for wearable, compliantly-driven robots, as illustrated in Fig. 3.5, i.e., the robot estimates human motion intention then outputs the varying impedance to suit it. The proposed formulation can guarantee safe interaction in both the hardware and software. In the hardware, the compliant actuator is deployed to make the robot tolerant to impacts and to store excessive energy. In the software, impedance control is implemented to deal with the physical interaction, instead of purely position or force.

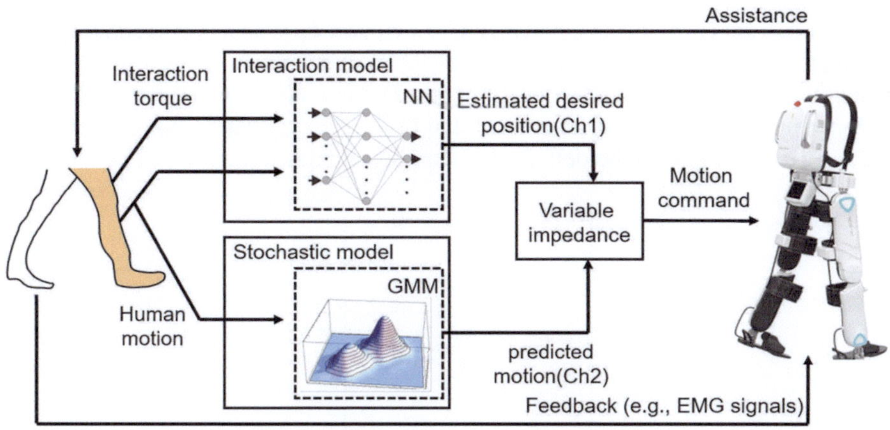

Fig. 3.5 The overall structure of an intention-driven exoskeleton robot. The dual-channel estimator receives the body information (e.g., interaction torque, human motion, etc.) then outputs the estimated human motion intention, which contains short-term and long-term intention (represented by $\hat{\boldsymbol{q}}_d$ and $\hat{\boldsymbol{q}}_h$ respectively). The control algorithm receives the estimated intention and then varies the robot impedance accordingly. The performance of the robot is assessed by referring to the feedback from human subjects, for example, EMG signals from the human body

First, as seen from (3.2), (3.3), and (3.7), the estimation of human motion intention should address: the unknown parameters of the human limb, stochastic distribution, and random disturbance. The first two issues are to be addressed under the dual-channel structure of this section, while the last one is to be addressed in the development of the controller.

Channel 1: This channel is based on the interaction model and then outputs the estimated desired position using an adaptive NNs. Note that (3.1) can be written as

$$\boldsymbol{A}\boldsymbol{y} = \boldsymbol{\tau}_e, \tag{3.8}$$

where

$$\boldsymbol{y} \triangleq \begin{bmatrix} \dot{\boldsymbol{q}}_h - \dot{\boldsymbol{q}}_d \\ \boldsymbol{q}_h - \boldsymbol{q}_d \end{bmatrix} \in \Re^{2n}, \tag{3.9}$$

$$\boldsymbol{A} \triangleq [\boldsymbol{D}_h, \boldsymbol{K}_h] \in \Re^{n\times 2n}, \tag{3.10}$$

That is, $\boldsymbol{y}$ is an extended vector and $\boldsymbol{A}$ is a matrix including the dynamic parameters of human limb. Because the dynamic parameters vary between different human beings with distinct gender, height, and weight, the exact information about $\boldsymbol{A}$ is hard to obtain.

From (3.8), we have

$$\boldsymbol{y} = \boldsymbol{A}^{+}\boldsymbol{\tau}_e, \tag{3.11}$$

where $A^{+} \triangleq A^{T}(AA^{T})^{-1}$, such that $AA^{+} = I_n$ and I_n is an $n \times n$ identity matrix. When the dynamic parameters of the human limb are unknown, the exact information about A^{+} is also unknown. Note that both D_h and K_h in (3.1) can be denoted as $D_h = \text{diag}(d_{h1}, \ldots, d_{hn})$, and $K_h = \text{diag}(k_{h1}, \ldots, k_{hn})$ where $d_{hi}, k_{hi} > 0$, $(i = 1, 2, \ldots, n)$. Then, from the definition, it can be calculated that

$$A^{+} = \begin{bmatrix} \frac{d_{h1}}{d_{h1}^2+k_{h1}^2} & 0 & \cdots & 0 \\ \vdots & \vdots & \ddots & \vdots \\ 0 & 0 & \cdots & \frac{d_{hn}}{d_{hn}^2+k_{hn}^2} \\ \frac{k_{h1}}{d_{h1}^2+k_{h1}^2} & 0 & \cdots & 0 \\ \vdots & \vdots & \ddots & \vdots \\ 0 & 0 & \cdots & \frac{k_{hn}}{d_{hn}^2+k_{hn}^2} \end{bmatrix} \in \Re^{2n \times n},$$

which always exists.

Then, the signal y is estimated as

$$y = \hat{W}\phi(\dot{q}_h, q_h, \tau_e) + E, \tag{3.12}$$

where $\hat{W} \in \Re^{2n \times n_w}$ denotes the estimated weight of NNs; $\phi(\cdot) \in \Re^{n_w}$ denotes the vector of neurons, where the output of the neuron function is bounded, n_w is the number of neurons; and E denotes the approximation error, which decreases when the number of neurons increases.

In this chapter, the exoskeleton robot is controlled to assist humans in such a way that the interaction torques are minimized. Hence, the cost function can be proposed as:

$$F = \frac{1}{2}\sum_{i=1}^{n} \tau_{ei}^2, \tag{3.13}$$

where τ_{ei} is the ith element of τ_e. A minimized cost function implies the robot effectively assists the human and also $q_h \to q_d$ (i.e., the current human limb tracks the intended motion). From (3.8), it is obtained that $\tau_{ei} = A_i y = \sum_{j=1}^{2n} A_{ij} y_j$, where A_i is the ith row of the matrix A, A_{ij} is an element, and y_j is the jth element. From (3.12), it is seen that the cost function is also related to the NN's weight $\hat{W}$.

Then, the weight can be updated using the gradient descent algorithm [21]. Specifically, the jth column of the weight matrix is updated as

$$
\begin{aligned}
\dot{\hat{\mathbf{W}}}_j^T &= -\gamma_j \frac{\partial F}{\partial \hat{\mathbf{W}}_j^T} \\
&= -\sum_{i=1}^{n} \gamma_j \frac{\partial F}{\partial \tau_{ei}} \frac{\partial \tau_{ei}}{\partial y_j} \frac{\partial y_j}{\partial \hat{\mathbf{W}}_j^T} \\
&= -\sum_{i=1}^{n} \gamma_j \tau_{ei} \frac{\partial \tau_{ei}}{\partial y_j} \phi(\dot{\boldsymbol{q}}_h, \boldsymbol{q}_h, \boldsymbol{\tau}_e) \\
&= -\sum_{i=1}^{n} \underbrace{\gamma_j \frac{\partial \tau_{ei}}{\partial y_j}}_{\boldsymbol{L}_j} \phi(\dot{\boldsymbol{q}}_h, \boldsymbol{q}_h, \boldsymbol{\tau}_e) \tau_{ei},
\end{aligned} \tag{3.14}
$$

where γ_j is a positive constant, and γ_j and $\frac{\partial \tau_{ei}}{\partial y_j}$ are combined as the matrix $\boldsymbol{L}_j \in \Re^{n_w \times n_w}$, which is positive-definite. Thus, the update law can be given as

$$
\dot{\hat{\mathbf{W}}}_j^T = -\sum_{i=1}^{n} \boldsymbol{L}_j \phi(\dot{\boldsymbol{q}}_h, \boldsymbol{q}_h, \boldsymbol{\tau}_e) \tau_{ei}. \tag{3.15}
$$

The update of NN's weight using (3.15) does not require the pre-training phase. The intention estimation using NNs is independent of the dynamic parameters of the human limb; thus, this method can be applied to different human subjects.

Channel 2: This channel is based on the GMM and outputs the predicted position of the human limb. This is achieved by representing the human motion in each time step (t) as one cluster and updating the model (3.3) as follows:

(1) Using the total amount of data stored (i.e., $\boldsymbol{q}_h$) in one time period as the mixture number (M). The initial values of GMM (i.e., $\boldsymbol{\mu}_m$ and $\boldsymbol{\sigma}_m$) are set based on the data collected. The initial value of $\boldsymbol{\pi}_m$ is the same for different normal distributions, meaning that all Gaussian distributions are evenly distributed at the start.
(2) Using the subject's history and current joint position, the PDF of the joint's position in the future is estimated based on the initialized GMM.
(3) The GMM is updated using the online EM algorithm [29] by exploring the stored sample data after updating.
(4) Repeat steps (2) and (3).

In fact, the above steps describe auto-regression. Specifically, the position of the human limb $\boldsymbol{q}_h$ at time $t+1$ can be given by the end of each iteration as

$$
\hat{\boldsymbol{q}}_h^{t+1} = \begin{cases} \mathbb{E}[\hat{\boldsymbol{q}}_h^{t+1} | \boldsymbol{q}_h^0, \ldots, \boldsymbol{q}_h^t] & t \geq M \\ \boldsymbol{q}_h^t & 0 \leq t < M \end{cases} \tag{3.16}
$$

where $\mathbb{E}\{\cdot\}$ represents the expectation of the stochastic model. The proposed method for predicting the long-term effect of human motion intention is robust to sensor noise

Algorithm 3.1 Online Estimation with a Modified EM Algorithm

Require: Sequentially collected data $\boldsymbol{q}_h$, initialized parameters $\boldsymbol{\mu}_m$, $\boldsymbol{\sigma}_m$, and π_m.
Ensure: Estimated joint position $\hat{\boldsymbol{q}}_h$.
  $\boldsymbol{q}_h^t$ is the current position, $\hat{\boldsymbol{q}}_h^{t+1}$ is the predicted position
  while $N_i = M$ **do**
    N_i: Number of collected data points $\boldsymbol{q}_h$
    E-step: evaluate the PDF using the GMM in Eq. (3.3) under current parameter values.
    M-step: find the maximum PDF value and assign the corresponding position as $\hat{\boldsymbol{q}}_h^{t+1}$.
    Update the mean $\boldsymbol{\mu}_m$ and variance $\boldsymbol{\sigma}_m$.
    Update the input $\boldsymbol{q}_h^{t+1}$ using the FIFO rule.
  end while

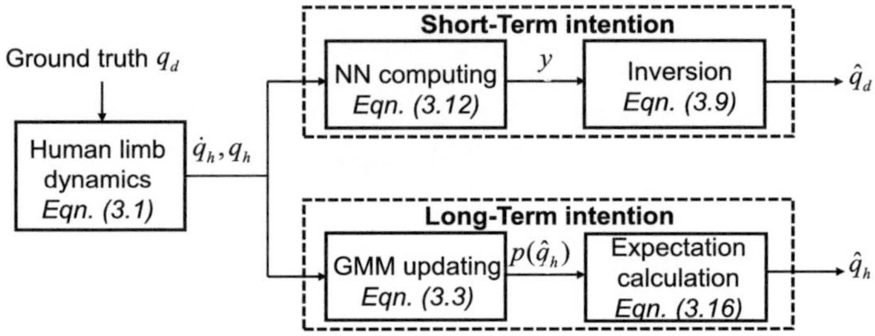

Fig. 3.6 A dual-channel structure for estimation of human motion intention

and uncertainty of human intent [19, 23]. In actual implementations, the proposed method can be realized by referring to Algorithm 3.1.

The output of both channels is used to formulate the variable impedance model. Such a formulation not only enables the robot to respond quickly to the human's immediate intention but also guarantees prediction accuracy subject to noise and uncertainty. The workflow of the proposed structure is shown in Fig. 3.6.

3.2.3 Task Planning

In impedance control, the control objective is specified as a desired impedance model, as in [30]

$$\boldsymbol{M}_d(\ddot{\boldsymbol{q}}-\ddot{\boldsymbol{q}}_r)+\boldsymbol{C}_d(\dot{\boldsymbol{q}}-\dot{\boldsymbol{q}}_r)+\boldsymbol{K}_d(\boldsymbol{q}-\boldsymbol{q}_r)=\boldsymbol{\tau}_e, \tag{3.17}$$

where $\boldsymbol{M}_d$, $\boldsymbol{C}_d$, and $\boldsymbol{K}_d \in \Re^{n\times n}$ denote the desired inertia, damping, and stiffness matrices, respectively, and $\boldsymbol{q}_r \in \Re^n$ is a reference trajectory for the robot in joint space. The parameters of $\boldsymbol{M}_d$, $\boldsymbol{C}_d$, and $\boldsymbol{K}_d$ in conventional impedance control are fixed.

This section introduces a variable impedance model as

$$\boldsymbol{M}_d(\ddot{\boldsymbol{q}}-\ddot{\hat{\boldsymbol{q}}}_d)+\boldsymbol{C}_d(\dot{\boldsymbol{q}}-\dot{\hat{\boldsymbol{q}}}_d)+\boldsymbol{K}_d(\boldsymbol{q}-\hat{\boldsymbol{q}}_d)=\frac{1}{w(\boldsymbol{q},\hat{\boldsymbol{q}}_h)}\boldsymbol{\tau}_e, \tag{3.18}$$

where $w(\boldsymbol{q}, \hat{\boldsymbol{q}}_h)$ is a weighting function. Note that (3.18) can be rewritten as

$$\boldsymbol{M}_d(t)(\ddot{\boldsymbol{q}} - \ddot{\hat{\boldsymbol{q}}}_d) + \boldsymbol{C}_d(t)(\dot{\boldsymbol{q}} - \dot{\hat{\boldsymbol{q}}}_d) + \boldsymbol{K}_d(t)(\boldsymbol{q} - \hat{\boldsymbol{q}}_d) = \boldsymbol{\tau}_e, \tag{3.19}$$

where $\boldsymbol{M}_d(t) \triangleq w(\boldsymbol{q}, \hat{\boldsymbol{q}}_h)\boldsymbol{M}_d, \boldsymbol{C}_d(t) \triangleq w(\boldsymbol{q}, \hat{\boldsymbol{q}}_h)\boldsymbol{C}_d, \boldsymbol{K}_d(t) \triangleq w(\boldsymbol{q}, \hat{\boldsymbol{q}}_h)\boldsymbol{K}_d$, represents the variable impedance parameters, which increase as the weighting function becomes larger, and vice versa.

An example of the weighting function can be given with reference to the difference between the current joint position $\boldsymbol{q}$ and the long-term estimated joint position $\hat{\boldsymbol{q}}_h$ as

$$\begin{aligned} w(\boldsymbol{q}, \hat{\boldsymbol{q}}_h) =& w_L + (w_H - w_L) \\ &\times \frac{(\min(0, (\min(0, h(\boldsymbol{q}, \hat{\boldsymbol{q}}_h)))^N - ((\kappa R)^2 - R^2)^N))^N}{((\kappa R)^2 - R^2)^{N^2}}, \end{aligned} \tag{3.20}$$

where w_H, w_L are positive constants representing the upper and lower bounds of $w(\boldsymbol{q}, \hat{\boldsymbol{q}}_h)$, respectively ($w_H > w_L$), $N \geq 4$ is an even integer, $0 < \kappa < 1$ is a constant which is set small, such that the variation of $w(\boldsymbol{q}, \hat{\boldsymbol{q}}_h)$ is not significant, R is a positive constant which specifies the radius of the region, and $h(\cdot)$ is the function of a region, which is specified as

$$h(\boldsymbol{q}, \hat{\boldsymbol{q}}_h) = ||\boldsymbol{q} - \hat{\boldsymbol{q}}_h||^2 - R^2. \tag{3.21}$$

That is, $w(\boldsymbol{q}, \hat{\boldsymbol{q}}_h) = w_H$ (maximum value) where $h(\boldsymbol{q}, \hat{\boldsymbol{q}}_h) \leq 0$, $w(\boldsymbol{q}, \hat{\boldsymbol{q}}_h) = w_L$ (minimum value) and where $h(\boldsymbol{q}, \hat{\boldsymbol{q}}_h) > 0$. The robot impedance varies according to the weighting function as follows:

(1) When the robot well matches the predicted human motion, $||\boldsymbol{q} - \hat{\boldsymbol{q}}_h|| < R$, the weighting function outputs a high value, such that the robot maintains a high impedance to enhance assistance.
(2) When the robot is far from the predicted human motion (indicating a conflict between the human and the robot), $||\boldsymbol{q} - \hat{\boldsymbol{q}}_h|| > R$, the robot becomes passive with a lower impedance to minimize ineffective assistance.

Note that the vector $\hat{\boldsymbol{q}}_d$ describes the short-term intention, which is treated as the reference trajectory in the variable impedance model (3.19). The vector $\hat{\boldsymbol{q}}_h$ denotes the long-term intention, which is used to construct the weighting function (3.20) in order to adjust the impedance parameters. Such a formulation combines the advantages of the short-term (i.e., faster response) and the long-term intention (i.e., higher accuracy), allowing them to complement one another.

3.2.4 *Intention-Driven Control*

In this section, a new adaptive impedance controller is proposed to control the compliantly-driven robot under the variable impedance model. First, multiplying both sides of (3.18) with $\boldsymbol{M}_d^{-1}$ and rewriting it yields

$$(\ddot{q}-\ddot{\hat{q}}_d)+M_d^{-1}C_d(\dot{q}-\dot{\hat{q}}_d)+M_d^{-1}K_d(q-\hat{q}_d)-M_d^{-1}\frac{1}{w(q,\hat{q}_h)}\tau_e=\mathbf{0}. \tag{3.22}$$

Then, the left side of (3.22) can be expressed in terms of an impedance vector and its time derivative as

$$\begin{aligned}&(\ddot{q}-\ddot{\hat{q}}_d)+M_d^{-1}C_d(\dot{q}-\dot{\hat{q}}_d)+M_d^{-1}K_d(q-\hat{q}_d)-M_d^{-1}\frac{1}{w(q,\hat{q}_h)}\tau_e\\&=\dot{z}+\Gamma z.\end{aligned} \tag{3.23}$$

where z is the impedance vector and can be treated as a low-pass filtered signal [30] with a cut-off frequency Γ, whose boundedness implies the boundedness of the left side of (3.23) at low frequency.

Specifically, the impedance vector is given as [31, 32]

$$\begin{aligned}z&=\dot{q}-\dot{q}_z\\&=\dot{q}-\dot{\hat{q}}_d+\Lambda(q-\hat{q}_d)-\tau_l,\end{aligned} \tag{3.24}$$

where $\dot{q}_z=\dot{\hat{q}}_d-\Lambda(q-\hat{q}_d)+\tau_l$ is a reference vector, and

$$\Lambda+\Gamma=M_d^{-1}C_d, \tag{3.25}$$

$$\Gamma\Lambda=M_d^{-1}K_d, \tag{3.26}$$

$$\dot{\tau}_l+\Gamma\tau_l=M_d^{-1}\frac{1}{w(q,\hat{q}_h)}\tau_e, \tag{3.27}$$

that is, $\Lambda,\Gamma\in\Re^{n\times n}$ are two positive-definite and diagonal matrices, and $\tau_l\in\Re^n$ is the low-pass filtered signal of $\frac{1}{w(q,\hat{q}_h)}\tau_e$.

Next, considering that the dynamic model of the compliantly-driven robot consists of two coupled subsystems, the backstepping approach [3, 31] is proposed for controller development. Specifically, a virtual desired position input θ_d is first proposed to stabilize the rigid-joint subsystem (3.4), then an actual control input u is proposed to stabilize the compliant-actuator subsystem. The backstepping approach is detailed as follows.

The dynamic model of the subsystem (3.4) can be rewritten by taking θ_d into consideration as

$$\begin{aligned}&M(q)\ddot{q}+C(\dot{q},q)\dot{q}+g(q)+Kq\\&=K\theta_d+K\Delta\theta+\tau_e+\tau_s,\end{aligned} \tag{3.28}$$

where $\Delta\theta=\theta-\theta_d$. By using (3.24) and Property (4) in Sect. 3.2.1, Eq. (3.28) can be written as

$$\begin{aligned} &\boldsymbol{M}(\boldsymbol{q})\dot{\boldsymbol{z}}+\boldsymbol{C}(\dot{\boldsymbol{q}},\boldsymbol{q})\boldsymbol{z}+\boldsymbol{Y}_q(\ddot{\boldsymbol{q}}_z,\dot{\boldsymbol{q}}_z,\dot{\boldsymbol{q}},\boldsymbol{q})\boldsymbol{\psi}_q+\boldsymbol{K}\boldsymbol{q}\\ &=\boldsymbol{K}\boldsymbol{\theta}_d+\boldsymbol{K}\Delta\boldsymbol{\theta}+\boldsymbol{\tau}_e+\boldsymbol{\tau}_s. \end{aligned} \tag{3.29}$$

Then, the virtual desired position input can be designed as

$$\boldsymbol{\theta}_d=\boldsymbol{q}-\boldsymbol{K}^{-1}(\boldsymbol{\tau}_e+\boldsymbol{K}_z\boldsymbol{z}-\boldsymbol{Y}_q(\ddot{\boldsymbol{q}}_z,\dot{\boldsymbol{q}}_z,\dot{\boldsymbol{q}},\boldsymbol{q})\hat{\boldsymbol{\psi}}_q), \tag{3.30}$$

where $\boldsymbol{K}_z\in\Re^{n\times n}$ is diagonal and positive-definite, and $\hat{\boldsymbol{\psi}}_q$ represents the estimate of $\boldsymbol{\psi}_q$, which is updated using the following adaptation law

$$\dot{\hat{\boldsymbol{\psi}}}_q = k_1\hat{\boldsymbol{\psi}}_q - \boldsymbol{L}_q\boldsymbol{Y}_q^T(\ddot{\boldsymbol{q}}_z,\dot{\boldsymbol{q}}_z,\dot{\boldsymbol{q}},\boldsymbol{q})\hat{\boldsymbol{\psi}}_q, \tag{3.31}$$

where k_1 is a positive constant and $\boldsymbol{L}_q\in\Re^{n_q\times n_q}$ is a diagonal and positive-definite matrix that regulates the convergence of the estimated parameters.

Substituting (3.30) into (3.29), we obtain the closed-loop equation for the rigid-link subsystem as

$$\boldsymbol{M}(\boldsymbol{q})\dot{\boldsymbol{z}}+(\boldsymbol{C}(\dot{\boldsymbol{q}},\boldsymbol{q})+\boldsymbol{K}_z)\boldsymbol{z}+\boldsymbol{Y}_q(\ddot{\boldsymbol{q}}_z,\dot{\boldsymbol{q}}_z,\dot{\boldsymbol{q}},\boldsymbol{q})\Delta\boldsymbol{\psi}_q = \boldsymbol{K}\Delta\boldsymbol{\theta}+\boldsymbol{\tau}_s, \tag{3.32}$$

where $\Delta\boldsymbol{\psi}_q=\boldsymbol{\psi}_q-\hat{\boldsymbol{\psi}}_q$.

Next, a sliding vector is given for the actuator subsystem as

$$\begin{aligned} \boldsymbol{s} &=\dot{\boldsymbol{\theta}}-\dot{\boldsymbol{\theta}}_r\\ &=\dot{\boldsymbol{\theta}}-\dot{\boldsymbol{\theta}}_d+\alpha(\boldsymbol{\theta}-\boldsymbol{\theta}_d)\\ &=\Delta\dot{\boldsymbol{\theta}}+\alpha\Delta\boldsymbol{\theta}. \end{aligned} \tag{3.33}$$

where α is a positive constant, and $\dot{\boldsymbol{\theta}}_r$ is a reference vector that is defined as

$$\dot{\boldsymbol{\theta}}_r = \dot{\boldsymbol{\theta}}_d - \alpha\Delta\boldsymbol{\theta}. \tag{3.34}$$

Using the sliding vector $\boldsymbol{s}$, the dynamic model of the subsystem (3.5) is expressed as

$$\boldsymbol{B}\dot{\boldsymbol{s}}+\boldsymbol{K}(\boldsymbol{\theta}-\boldsymbol{q})+\boldsymbol{B}\ddot{\boldsymbol{\theta}}_r = \boldsymbol{u}. \tag{3.35}$$

Then, the actual control input can be designed as

$$\boldsymbol{u}=\boldsymbol{K}(\boldsymbol{\theta}-\boldsymbol{q})+\boldsymbol{B}\ddot{\boldsymbol{\theta}}_r-\boldsymbol{K}_s\boldsymbol{s}-\boldsymbol{K}_\theta\Delta\boldsymbol{\theta}, \tag{3.36}$$

where $\boldsymbol{K}_s$, $\boldsymbol{K}_\theta\in\Re^{n\times n}$ are positive-definite matrices.

Substituting (3.36) into (3.35), the closed-loop equation of the compliantly-driven robot is obtained as

$$B\dot{s}+K_s s+K_\theta \Delta\theta=\mathbf{0}. \tag{3.37}$$

While the impedance model varies according to human motion intention, such a transition will not induce impulses in the control input. This is because (1) the slope between the transition is dependent on the parameter κ (3.20), which can be set small to ensure a smooth transition; (2) the impedance model is represented by a low-pass filtered impedance vector z (3.23), and only z (instead of the whole impedance model) is included in the virtual desired position input (3.30), therefore, potential large impulses during the transition stage will not be transferred to the control input; (3) the robot is driven with compliant actuators, which is capable of energy storage and tolerance to large impacts.

3.2.5 Stability Analysis

We can now state the following theorem.

Theorem 3.2.1 *When the proposed adaptive impedance control scheme described by (3.30), (3.31), and (3.36) is applied to the wearable, compliantly-driven robot described by (3.4) and (3.5), the closed-loop system is stable and the boundedness of the impedance vector is guaranteed at a steady state if the control parameters are chosen such that:*

$$z^T(K_z-\frac{1}{2}K)z \geq \eta_1 z^T M(\cdot)z, \tag{3.38}$$

$$\frac{1}{2}k_1 l_q^{-1}||\Delta\psi_q||^2 \geq \eta_2 \Delta\psi_q^T L_q^{-1}\Delta\psi_q, \tag{3.39}$$

$$\Delta\theta^T(\alpha K_\theta-\frac{1}{2}K)\Delta\theta \geq \eta_3 \Delta\theta^T K_\theta \Delta\theta, \tag{3.40}$$

where η_1, η_2, η_3 are positive constants whose selection is related to the control parameters of K_z, k_1, and α, respectively.

Proof The stability of the closed-loop system is analyzed from the stochastic perspective, such that the random disturbance is attenuated and the boundedness of the impedance error is guaranteed. First, by considering (3.32), (3.31), (3.33), and (3.37) together, an augmented system with four discussed systems and one-dimensional Brownian motion is constructed as

$$\begin{bmatrix} \mathrm{d}z \\ \mathrm{d}\Delta\psi_q \\ \mathrm{d}\Delta\theta \\ \mathrm{d}s \end{bmatrix} = \begin{bmatrix} f_z \\ f_\psi \\ f_\theta \\ f_s \end{bmatrix}\mathrm{d}t+\begin{bmatrix} M^{-1}(\cdot) & \mathbf{0}\,\mathbf{0}\,\mathbf{0} \\ \mathbf{0} & \mathbf{0}\,\mathbf{0}\,\mathbf{0} \\ \mathbf{0} & \mathbf{0}\,\mathbf{0}\,\mathbf{0} \\ \mathbf{0} & \mathbf{0}\,\mathbf{0}\,\mathbf{0} \end{bmatrix}\begin{bmatrix} \mathrm{d}d_s \\ \mathbf{0} \\ \mathbf{0} \\ \mathbf{0} \end{bmatrix}, \tag{3.41}$$

where $d_s=[d_1,\ldots,d_n]^T$ and

$$\begin{aligned}
\boldsymbol{f}_z =& -\boldsymbol{M}^{-1}(\cdot)((\boldsymbol{C}(\cdot)+\boldsymbol{K}_z)\boldsymbol{z}+\boldsymbol{Y}_q(\cdot)\Delta\boldsymbol{\psi}_q-\boldsymbol{K}\Delta\boldsymbol{\theta}),\\
\boldsymbol{f}_\psi =& -k_1\hat{\boldsymbol{\psi}}_q+\boldsymbol{L}_q\boldsymbol{Y}_q^T(\cdot)\boldsymbol{z},\\
\boldsymbol{f}_\theta =& \boldsymbol{s}-\alpha\Delta\boldsymbol{\theta},\\
\boldsymbol{f}_s =& -\boldsymbol{B}^{-1}(\boldsymbol{K}_s\boldsymbol{s}+\boldsymbol{K}_\theta\Delta\boldsymbol{\theta}),
\end{aligned} \tag{3.42}$$

and (3.41) is then expressed as

$$\mathrm{d}\boldsymbol{r} = \boldsymbol{f}_r\mathrm{d}t + \boldsymbol{\sigma}_r\mathrm{d}\boldsymbol{d}_r, \tag{3.43}$$

where

$$\begin{aligned}
\boldsymbol{r} =& [\boldsymbol{z};\ \Delta\boldsymbol{\psi}_q;\ \Delta\boldsymbol{\theta};\ \boldsymbol{s}],\\
\boldsymbol{f}_r =& [\boldsymbol{f}_z;\ \boldsymbol{f}_\psi;\ \boldsymbol{f}_\theta;\ \boldsymbol{f}_s],\\
\mathrm{d}\boldsymbol{d}_r =& [\mathrm{d}\boldsymbol{d}_s;\ \boldsymbol{0};\ \boldsymbol{0};\ \boldsymbol{0}],\\
\boldsymbol{\sigma}_r =& \mathrm{diag}(\boldsymbol{M}^{-1}(\cdot), \boldsymbol{0}, \boldsymbol{0}, \boldsymbol{0}).
\end{aligned} \tag{3.44}$$

Next, a Lyapunov-like function is defined as

$$V = \frac{1}{2}\boldsymbol{r}^T\boldsymbol{M}_r\boldsymbol{r}, \tag{3.45}$$

where $\boldsymbol{M}_r$=diag($\boldsymbol{M}(\cdot)$, $\boldsymbol{L}_q^{-1}$, $\boldsymbol{K}_\theta$, $\boldsymbol{B}$). The differential generator of the Itô process [33] is given as

$$\mathscr{L}V(t) = \frac{\partial V}{\partial t} + \frac{\partial V}{\partial \boldsymbol{r}}\boldsymbol{f}_r + \frac{1}{2}\mathrm{tr}\{\boldsymbol{\sigma}_r^T\frac{\partial^2 V}{\partial \boldsymbol{r}^2}\boldsymbol{\sigma}_r\}. \tag{3.46}$$

where tr{·} denotes the trace of matrix, and $\mathscr{L}$ represents the differential generator. Note that (3.46) can be rewritten as

$$\begin{aligned}
\mathscr{L}V(t) =& \frac{1}{2}\boldsymbol{z}^T\dot{\boldsymbol{M}}(\cdot)\boldsymbol{z} + [\boldsymbol{z};\ \Delta\boldsymbol{\psi}_q;\ \Delta\boldsymbol{\theta};\ \boldsymbol{s}]^T\times\\
& \begin{bmatrix} \boldsymbol{M}(\cdot) & \boldsymbol{0} & \boldsymbol{0} & \boldsymbol{0} \\ \boldsymbol{0} & \boldsymbol{L}_q^{-1} & \boldsymbol{0} & \boldsymbol{0} \\ \boldsymbol{0} & \boldsymbol{0} & \boldsymbol{K}_\theta & \boldsymbol{0} \\ \boldsymbol{0} & \boldsymbol{0} & \boldsymbol{0} & \boldsymbol{B} \end{bmatrix}\begin{bmatrix} \boldsymbol{f}_z \\ \boldsymbol{f}_\psi \\ \boldsymbol{f}_\theta \\ \boldsymbol{f}_s \end{bmatrix} + \frac{1}{2}\mathrm{tr}\{\boldsymbol{M}^{-T}(\cdot)\}.
\end{aligned} \tag{3.47}$$

By substituting (3.42) into (3.47), we have

$$\begin{aligned}\mathscr{L}V(t)=&\frac{1}{2}z^T\dot{\boldsymbol{M}}(\cdot)z-z^T((\boldsymbol{C}(\cdot)+\boldsymbol{K}_z)z+\boldsymbol{Y}_q(\cdot)\Delta\boldsymbol{\psi}_q-\boldsymbol{K}\Delta\boldsymbol{\theta})\\&+\Delta\boldsymbol{\psi}_q^T\boldsymbol{L}_q^{-1}(-k_1\hat{\boldsymbol{\psi}}_q+\boldsymbol{L}_q\boldsymbol{Y}_q^T(\cdot)z)+\Delta\boldsymbol{\theta}^T\boldsymbol{K}_\theta(\boldsymbol{s}-\alpha\Delta\boldsymbol{\theta})\\&-\boldsymbol{s}^T(\boldsymbol{K}_s\boldsymbol{s}+\boldsymbol{K}_\theta\Delta\boldsymbol{\theta})+\frac{1}{2}\mathrm{tr}\{\boldsymbol{M}^{-T}(\cdot)\}.\end{aligned}\tag{3.48}$$

Note that

$$\begin{aligned}\Delta\boldsymbol{\psi}_q^T\boldsymbol{L}_q^{-1}\hat{\boldsymbol{\psi}}_q\leq&\frac{l_q^{-1}}{2}(||\boldsymbol{\psi}_q||^2-||\Delta\boldsymbol{\psi}_q||^2),\\ z^T\boldsymbol{K}\Delta\boldsymbol{\theta}\leq&\frac{1}{2}(z^T\boldsymbol{K}z+\Delta\boldsymbol{\theta}^T\boldsymbol{K}\Delta\boldsymbol{\theta}),\end{aligned}\tag{3.49}$$

where l_q is a positive constant. Substituting the above inequalities into (3.48) and using Property (2) in Sect. 3.2.1 yields

$$\begin{aligned}\mathscr{L}V(t)\leq&-z^T(\boldsymbol{K}_z-\frac{1}{2}\boldsymbol{K})z-\Delta\boldsymbol{\theta}^T(\alpha\boldsymbol{K}_\theta-\frac{1}{2}\boldsymbol{K})\Delta\boldsymbol{\theta}\\&-\boldsymbol{s}^T\boldsymbol{K}_s\boldsymbol{s}-\frac{1}{2}k_1l_q^{-1}||\Delta\boldsymbol{\psi}_q||^2\\&+\frac{1}{2}k_1l_q^{-1}||\boldsymbol{\psi}_q||^2+\frac{1}{2}\mathrm{tr}\{\boldsymbol{M}^{-T}(\cdot)\}.\end{aligned}\tag{3.50}$$

When the control parameters are set large, such that the conditions (3.38)–(3.40) are satisfied, it is obtained that

$$\begin{aligned}\mathscr{L}V(t)\leq&-2\eta_1(\frac{1}{2}z^T\boldsymbol{M}(\cdot)z)-2\frac{\lambda_1}{\lambda_2}(\frac{1}{2}\boldsymbol{s}^T\boldsymbol{B}\boldsymbol{s})\\&-2\eta_3(\frac{1}{2}\Delta\boldsymbol{\theta}^T\boldsymbol{K}_\theta\Delta\boldsymbol{\theta})-2\eta_2(\frac{1}{2}\Delta\boldsymbol{\psi}_q^T\boldsymbol{L}_q^{-1}\Delta\boldsymbol{\psi}_q)+\mu\\\leq&-2\beta V(t)+\mu,\end{aligned}\tag{3.51}$$

where

$$\mu\triangleq\frac{1}{2}k_1l_q^{-1}||\boldsymbol{\psi}_q||^2+\frac{1}{2}|\mathrm{tr}\{\boldsymbol{M}^{-T}(\cdot)\}|,\tag{3.52}$$

$$\beta\triangleq\min\{\eta_1,\frac{\lambda_1}{\lambda_2},\eta_2,\eta_3\},\tag{3.53}$$

and $\lambda_1\triangleq\lambda_{\max}\{\boldsymbol{K}_s\}$, $\lambda_2\triangleq\lambda_{\min}\{\boldsymbol{B}\}$, where $\lambda_{\min}\{\cdot\}$ and $\lambda_{\max}\{\cdot\}$ denote the minimum and maximum eigenvalues of a matrix, respectively.

Next, the expected value of $V(t)$ is calculated as [33]

$$\mathbb{E}\{V(t)\}-V(0)=\mathbb{E}\{\int_0^t\mathscr{L}V(s)\mathrm{d}s\}.\tag{3.54}$$

With the condition of Eq. (3.51), the above expected value can be further calculated by referring to the Gronwall-type lemma [34]

$$\forall t \geq 0, \mathbb{E}\{V(t)\} \leq \frac{\mu}{2\beta} + B_{\max}\{V(0) - \frac{\mu}{2\beta}\}\mathrm{e}^{-2\beta t}, \tag{3.55}$$

that is, the scalar function is bounded, which also implies that the impedance errors are bounded from (3.45).

Furthermore, note that the random disturbance is related to the term μ from (3.52). Therefore, if the control parameters are chosen to be larger according to (3.38)–(3.40), the term β in (3.53) increases, hence $\frac{\mu}{2\beta}$ in (3.55) is smaller and thus the random disturbance is attenuated. □

3.2.6 Results

In this section, simulation studies are conducted to validate the effectiveness of the intention estimation under the proposed dual-channel structure. Specifically, the knee and hip joints are considered, hence $\boldsymbol{q}_h = [q_{h1}, q_{h2}]^T$, representing the joint angles, respectively. The parameters in the dynamic model are set as [35]

$$\boldsymbol{D}_h = \begin{bmatrix} 0.1+0.04\sin(\omega t) & 0 \\ 0 & 0.13+0.04\sin(\omega t) \end{bmatrix}$$
$$\boldsymbol{K}_h = \begin{bmatrix} 3.27+0.8\cos(\omega t) & 0 \\ 0 & 4.3+0.8\cos(\omega t) \end{bmatrix} \tag{3.56}$$

where $\omega = 1.256$ denotes the angular velocity.

In addition, the interaction torque is set as

$$\boldsymbol{\tau}_e = \begin{bmatrix} 0.2\cos(2\omega t) \\ 0.2\sin(2\omega t) \end{bmatrix} \tag{3.57}$$

The estimation of human motion intention follows the structure shown in Fig. 3.6, which comprises the estimation of the desired joint position and the prediction of the future position.

The ground truth $\boldsymbol{q}_d = [q_{d1}, q_{d2}]^T$ is set as

$$\begin{aligned} q_{d1} =& -30 + 2\cos(0.4\pi t) + 4\cos(0.8\pi t) + 6\cos(2\pi t), \\ q_{d2} =& 20 + 20\sin(0.4\pi t + 1.5\pi) \\ &+ 10\sin(0.8\pi t + 4.5\pi) + 10\sin(2\pi t + 7.5\pi), \end{aligned} \tag{3.58}$$

representing the desired positions for the knee and the hip joints, respectively. Such a setting is to simulate the diverse frequencies and time-varying aptitude of knee and

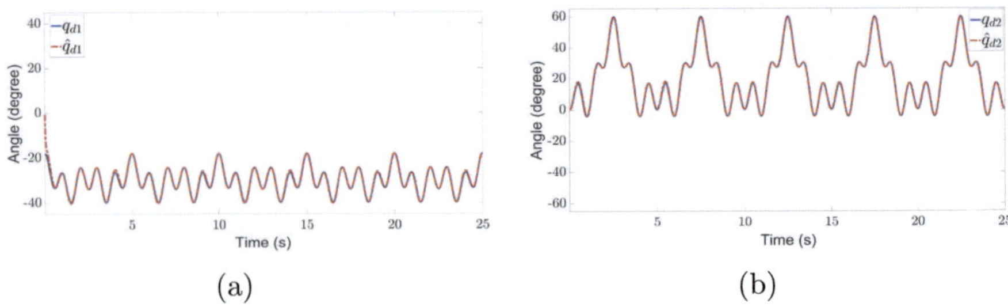

Fig. 3.7 Estimation of the desired position (i.e., the short-term intention): **a** knee joint; **b** hip joint

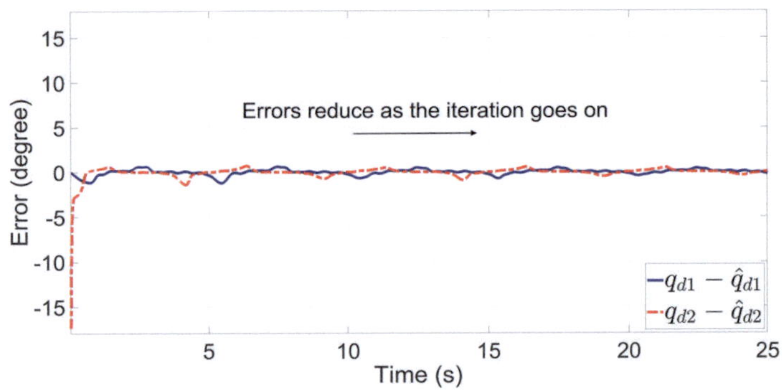

Fig. 3.8 Estimation of the desired position (i.e., the short-term intention): tracking errors

hip joint motion in a real-life situation. Note that the ground truth is not measurable or available in the experiments. The current position of human limb $\boldsymbol{q}_h$ is computed by referring to (3.1) and (3.58).

First, the desired position of the human limb is estimated using the adaptive NNs and represented as the short-term human motion intention estimation. The hidden layer of the NN is constructed using 10 neurons, where its activation function is specified with the kernel of the radial basis function (RBF), where its center is a 3×10 random vector with a range of $[-2,2]$, the bias is set as 10, and the initial value $\hat{\boldsymbol{W}}_j(0)$ is set as $0.1\boldsymbol{J}_{10}$ ($\boldsymbol{J}_{10} \in \Re^{10}$ is a matrix of ones). In addition, the learning rate in (3.15) is set as $\boldsymbol{L}_j = 0.25\boldsymbol{I}_{10}$ where $\boldsymbol{I}_{10} \in \Re^{10\times 10}$ is an identity matrix.

The results, shown in Fig. 3.7, reveal that the estimated desired position $\hat{\boldsymbol{q}}_d$ converges to the ground truth in several seconds. The tracking errors are within $[-0.5, 0.5]$ (see Fig. 3.8), guaranteeing estimation efficiency and accuracy in real-time implementation; It was also seen that the tracking errors kept reducing as the iteration was going on. The proposed method can better explore the intentional information hidden in the interaction torque, which is included in the NN input.

Second, the future position of the human limb is predicted by referring to the probabilistic model comprising past data and represented as the long-term human motion intention estimation. In the simulation, random noise is added to the joint

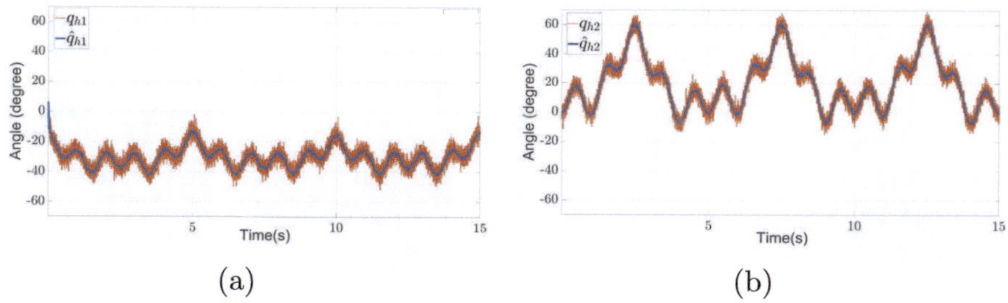

Fig. 3.9 Prediction of the future position (i.e., the long-term intention): **a** knee joint; **b** hip joint

positions of the human limb $\boldsymbol{q}_h$ to mimic the uncertainty in human motion intention and sensor noise; The added noise is white Gaussian noise. The mean $\boldsymbol{\mu}_m$ in Algorithm 3.1 is set as the average of the first $M = 20$ sampled data, and the variance $\boldsymbol{\sigma}_m$ is set as the standard deviation of the first $M = 20$ sampled data. Each $\boldsymbol{q}_h$ is specified as one cluster, i.e., $\pi_m = 1$.

The results in Fig. 3.9 show that the predicted position $\hat{\boldsymbol{q}}_h$ is smoother than the original thus it improves the accuracy compared to the consideration of the current dynamic state only because the current state may be significantly affected by noise and uncertainty. When such results are included in the weighting function in order to regulate robot impedance, the transition between different interaction modes is more stable and appropriate.

The proposed control scheme has also been implemented in a four-joint, lower-limb exoskeleton robot. The experimental results, including a series of comparative studies, ablation studies, and the overall assessment, are presented in this section as follows:

(1) First, the experimental setup is introduced;
(2) Second, a comparative study is presented, between a PD controller, an impedance controller with fixed parameters, and the proposed variable impedance controller, to highlight the improved performance of the proposed model;
(3) Third, the ablation study of the proposed dual channel structure is presented to show the advantages of using both short-term and long-term human motion intention; and
(4) Fourth, the results of an overground walking test are presented to validate the overall performance of the proposed controller with respect to other schemes.

Experimental Setup: Rather than a clinical or medical trial to test the biological response of patients, the experiments aimed to test the functionality of the controller. Thus, only healthy subjects were involved. Corresponding informed consent was granted by the human subjects involved in these experiments.

The experimental setup shown in Fig. 3.10 comprises the mechanical structure, control module, sensors, and assessment module. The overall exoskeleton robot has four degrees of freedom, that is, two hip joints and two knee joints, and all joints are driven by serial elastic actuators (SEAs). The stiffness of the springs in the SEAs is

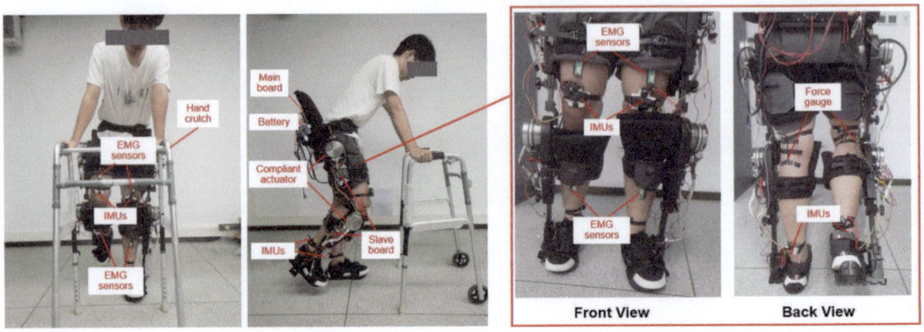

(a) (b)

Fig. 3.10 The developed lower-limb exoskeleton robot, where all four joints are driven with the compliant joints, and a series of sensors are installed on the robot or mounted on the human subject for measurement or assessment; **a** overview; **b** more details

selected as 635 N/m, and their rotary motion is measured by encoders (2048 lines) at both ends of the springs. Note that the length of the legs and the width of the waist of the robot can be adjusted to suit human subjects with different body parameters. A battery is installed inside the backpack.

The control module is hierarchical in structure, comprising one main board at the high level and four slave boards at the low level. The control algorithm is programmed into the main board, where commands are sent to the buffer as pulse-width modulation (PWM) signals and finally received by the four slave boards, thus driving the motion of the four joints.

A series of sensors are mounted on the human body during the experiment, including inertial measurement units (IMUs) (JY901) and force gauges (SICHIRAY) to measure the motion of the human limb and the interaction torque, respectively. In addition, the assessment module comprises four wireless surface EMG sensors (WS450, Biometrics Ltd.) that are mounted on human leg muscles, namely, the quadriceps femoris (QF) and tibialis anterior (TA) muscles of the left and right leg, as shown in Fig. 3.10. EMG signals will be used to assess the robot's performance in different scenarios. All aforementioned sensory feedback is collected by the slave boards and fed back to the main board via the RS-485 bus.

The reasons for using surface EMG sensors on the selected muscles for assessment can be summarized as follows:

(1) Quadriceps femoris (QF) and tibialis anterior (TA) muscles are able to represent muscle fatigue during walking. They are thus considered an effective measure of muscle strength research for medical applications [36–38]. Such selections can also be found in the assessment of other lower-limb exoskeletons [4].
(2) Existing work shows that surface EMG sensors are effective in evaluating a robot's assistance [39, 40]. To guarantee the fair comparison of different methods, it is more important to maintain the same sensor settings (e.g., the same placement on a human subject) than absolute accuracy.

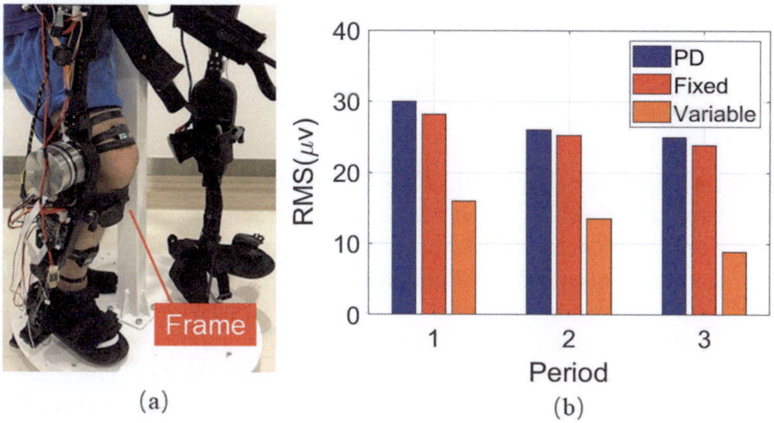

Fig. 3.11 Comparative study of the PD controller and fixed/variable impedance controller: **a** experimental setup; **b** mean of the RMS of the EMG data

Table 3.1 Impedance model parameters

Fixed	Variable
$w(\cdot) \equiv 10$	$w_L = 1$, $w_H = 10$, $\kappa = 0.6$
$M_d = 1$, $C_d = 21$, $K_d = 90$	

(3) To assess the robot's performance during a given period, the readings of EMG sensors within that period are recorded for each QF or TA muscle on both legs. Then, the root mean square (RMS) and the standard derivation are computed and compared. In general, smaller RMS implies more comfort and hence better assistance from the robot, and vice versa. Such a conclusion can also be the human subject found in other exoskeleton robots [41].

Variable Impedance Model: A comparative study was performed to validate the effectiveness of the variable impedance model. For the sake of highlighting the difference, the exoskeleton robot was hung on a frame, and Subject B wore the robot and moved his leg in the air, as shown in Fig. 3.11a; the detailed exoskeleton robot hardware settings are shown in Fig. 3.10. In this experiment, only a single joint of the robot was activated (i.e., the right knee joint), and its reference trajectory was set as

$$q_r = -0.3 + 0.3\sin(1.256\,t). \tag{3.59}$$

The reference trajectory was included in fixed and variable impedance models, and the impedance parameters are given in Table 3.1. The parameters in the fixed impedance model were invariant, and those in the variable impedance model are configured as: $1 \le M_d(t) \le 10$, $21 \le C_d(t) \le 210$, and $90 \le K_d(t) \le 900$. The weighting function of the variable impedance model was set to vary according to the offset of $\boldsymbol{q} - \boldsymbol{q}_r$ (that is, the difference between the robot joint and the reference trajectory) in this experiment. The controller parameters are set in Table 3.2. The compliant actuator parameters were obtained based on the mechanical design and set as $K = 635$, $B = 0.000317$.

Table 3.2 Controller parameters settings

Parameter	K_z	k_1	L_q	α	K_s	K_θ
Value	25	0.1	0.1	15	11	18

During the experiment, the subject intentionally held his leg (i.e., exerted force against the robot) several times to mimic the situation that some conflicts arose between the human and the robot; A total of three control schemes were implemented, that is, a PD controller (with P and D gains of 450 and 35 respectively), an impedance controller with fixed parameters, and the proposed variable impedance controller; Note that the PD has been extensively used as baselines in existing works [4, 42], which has the advantage of high robustness to modeling errors; The mean of the RMS of the EMG signals from four channels were collected and compared between different controllers, and the results are shown in Fig. 3.11b. Specifically, a lower RMS indicated that the subject was more relaxed, proving better assistance was generated from that control scheme, and vice versa.

From Fig. 3.11b, it was seen that the PD controller resulted in the largest RMS of EMG data (blue bar), implying less comfort than the fixed impedance controller (red bar) or (orange bar) the proposed variable impedance controller; This was mainly because that the PD controller treated the interaction torque/force as disturbance and hence rejects it, instead of suiting it as impedance controllers.

The mean of the RMS under the variable impedance model was smaller than that under the fixed impedance model, and the difference increased as the holding actions were repeated. This was mainly because of the proposed model's capability of dealing with conflicts; Specifically, when the mock conflict arose (i.e., a large offset between $\boldsymbol{q}$ and $\boldsymbol{q}_r$), the weighting function (see Fig. 3.12a) in the variable impedance model was scaled down, driving the robot to become passive (see Fig. 3.12b); In comparison, the fixed impedance model always had the same impedance parameters, i.e., the same weight (see Fig. 3.12c), and hence could not effectively respond to the conflict (see Fig. 3.12d). For the fixed and variable impedance models, the impedance errors z (see Fig. 3.12b, d) in the steady state were approximately $[-0.01, 0.01]$, proving the realization of the impedance control task.

Human Motion Intention: Ablation experiments were conducted to validate the proposed method for the estimation of human motion intention. The subject wore the robot and walked on the ground. The robot assessed human motion intention and provided assistance accordingly. Specifically, five scenarios were considered, as shown in Table 3.3. Note that the reference trajectory in setting I was set as a series of predefined gait trajectories, which were recorded on healthy subjects beforehand. In addition, to minimize potential conflicts subject to the predefined trajectory, only two knee joints were activated in this scenario.

The EMG data of the human subject in different scenarios is shown in Fig. 3.13 (each bar chart corresponds to one scenario in Table 3.3). It can be seen that the RMS of the EMG data in setting I is often higher than in the other scenarios, indicating that the subject exerts more effort without predicting human motion intention. The results

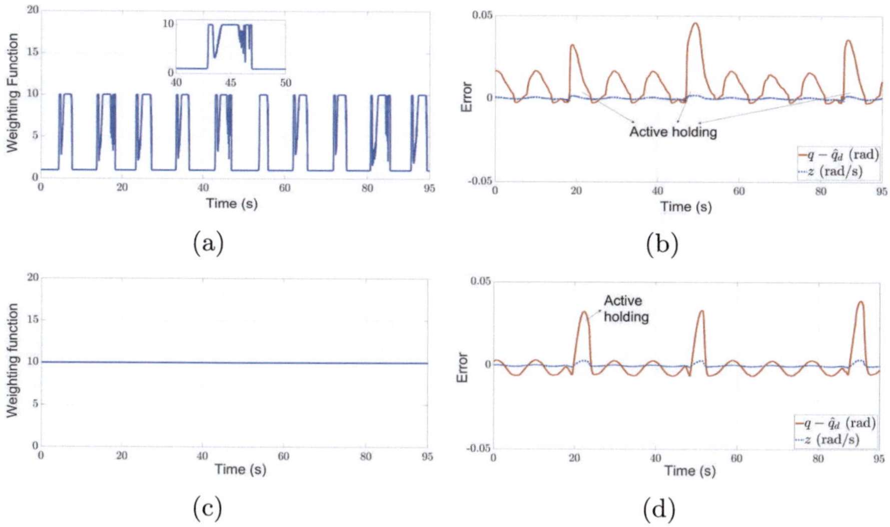

Fig. 3.12 Experimental results: **a** weighting function under the variable impedance model; **b** position error and impedance error under the variable impedance model; **c** weighting function under the fixed impedance model; **d** position error and impedance error under the fixed impedance model

Table 3.3 Comparative study of human motion intention

Setting	Joints	Trajectory	Impedance	Remark
I	Two knees	$\boldsymbol{q}_p$	Fixed	No intention
II	Four	$\boldsymbol{q}_h$	$w(\boldsymbol{q} - \boldsymbol{q}_h)$	Limb position
III	Four	$\hat{\boldsymbol{q}}_d$	$w(\boldsymbol{q} - \boldsymbol{q}_h)$	Short-term
IV	Four	$\boldsymbol{q}_h$	$w(\boldsymbol{q} - \hat{\boldsymbol{q}}_h)$	Long-term
V	Four	$\hat{\boldsymbol{q}}_d$	$w(\boldsymbol{q} - \hat{\boldsymbol{q}}_h)$	Short and long

of both the short-term (setting III) and the long-term intention (setting IV) are better than in the case of simply treating the current human limb as the intention (setting II), proving the effectiveness of the proposed estimation method. The best performance is achieved in setting V. In setting V, dual-channel estimation is fully deployed, combining fast response with immediate human intention for better accuracy in the long term. The statistical results based on Fig. 3.13 are shown in Table 3.4, where $R =$ right Leg, $L =$ left Leg, MEAN = mean of the RMS, and STD =standard derivation.

The proposed adaptive impedance control scheme is implemented in the robot during the walking experiment, where the parameters are set to be the same as the parameters in the previous section. The impedance errors in Table 3.3 setting V are shown in Fig. 3.14. In Fig. 3.14b, the red line denotes the position error of $\boldsymbol{q} - \hat{\boldsymbol{q}}_d$, and the blue line represents the impedance error z. Figure 3.14a illustrates the change of the scaled weighting function. It can be seen that the weighting function varies

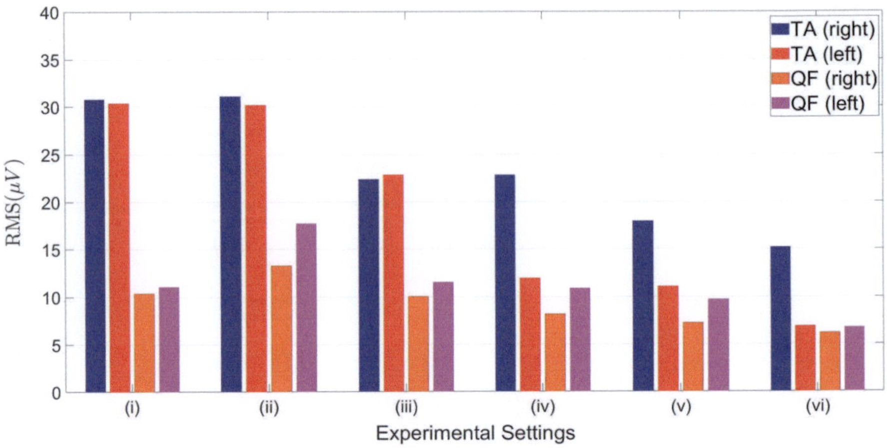

Fig. 3.13 The RMS of the EMG data in the comparative study of human motion intention (Subject A), where QF and TA denote the quadriceps femoris and tibialis anterior muscles, respectively

Table 3.4 Statistic results (human motion intention)

Setting	RMS OF EMG DATA				MEAN	STD
	knee (*R*)	knee (*L*)	hip (*R*)	hip (*L*)		
I	14.60	15.01	12.14	16.27	14.51	1.50
II	17.04	17.85	7.09	8.57	12.64	4.84
III	16.84	14.68	8.01	9.41	12.24	3.64
IV	15.84	11.68	7.71	8.71	10.99	3.16
V	15.17	6.86	6.19	6.74	8.74	3.72

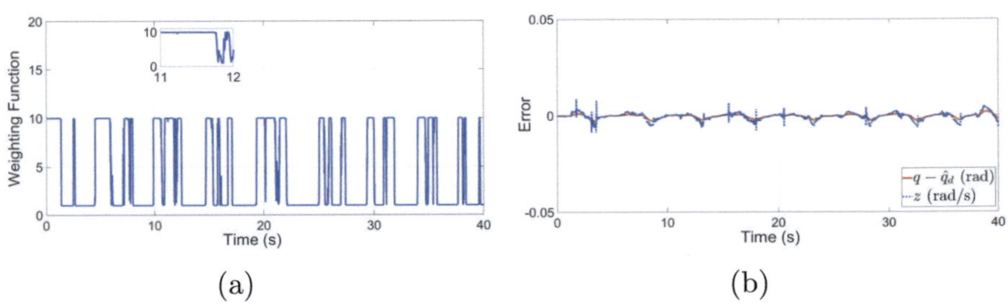

Fig. 3.14 Walking experiment under the adaptive intention-driven variable impedance control: **a** weighting function; **b** position error and impedance error

smoothly during walking in order to vary robot impedance and match human motion intention. The impedance and position errors are within $[-0.01, 0.01]$, which ensures the impedance control task.

Table 3.5 Comparative study of overall assessment

Setting	Joints	Intention	Control	Remark
I	N.A.	N.A.	N.A.	Without exoskeleton
II	None	N.A.	N.A.	Disabled exoskeleton
III	Two hips	None	Position	Predefined trajectory
IV	Two hips	Short-term	Fixed-impedance	Intention-driven
V	Two hips	Dual-channel	Variable-impedance	Intention-driven
VI	Four	Dual-channel	Variable-impedance	Intention-driven

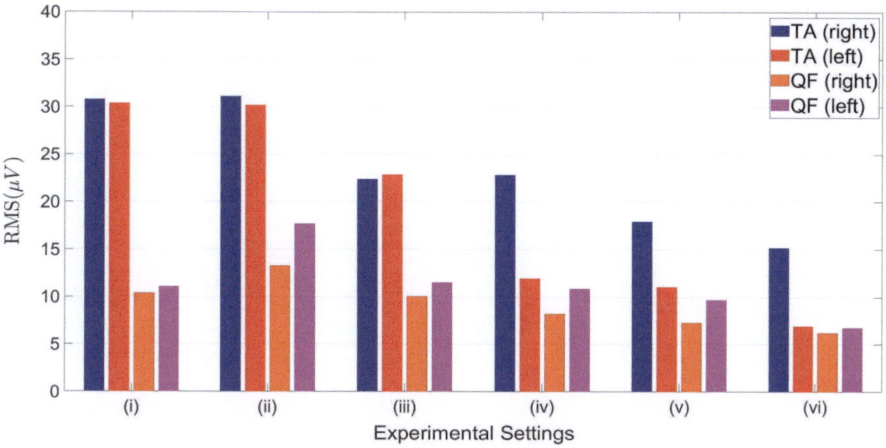

Fig. 3.15 The RMS of the EMG data in the comparative study of the overall assessment (Subject A), where QF and TA denote the quadriceps femoris and tibialis anterior muscles, respectively

Overall Assessment: Third, comparative studies are conducted to evaluate overall performance in walking experiments for different subjects. Specifically, the following scenarios in Table 3.5 (N.A. stands for not available) are considered. Two human subjects are involved in the experiment, Subject A and Subject B. Similarly, the EMG data of the human limb during the experiment is collected and analyzed to assess the robot's performance. The details of the EMG data for Subject A are shown in Fig. 3.15. Similar results could be obtained for Subject B. The statistical results for Subjects A and B are summarized in Tables 3.6 and 3.7.

By analyzing the results, the following conclusion can be made.

(1) The results in setting I and II are treated as the baseline of the experiment. The subject must exert more effort in II than in I, which is natural because the disabled robot results in additional load on the human.
(2) The intention-driven mode (i.e., setting IV, V, and VI) is better than purely following the predefined trajectory (i.e., setting III) because the robot can adjust

Table 3.6 Statistic results with Subject A

Setting	RMS OF EMG DATA				MEAN	STD
	knee (*R*)	knee (*L*)	hip (*R*)	hip (*L*)		
I	30.74	30.36	10.34	11.05	20.62	9.93
II	31.06	30.17	13.27	17.67	23.04	7.74
III	22.35	22.81	10.03	11.52	16.68	5.93
IV	22.78	11.91	8.15	10.87	13.43	5.57
V	17.91	11.04	7.25	9.69	11.47	3.96
VI	15.17	6.86	6.19	6.74	9.49	5.02

Table 3.7 Statistic results with Subject B

Setting	RMS OF EMG DATA				MEAN	STD
	knee (*R*)	knee (*L*)	hip (*R*)	hip (*L*)		
I	6.34	17.14	36.81	86.81	36.78	35.66
II	17.24	32.16	60.80	50.59	40.20	19.36
III	7.89	13.49	30.27	60.11	27.94	23.46
IV	10.89	11.49	13.71	32.56	17.16	8.95
V	8.42	22.06	9.42	20.5	15.1	6.21
VI	7.78	19.16	6.35	16.12	12.35	5.41

the reference trajectory in the impedance model by assessing human motion intention.

(3) By comparing the results of setting IV and V, respectively, it can be determined that the use of dual-channel estimation is better than only the estimation of short-term intention, and the use of variable impedance is better than fixed impedance.

(4) When the four joints are fully deployed under the proposed control scheme (i.e., setting VI), the robot achieves the best performance, where the mean of the RMS of the EMG data is the smallest.

The aforementioned results prove the effectiveness and universality of the proposed control scheme. In summary,

(1) The simulation model simulates the change of "real" short-term intention (i.e., $\boldsymbol{q}_d$) regulated by the dynamic model of the human limb (3.56), which is difficult to measure with sensors in the real world. With the parameters updated by the adaptive NN, the estimated desired position quickly converges to the ground truth within 0.1s. In addition, the tracking errors are also kept as small as $0.5°$ at the steady state, which is 8% of the trajectory's amplitude (see Fig. 3.7). For the estimation of long-term intention, the proposed EM-based algorithm also successfully suppresses random noise; Specifically, variation is reduced (see Fig. 3.9). Such improvement also guarantees a more accurate and stable impedance model, where both the short-term and long-term human motion intentions are incorporated.

(2) The variable impedance model, without intention estimation, is first compared with a fixed impedance model under the same reference trajectory (3.59); The EMG results indirectly prove better assistance using the variable impedance model, whose RMS of the EMG signals are significantly lower than the fixed model (see Fig. 3.11b; roughly reduced by half). Then, the intention-driven variable impedance model is tested in a series of ablation studies. Specifically, the RMS of the EMG signals is recorded in the cases of no intention, fixed intention, short-term intention only, long-term intention only, and both forms of intention. The lowest EMG (or the best assistance) is due to a combination of the advantages of short-term intention (i.e., fast response) and long-term intention (i.e., higher accuracy). The experimental results also prove that the proposed model of human motion intention well suits its underlying principle in a wearable robotic system. That is, both the nature of the physical interaction and the probabilistic properties are considered.

(3) Compared with other control algorithms, the proposed intention-driven controller realizes the lowest RMS in the overground walking experiment(e.g., a reduction by half in the case of no exoskeleton). This is because the controlled robot can better understand the human subject's intention during the physical interaction; it also progressively adjusts the output impedance to match the intention and hence provides effective assistance. In addition, the implemented adaptive NN fulfills the real-time requirements of an exoskeleton robot 100 Hz updating frequency. This ensures that the intention-driven controller is applicable to human subjects with different body parameters (see Tables 3.6 and 3.7).

3.3 Interaction Control

This section proposes a multi-mode control scheme for compliantly-driven exoskeleton robots, where the control objective is specified as a series of regions. Different from the previous approaches, the development of the proposed controller follows the singular perturbation approach, and the stability of the overall system is rigorously proved by using Tikhonov's theorem. Experimental results in both upper-limb and lower-limb exoskeleton systems are presented to demonstrate the effectiveness of the proposed control method.

3.3.1 Singular Perturbation Approach

The dynamic model of a compliantly driven exoskeleton is given as

$$\boldsymbol{M}(\boldsymbol{q})\ddot{\boldsymbol{q}}+\boldsymbol{C}(\dot{\boldsymbol{q}},\boldsymbol{q})\dot{\boldsymbol{q}}+\boldsymbol{g}(\boldsymbol{q}) =\boldsymbol{K}(\boldsymbol{\theta}-\boldsymbol{q})+\boldsymbol{\tau}_e, \tag{3.60}$$

$$\boldsymbol{B}\ddot{\boldsymbol{\theta}}+\boldsymbol{K}(\boldsymbol{\theta}-\boldsymbol{q}) =\boldsymbol{u}, \tag{3.61}$$

which has slower mechanical dynamics (3.60) and faster actuator dynamics (3.61), thus exhibiting the features of two time scales [43]. Therefore, the singular perturba-

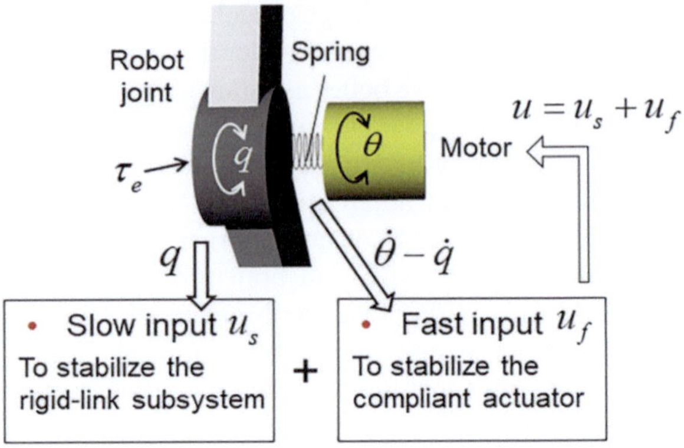

Fig. 3.16 Illustration of singular perturbation approach

tion theory [44] can be introduced to design controllers for two subsystems separately by treating the faster dynamics (3.60) as a perturbation of the slower dynamics (3.61).

By following the singular perturbation approach, the overall controller is proposed as:

$$\boldsymbol{u} = \boldsymbol{u}_r - \boldsymbol{K}_v(\dot{\boldsymbol{\theta}} - \dot{\boldsymbol{q}}), \tag{3.62}$$

where $\boldsymbol{K}_v \in \Re^{n \times n}$ is a positive definite and diagonal matrix, and $\boldsymbol{u}_r$ is the slow time-scale control, which is detailed later. It is seen that the controller in Eq. (3.62) consists of two parts, where the slow time-scale control part $\boldsymbol{u}_r$ is to stabilize the slower robot dynamics (3.60), and the control term $-\boldsymbol{K}_v(\dot{\boldsymbol{\theta}} - \dot{\boldsymbol{q}})$ is to stabilize the faster actuator dynamics (3.61). While the overall system is a fourth-order system described by (3.60) and (3.61), the control complexity is the same as that of the slow time-scale controller $\boldsymbol{u}_r$ (Fig. 3.16).

Substituting Eq. (3.62) into Eq. (3.61) yields:

$$\boldsymbol{B}\ddot{\boldsymbol{\theta}} + \boldsymbol{K}(\boldsymbol{\theta} - \boldsymbol{q}) + \boldsymbol{K}_v(\dot{\boldsymbol{\theta}} - \dot{\boldsymbol{q}}) = \boldsymbol{u}_r, \tag{3.63}$$

which can be written as:

$$\boldsymbol{B}(\ddot{\boldsymbol{\theta}} - \ddot{\boldsymbol{q}}) + \boldsymbol{K}_v(\dot{\boldsymbol{\theta}} - \dot{\boldsymbol{q}}) + \boldsymbol{K}(\boldsymbol{\theta} - \boldsymbol{q}) = \boldsymbol{u}_r - \boldsymbol{B}\ddot{\boldsymbol{q}}. \tag{3.64}$$

Next, a variable $\boldsymbol{\tau}_o$ is defined as the output torque of SEA as: $\boldsymbol{\tau}_o = \boldsymbol{K}(\boldsymbol{\theta} - \boldsymbol{q})$, such that equations (3.60) and (3.64) is written as:

$$\boldsymbol{M}(\boldsymbol{q})\ddot{\boldsymbol{q}} + \boldsymbol{C}(\dot{\boldsymbol{q}}, \boldsymbol{q})\dot{\boldsymbol{q}} + \boldsymbol{g}(\boldsymbol{q}) = \boldsymbol{\tau}_o + \boldsymbol{\tau}_e, \tag{3.65}$$

$$\boldsymbol{B}\ddot{\boldsymbol{\tau}}_o + \boldsymbol{K}_v\dot{\boldsymbol{\tau}}_o + \boldsymbol{K}\boldsymbol{\tau}_o = \boldsymbol{K}(\boldsymbol{u}_r - \boldsymbol{B}\ddot{\boldsymbol{q}}). \tag{3.66}$$

Next, $\boldsymbol{K}$ and $\boldsymbol{K}_v$ are expressed as [43]:

$$K = \frac{1}{\epsilon^2} K_1, \tag{3.67}$$

$$K_v = \frac{1}{\epsilon} K_2, \tag{3.68}$$

where ϵ is a small parameter. Then, substituting them into Eq. (3.66) yields:

$$\epsilon^2 B \ddot{\tau}_o + \epsilon K_2 \dot{\tau}_o + K_1 \tau_o = K_1 (u_r - B\ddot{q}). \tag{3.69}$$

The system described by (3.65) and (3.69) is singularly perturbed. The variables τ_o and $\dot{\tau}_o$ can be viewed as fast time-scale variables, while the robot state variables q and $\dot{q}$ denote slow time-scale variables.

With $\epsilon = 0$, Eq. (3.69) becomes:

$$\bar{\tau}_o = u_r - B\ddot{q}, \tag{3.70}$$

in which, the overbar variable denotes the value of the variable at ϵ=0. Then, substituting Eq. (3.70) into Eq. (3.65) at ϵ=0 yields the *quasi-steady-state* model:

$$(M(q) + B)\ddot{q} + C(\dot{q}, q)\dot{q} + g(q) = u_r + \tau_e. \tag{3.71}$$

Next, a fast time scale is introduced as: $\tau = \frac{t}{\epsilon}$. Note that $\bar{\tau}_o$ is obtained at $\tau = \frac{t}{\epsilon} \to \infty$. Therefore, $\bar{\tau}_o$ is constant on the fast time-scale τ. Then, a new variable is introduced as: $\eta = \tau_o - \bar{\tau}_o$. Substituting the fast time-scale τ and $\tau_o = \eta + \bar{\tau}_o$ into (3.69), with

$$\frac{d\bar{\tau}_o}{d\tau} = \frac{d^2\bar{\tau}_o}{d\tau^2} = \mathbf{0}, \tag{3.72}$$

and $\epsilon = 0$, we have:

$$B\left(\frac{d^2\eta}{d\tau^2}\right) + K_2\left(\frac{d\eta}{d\tau}\right) + K_1 \eta = \mathbf{0}, \tag{3.73}$$

which is referred to as *boundary-layer system.*

Note that both the setting $\epsilon = 0$ and the value $\bar{z}$ are used to separate the *quasi-steady-state* model for the stability analysis only. The development of the proposed control scheme neither requires it nor assumes that the stiffness of SEA is infinitely large. The stability of the overall system can be concluded by using Tikhonov's theorem [44] as:

Theorem 3.3.1 *The quasi-steady-state system (3.71) has a unique solution defined on an infinite time interval for sufficiently small* ϵ*, if both the boundary-layer system (3.73) and the quasi-steady-state system (3.71) are exponentially stable. That is, there exists* ϵ^* *such that, for all* $\epsilon < \epsilon^*$*:*

$$\begin{aligned} \tau_o(t) &= \bar{\tau}_o(t) + \eta(\tau) + O(\epsilon), \\ q(t) &= \bar{q}(t) + O(\epsilon), \end{aligned} \tag{3.74}$$

holds uniformly for $t > 0$, where $\bar{\boldsymbol{q}}$ is the value of $\boldsymbol{q}$ at $\epsilon = 0$, and $O(\epsilon)$ denotes terms of order ϵ and higher.

The exponential stability of the *boundary-layer system* can be guaranteed by choosing some $\epsilon < \epsilon^*$, and appropriate $\boldsymbol{K}_1$ and $\boldsymbol{K}_2$ according to $\boldsymbol{K}$ and $\boldsymbol{K}_v$ from Eqs. (3.67) and (3.68). Next, a controller $\boldsymbol{u}_r$ is to be developed to guarantee the exponential stability of the *quasi-steady-state system* (3.71).

3.3.2 Multiple Interaction Modes

The slow time scale controller $\boldsymbol{u}_r$ is developed to guarantee the exponential stability of the *quasi-steady-state system* (3.71) and also enables the robot to safely and effectively interact with the wearer. To suit the scenario of rehabilitation training, three interaction modes are introduced, i.e., *robot-assisted*, *robot-dominant*, and *safety-stop* are specified in the proposed controller, which is illustrated in Fig. 3.17.

In the *robot-assisted* mode (see Fig. 3.17a), the human initiates a movement, while the robot acts as a follower and provides assistive forces. The assistive forces are generated by controlling the robot to follow a moving region around the reference trajectory (e.g., gait trajectory of healthy subjects for lower-limb exoskeletons), such that a certain level of flexibility is provided for the free movement of humans to allow him/her to gain enough exercises.

In the *robot-dominant* mode (see Fig. 3.17b), the robot is outside the region due to interactions between the human and the robot, and it is concluded that the movement of the human is either unreliable or insufficient. Therefore, the robot should take the lead to move back to the region, so as to correct the improper movement of the human.

In the *robot-assisted* mode and the *robot-dominant* mode, the interaction force is not very large, and it is safe for the human to interact with the robot. If the interaction force is very large due to the conflicts of movement between the human and the robot, the safety of the human may be compromised, and hence the *safety-stop* mode (see Fig. 3.17c) is activated to stop the robot.

Therefore, two types of feedback information: (1) the robot's position with respect to the desired region, and (2) the interaction force between the human and the robot, are employed to evaluate the performance of rehabilitation tasks and also guarantee the safety of the human. Next, two weighting functions are formulated as the quantitative measures for rehabilitation control tasks and safety respectively.

First, a series of joint-space regions are specified to enclose the reference trajectory $\boldsymbol{q}_d$ as:

$$h_{qi}(q_i) = \frac{(q_i - q_{di})^2}{b_{qi}^2} - 1 \leq 0, \tag{3.75}$$

where $\boldsymbol{q}_d = [q_{d1}, \ldots, q_{dn}] \in \Re^n$ represents the reference trajectory, and b_{qi} $(i = 1, \ldots, n)$ denotes the bound of region for the ith degree of freedom. When

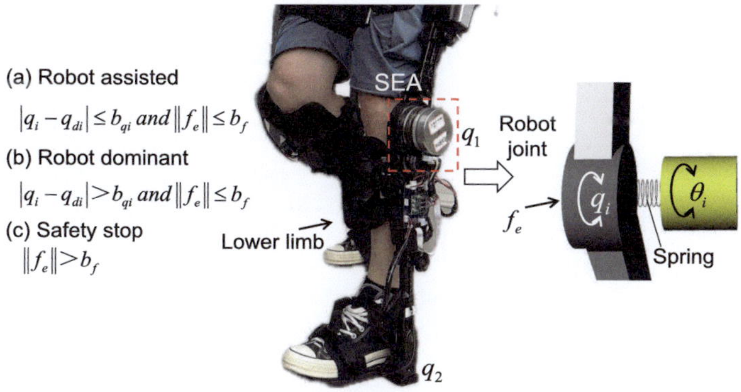

Fig. 3.17 An illustration of different control modes, where $\boldsymbol{q} = [q_1, q_2]^T$ is the vector of joint angles, $\boldsymbol{q}_d = [q_{d1}, q_{d2}]^T$ denotes the reference trajectory in joint space, $\boldsymbol{f}_e$ is the external force, and b_{qi} $(i = 1, \ldots, n)$, b_f are positive constants. **a** *robot-assisted*: the joint angles are around the reference trajectory such that $|q_i - q_{di}| \le b_{qi}$, and the interaction force is not large such that $||\boldsymbol{f}_e|| \le b_f$; **b** *robot-dominant*: the joint angles are away from the reference trajectory such that $|q_i - q_{di}| > b_{qi}$, and the interaction force is not large such that $||\boldsymbol{f}_e|| \le b_f$; **c** *safety-stop*: the force is large such that $||\boldsymbol{f}_e|| > b_f$

the joint angles are near the reference trajectory such that $h_{qi}(q_i) \le 0$, the joint angles are inside the joint-space regions, and vice versa.

Next, a force region is specified to monitor the variation of interaction forces as:

$$h_f(\boldsymbol{f}_e) = \frac{f_{e1}^2}{b_{f1}^2} + \cdots + \frac{f_{em}^2}{b_{fm}^2} - 1 \le 0, \tag{3.76}$$

where $\boldsymbol{f}_e = [f_{e1}, \ldots, f_{em}] \in \Re^m$, and b_{fi} $(i = 1, \ldots, m)$ denotes the bound of region in the ith coordinate of task space. If the interaction force is not large such that $h_f(\boldsymbol{f}_e) \le 0$, the force is inside the region, and vice versa.

In actual implementations, the region parameters should be initialized according to the assessment of patients' specific conditions. For example, if the patient is at the early stage of the therapy and has minimal motion control, the joint-space regions should be set very small to amplify robot assistance; as the patient's motion control ability is being restored, the joint-space regions could be enlarged progressively to encourage patient's active motions [45]. The force region should be set small for small joint-space regions so as to guarantee safety when the robot is playing a more dominant role, and it could be set larger as the joint-space regions are enlarged to avoid over-frequent activation of *safety-stop* mode when the patient is willing to exert more voluntary efforts.

The joint-space regions can also be specified as dynamic regions [46]. The regions can be scaled up or scaled down through the transformation matrix, and the variations can be dependent on the tracking errors, the interaction force, or other sensory feedback information. Based on the concept of dynamic region, the proposed con-

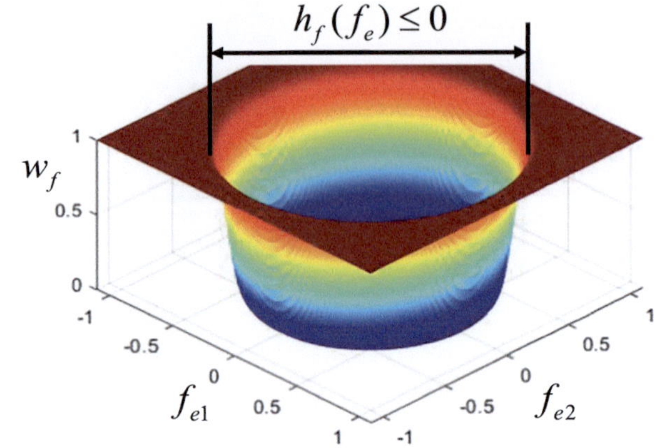

Fig. 3.18 An illustration of weighting function $w_f(\boldsymbol{f}_e)$

trol scheme is able to adjust the size of joint-space regions, and thus the level of assistance, according to the requirements of specific subjects online.

Then, weighting functions are specified based on the joint-space regions and the force region as:

$$w_{qi}(q_i) = \frac{(\min(0, (\min(0, h_{qi}(q_i)))^4 - (\kappa_{qi}^2 - 1)^4))^4}{(\kappa_{qi}^2 - 1)^{16}}$$
$$= \begin{cases} 0, & |q_i - q_{di}| \leq \kappa_q b_{qi} \\ \frac{((h_{qi}(q_i))^4 - (\kappa_{qi}^2 - 1)^4)^4}{(\kappa_{qi}^2 - 1)^{16}}, & \kappa_{qi} b_{qi} < |q_i - q_{di}| \leq b_{qi} \\ 1, & |q_i - q_{di}| > b_{qi} \end{cases} \tag{3.77}$$

and

$$w_f(\boldsymbol{f}_e) = \frac{(\min(0, (\min(0, h_f(\boldsymbol{f}_e)))^4 - (\kappa_f^2 - 1)^4))^4}{(\kappa_f^2 - 1)^{16}}$$
$$= \begin{cases} 0, & |f_{ei}| \leq \kappa_f b_{fi} \\ \frac{((h_f(\boldsymbol{f}_e))^4 - (\kappa_f^2 - 1)^4)^4}{(\kappa_f^2 - 1)^{16}}, & \kappa_f b_{fi} < |f_{ei}| \leq b_{fi} \\ 1, & |f_{ei}| > b_{fi} \end{cases} \tag{3.78}$$

where $0 < \kappa_{qi}, \kappa_f < 1$ are constants. An illustration of $w_f(\boldsymbol{f}_e)$ is shown in Fig. 3.18. From (3.77) and (3.78), both $w_{qi}(q_i)$ and $w_f(\boldsymbol{f}_e)$ are second-order continuous. The weighting functions are equal to 1 when the corresponding region function $h_{qi}(q_i)$ or $h_f(\boldsymbol{f}_e)$ is positive, and the weighting functions reduce to zero when $h_{qi}(q_i)$ or $h_f(\boldsymbol{f}_e)$ is less than zero. The gradient of weighting functions between 1 and 0 is determined by the parameters κ_{qi} and κ_f. In particular, larger values lead to a steeper gradient of weighting functions and vice versa.

Different control modes can be described by using the combination of two types of weighting functions. When $w_{qi}(q_i) = 0$ and $w_f(\boldsymbol{f}_e) = 0$, the robot is inside the joint-space regions where $h_{qi}(q_i) \leq 0$ and the interaction force is not large, and thus the robot works in the *robot-assisted* mode. When $w_{qi}(q_i)=1$ and $w_f(\boldsymbol{f}_e)=0$, the robot is outside the joint-space regions where $h_{qi}(q_i) > 0$ and the interaction force is also not large, and the robot works in the *robot-dominant* mode. When $w_f(\boldsymbol{f}_e)=1$, the interaction force is too large and the task is terminated.

By using the weighting functions, a region error is defined as:

$$\Delta\varepsilon = \begin{bmatrix} \Delta\varepsilon_1 \\ \Delta\varepsilon_2 \\ \cdots \\ \Delta\varepsilon_n \end{bmatrix} \triangleq \begin{bmatrix} w_{q1}(q_1)(q_1 - q_{c1}) \\ w_{q2}(q_2)(q_2 - q_{c2}) \\ \cdots \\ w_{qn}(q_n)(q_n - q_{cn}) \end{bmatrix} \in \Re^n, \tag{3.79}$$

where $\boldsymbol{q}_c = [q_{c1}, q_{c2}, \ldots, q_{cn}]^T \in \Re^n$ is a composite vector, which is defined as:

$$\begin{aligned} \boldsymbol{q}_c &\triangleq w_f(\boldsymbol{f}_e)\boldsymbol{q}_s + (1 - w_f(\boldsymbol{f}_e))\boldsymbol{q}_d \\ &= \begin{cases} \boldsymbol{q}_d, & w_f(\boldsymbol{f}_e)=0 \\ \boldsymbol{q}_s, & w_f(\boldsymbol{f}_e)=1 \end{cases} \end{aligned} \tag{3.80}$$

and $\boldsymbol{q}_s$ denotes a "stop-position" which is defined as $\dot{\boldsymbol{q}}_s = \boldsymbol{0}$ when $w_f(\boldsymbol{f}_e)=1$, and $\boldsymbol{q}_s = \boldsymbol{q}$ when $w_f(\boldsymbol{f}_e) \neq 1$. Therefore, the "stop-position" is updated as the real-time position of robot in the *robot-assisted* and *robot-dominant* modes, and the updating is suspended in the *safety-stop* mode.

From (3.79) and (3.80), if the interaction force is very large such that $h_f(\boldsymbol{f}_e) > 0$, $w_f(\boldsymbol{f}_e)=1$, $\boldsymbol{q}_c = \boldsymbol{q}_s$, and hence $\Delta\varepsilon$ holds the robot at the stop position. On the other hand, if the interaction force is not large such that $h_f(\boldsymbol{f}_e) \leq 0$, $w_f(\boldsymbol{f}_e)=0$, $\boldsymbol{q}_c = \boldsymbol{q}_d$, and the region error becomes $\Delta\varepsilon_i = w_{qi}(q_i)(q_i - q_{di})$. Then, if the robot is outside the joint-space regions where $h_{qi}(q_i) > 0$, $w_{qi}(q_i)=1$, the region error further becomes $\Delta\varepsilon_i = q_i - q_{di}$, driving the robot toward the joint-space regions. After the robot enters the joint-space regions where $h_{qi}(q_i) \leq 0$, $\Delta\varepsilon_i$ is equal to zero. The aforementioned variations of the region error $\Delta\varepsilon$ correspond to different modes of the proposed controller.

As a safety measure, the weighting function $w_f(\boldsymbol{f}_e)$ has a higher priority over the weighting functions $w_{qi}(q_i)$. That is, the *safety-stop* mode is activated at any time when $w_f(\boldsymbol{f}_e)=1$. The safety measures can also be formulated by using other sensory feedback information instead of the interaction force, such as the electrical signal of human muscles evaluated and recorded by Electromyography (EMG) sensors, or the output torque of SEA.

3.3.3 Controller Development

We can now propose the controller $\boldsymbol{u}_r$ as:

$$\boldsymbol{u}_r = -\boldsymbol{K}_s\boldsymbol{s} + \boldsymbol{\phi}(\boldsymbol{f}_e, \boldsymbol{s}) + (\boldsymbol{M}(\boldsymbol{q}) + \boldsymbol{B})\ddot{\boldsymbol{q}}_r + \boldsymbol{C}(\dot{\boldsymbol{q}}, \boldsymbol{q})\dot{\boldsymbol{q}}_r + \boldsymbol{g}(\boldsymbol{q}), \tag{3.81}$$

where $\boldsymbol{K}_s \in \Re^{n\times n}$ is a positive-definite matrix, and the vector

$$\begin{aligned} \boldsymbol{s} &= \dot{\boldsymbol{q}} - \dot{\boldsymbol{q}}_r \\ &= \dot{\boldsymbol{q}} - \dot{\boldsymbol{q}}_c + \alpha\Delta\boldsymbol{\varepsilon} \end{aligned} \tag{3.82}$$

is a sliding vector where

$$\begin{aligned} \dot{\boldsymbol{q}}_r &= \dot{\boldsymbol{q}}_c - \alpha\Delta\boldsymbol{\varepsilon} \\ &= \begin{cases} \dot{\boldsymbol{q}}_d, & w_{qi}(q_i)=0 \text{ and } w_f(\boldsymbol{f}_e)=0 \\ \dot{\boldsymbol{q}}_d - \alpha(\boldsymbol{q}-\boldsymbol{q}_d), & w_{qi}(q_i)=1 \text{ and } w_f(\boldsymbol{f}_e)=0 \\ -\alpha\Delta\boldsymbol{\varepsilon}, & w_f(\boldsymbol{f}_e)=1 \end{cases} \end{aligned} \tag{3.83}$$

represents a reference vector, and α is a positive constant. In addition, the vector $\boldsymbol{\phi}(\boldsymbol{f}_e, \boldsymbol{s})$ in (3.81) denotes an interaction vector which is specified as:

$$\begin{aligned} \boldsymbol{\phi}(\boldsymbol{f}_e, \boldsymbol{s}) &= -w_f(\boldsymbol{f}_e)\boldsymbol{\tau}_e - (1-w_f(\boldsymbol{f}_e))\frac{\boldsymbol{s}^T\boldsymbol{\tau}_e+\delta}{|\boldsymbol{s}^T\boldsymbol{\tau}_e|+\delta}\boldsymbol{\tau}_e \\ &= \begin{cases} -\boldsymbol{\tau}_e, & w_f(\boldsymbol{f}_e)=1 \\ -\boldsymbol{\tau}_e, & w_f(\boldsymbol{f}_e)=0 \text{ and } \boldsymbol{s}^T\boldsymbol{\tau}_e \ge 0 \\ -\frac{\boldsymbol{s}^T\boldsymbol{\tau}_e+\delta}{|\boldsymbol{s}^T\boldsymbol{\tau}_e|+\delta}\boldsymbol{\tau}_e, & w_f(\boldsymbol{f}_e)=0 \text{ and } \boldsymbol{s}^T\boldsymbol{\tau}_e < 0 \end{cases} \end{aligned} \tag{3.84}$$

where δ is a very small positive constant. The role of δ is to avoid singularity when $\boldsymbol{s}^T\boldsymbol{\tau}_e=0$.

In the proposed control scheme for robotic rehabilitation (3.81), the sliding vector $\boldsymbol{s}$ includes the position and velocity control terms, the interaction vector $\boldsymbol{\phi}(\boldsymbol{f}_e, \boldsymbol{s})$ is used to handle interaction forces between the human and the robot, and the last three terms represent the dynamic compensation.

The interaction vector in (3.84) consists of two compensation terms, where the first term deals with the human-robot interaction when the force is very large (i.e., $w_f(\boldsymbol{f}_e)=1$) and the second term deals with the interaction when the force is not large (i.e., $w_f(\boldsymbol{f}_e)=0$). The scalar $\frac{\boldsymbol{s}^T\boldsymbol{\tau}_e+\delta}{|\boldsymbol{s}^T\boldsymbol{\tau}_e|+\delta}$ can be treated as the amplitude of the second compensation term, which is dependent on the sign of $\boldsymbol{s}^T\boldsymbol{\tau}_e$. An illustration of the amplitude $\frac{\boldsymbol{s}^T\boldsymbol{\tau}_e+\delta}{|\boldsymbol{s}^T\boldsymbol{\tau}_e|+\delta}$ is shown in Fig. 3.19.

When $\boldsymbol{s}^T\boldsymbol{\tau}_e \ge 0$, the interaction force is driving the robot away from the reference trajectory, which should be corrected, and hence $\boldsymbol{\phi}(\boldsymbol{f}_e, \boldsymbol{s}) = -\boldsymbol{\tau}_e$ directly canceling out the interaction force.

When $\boldsymbol{s}^T\boldsymbol{\tau}_e \le -\delta$, the force is driving the robot to follow the reference trajectory, which should be encouraged, and hence $\boldsymbol{\phi}(\boldsymbol{f}_e, \boldsymbol{s}) = -\frac{\boldsymbol{s}^T\boldsymbol{\tau}_e+\delta}{|\boldsymbol{s}^T\boldsymbol{\tau}_e|+\delta}\boldsymbol{\tau}_e$ which is in the same direction as $\boldsymbol{\tau}_e$. Therefore, the interaction vector is capable of rejecting incorrect forces and amplifying correct forces, and the use of the interaction vector realizes the paradigm of "AAN". The zone where $-\delta < \boldsymbol{s}^T\boldsymbol{\tau}_e < 0$ is treated as a transition stage guaranteeing the continuity of $\boldsymbol{\phi}(\boldsymbol{f}_e, \boldsymbol{s})$.

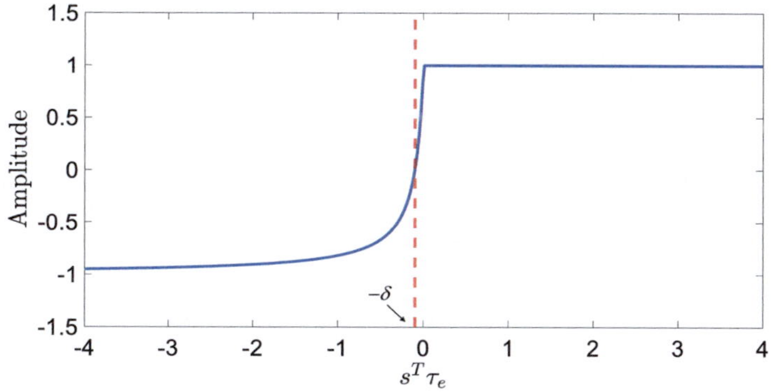

Fig. 3.19 An illustration of the amplitude $\frac{s^T\tau_e+\delta}{|s^T\tau_e|+\delta}$, which is 1 when $s^T\tau_e \geq 0$, and less than 0 when $s^T\tau_e \leq -\delta$. The zone where $-\delta < s^T\tau_e < 0$ is a transition stage

Both the sliding vector and the interaction vector can be varied to suit different control modes as follows.

(1) In the *robot-assisted* mode, the robot is inside the joint-space regions, and the interaction force is small such that $w_{qi}(q_i) = 0$ and $w_f(\boldsymbol{f}_e) = 0$. Hence we have

$$\begin{cases} \boldsymbol{s} = \dot{\boldsymbol{q}} - \dot{\boldsymbol{q}}_d, \\ \phi(\boldsymbol{f}_e, \boldsymbol{s}) = -\frac{\boldsymbol{s}^T\boldsymbol{\tau}_e+\delta}{|\boldsymbol{s}^T\boldsymbol{\tau}_e|+\delta}\boldsymbol{\tau}_e. \end{cases} \tag{3.85}$$

The sliding vector $\boldsymbol{s}$ only includes the velocity control term, to avoid abnormal motions that are too slow and too fast, while the position information is completely provided by the human such that the human is able to gain enough exercise inside the joint space regions. The interaction vector $\phi(\boldsymbol{f}_e, \boldsymbol{s})$ cancels or reinforces the interaction force according to the sign of $\boldsymbol{s}^T\boldsymbol{\tau}_e$, i.e., the relative direction between the interaction force and the reference trajectory.

(2) In the *robot-dominant* mode, the robot is outside the joint-space regions, and the interaction force is small such that $w_{qi}(q_i) = 1$ and $w_f(\boldsymbol{f}_e) = 0$. Then,

$$\begin{cases} \boldsymbol{s} = \dot{\boldsymbol{q}} - \dot{\boldsymbol{q}}_d + \alpha(\boldsymbol{q} - \boldsymbol{q}_d), \\ \phi(\boldsymbol{f}_e, \boldsymbol{s}) = -\frac{\boldsymbol{s}^T\boldsymbol{\tau}_e+\delta}{|\boldsymbol{s}^T\boldsymbol{\tau}_e|+\delta}\boldsymbol{\tau}_e. \end{cases} \tag{3.86}$$

Since the robot is outside the joint-space regions, the robot joints are away from the reference trajectory. Thus, it is concluded that human movement is either unreliable or inefficient. The position feedback is activated in the sliding vector to drive the robot back to the joint-space regions and thus correct human motions. The interaction vector plays a similar role.

(3) In the *safety-stop* mode, the interaction force is very large such that $w_f(\boldsymbol{f}_e) = 1$. Hence we have

Table 3.8 Control terms in different control modes

Term	Robot-assisted	Robot-dominant	Safety-stop
w_{qi}	0	1	0/1
w_f	0	0	1
Sliding vector	$s = \dot{q} - \dot{q}_d$	$s = \dot{q} - \dot{q}_d$ $+\alpha(q - q_d)$	$s = \dot{q} + \alpha\Delta\varepsilon$
Control scheme	$u_r = -K_s s$ $+\phi(f_e, s)$ $+(M(q) + B)\ddot{q}_d$ $+C(q, \dot{q})\dot{q}_d + g(q)$	$u_r = -K_s s$ $+\phi(f_e, s)$ $+(M(q) + B)\ddot{q}_r$ $+C(q, \dot{q})\dot{q}_r + g(q)$	$u_r = -K_s s$ $-\tau_e$ $+(M(q) + B)\ddot{q}_r$ $+C(q, \dot{q})\dot{q}_r + g(q)$

$$\begin{cases} s = \dot{q} + \alpha\Delta\varepsilon, \\ \phi(f_e, s) = -\tau_e. \end{cases} \tag{3.87}$$

The sliding vector holds the robot at the stop position q_s, while the interaction force is completely compensated by $\phi(f_e, s)$. Hence, the robot is stopped.

Therefore, different control modes are integrated into the proposed control scheme (3.81). The variations of control terms are summarized in Table 3.8. Using the sliding vector s, the *quasi-steady-state system* (3.71) can be represented as:

$$\begin{aligned} &(M(q) + B)\dot{s} + C(\dot{q}, q)s(M(q) + B)\ddot{q}_r + C(\dot{q}, q)\dot{q}_r + g(q) \\ &= u_r + \tau_e. \end{aligned} \tag{3.88}$$

Substituting u_r (3.81) into (3.88), we have:

$$(M(q) + B)\dot{s} + (C(\dot{q}, q) + K_s)s = \phi(f_e, s) + \tau_e. \tag{3.89}$$

Equation (3.89) describes the relationship between the position of the robot and the interaction torque. In the *robot-assisted* mode ($w_{qi}(q_i) = 0$, $w_f(f_e) = 0$), the position feedback reduces to zero while the velocity error $\dot{q} - \dot{q}_d$ is very small as the robot is inside the joint-space regions, and thus the sliding vector s is also very small which leads to a low impedance of controller as:

$$\begin{aligned} &(M(q) + B)(\ddot{q} - \ddot{q}_d) + (C(\dot{q}, q) + K_s)(\dot{q} - \dot{q}_d) \\ &= \phi(f_e, s) + \tau_e. \end{aligned} \tag{3.90}$$

The low impedance enables humans to move freely inside the joint-space regions and thus gain enough exercise. Once the robot is outside the joint-space regions, the controller works in the *robot-dominant* mode, and the region error $\Delta\varepsilon$ is nonzero, leading to:

$$\begin{aligned} &(M(q) + B)(\ddot{q} - \ddot{q}_d + \alpha\Delta\dot{\varepsilon}) + \\ &(C(\dot{q}, q) + K_s)(\dot{q} - \dot{q}_d + \alpha\Delta\varepsilon) \\ &= \phi(f_e, s) + \tau_e. \end{aligned} \tag{3.91}$$

That is, the impedance of the robot increases, preventing humans from moving further away from the joint-space regions and thus correcting improper human movements. When the controller transitions to the *safety-stop* mode, the closed-loop system can be given as

$$(\boldsymbol{M}(\boldsymbol{q})+\boldsymbol{B})(\ddot{\boldsymbol{q}}+\alpha\Delta\dot{\varepsilon}) + (\boldsymbol{C}(\dot{\boldsymbol{q}},\boldsymbol{q})+\boldsymbol{K}_s)(\dot{\boldsymbol{q}}+\alpha\Delta\varepsilon) = \boldsymbol{0}, \tag{3.92}$$

where $\Delta\varepsilon_i = w_{qi}(q_i)(q_i - q_{si})$. If $w_{qi}(q_i) \neq 0$ and $\boldsymbol{K}_s$ is sufficiently large, Eq. (3.92) implies that the robot is stabilized at the position $\boldsymbol{q}_s$. If $w_{qi}(q_i)=0$ and $\boldsymbol{K}_s$ is sufficiently large, the above equation can be further simplified as: $(\boldsymbol{M}(\boldsymbol{q})+\boldsymbol{B})\ddot{\boldsymbol{q}} + (\boldsymbol{C}(\dot{\boldsymbol{q}},\boldsymbol{q})+\boldsymbol{K}_s)\dot{\boldsymbol{q}}=\boldsymbol{0}$ indicating the robot is stabilized at its current position. In either case, the robot is stopped.

Note that the controller $\boldsymbol{u}_r$ in (3.81) is dependent on the derivatives of composite vector $\dot{\boldsymbol{q}}_c$ and $\ddot{\boldsymbol{q}}_c$, and hence the acceleration $\ddot{\boldsymbol{q}}$ and $\ddot{\boldsymbol{f}}_e$ are required. To eliminate the requirement of the acceleration, an observer is constructed to estimate the composite vector $\boldsymbol{q}_c$ as:

$$\begin{cases} \dot{\hat{\boldsymbol{q}}}_c = \boldsymbol{\eta} + \beta\boldsymbol{e}_o + \alpha\Delta\hat{\varepsilon} \\ \dot{\boldsymbol{\eta}} = \boldsymbol{M}^{-1}(\boldsymbol{q})(\boldsymbol{\tau}_e+\boldsymbol{K}(\boldsymbol{\theta}-\boldsymbol{q})-\boldsymbol{C}(\dot{\boldsymbol{q}},\boldsymbol{q})\boldsymbol{\eta}-\boldsymbol{g}(\boldsymbol{q}) \\ \quad +\boldsymbol{K}_y\boldsymbol{y}+\boldsymbol{K}_e\boldsymbol{e}_o), \end{cases} \tag{3.93}$$

where $\hat{\boldsymbol{q}}_c$ is the estimated signal of $\boldsymbol{q}_c$, $\boldsymbol{e}_o=\boldsymbol{q}_c-\hat{\boldsymbol{q}}_c$ represents the observation error, $\boldsymbol{\eta}$ is an auxiliary variable, $\boldsymbol{K}_e$, $\boldsymbol{K}_y$ are positive definite matrices, β is a positive constant, and $\boldsymbol{y}=\dot{\boldsymbol{q}}-\boldsymbol{\eta}$ is an auxiliary vector. In addition,

$$\Delta\hat{\varepsilon} = \begin{bmatrix} w_{q1}(q_1)(q_1 - \hat{q}_{c1}) \\ w_{q2}(q_2)(q_2 - \hat{q}_{c2}) \\ \cdots \\ w_{qn}(q_n)(q_n - \hat{q}_{cn}) \end{bmatrix}. \tag{3.94}$$

By using the estimated composite vector $\hat{\boldsymbol{q}}_c$, a new sliding vector is then defined as:

$$\begin{aligned} \hat{\boldsymbol{s}} &= \dot{\boldsymbol{q}} - \dot{\hat{\boldsymbol{q}}}_r \\ &= \dot{\boldsymbol{q}} - \dot{\hat{\boldsymbol{q}}}_c + \alpha\Delta\hat{\varepsilon}, \end{aligned} \tag{3.95}$$

where the reference vector $\dot{\hat{\boldsymbol{q}}}_r$ is defined as:

$$\dot{\hat{\boldsymbol{q}}}_r = \dot{\hat{\boldsymbol{q}}}_c - \alpha\Delta\hat{\varepsilon}. \tag{3.96}$$

The controller for robotic exoskeletons is now revised as:

$$\hat{\boldsymbol{u}}_r = -\boldsymbol{K}_s\hat{\boldsymbol{s}} + \hat{\phi}(\boldsymbol{f}_e, \hat{\boldsymbol{s}}) + (\boldsymbol{M}(\boldsymbol{q})+\boldsymbol{B})\ddot{\hat{\boldsymbol{q}}}_r + \boldsymbol{C}(\dot{\boldsymbol{q}},\boldsymbol{q})\dot{\hat{\boldsymbol{q}}}_r + \boldsymbol{g}(\boldsymbol{q}), \tag{3.97}$$

where

$$\hat{\phi}(\boldsymbol{f}_e, \hat{\boldsymbol{s}}) = -w_f(\boldsymbol{f}_e)\boldsymbol{\tau}_e - (1 - w_f(\boldsymbol{f}_e))\frac{\hat{\boldsymbol{s}}^T\boldsymbol{\tau}_e + \delta}{|\hat{\boldsymbol{s}}^T\boldsymbol{\tau}_e| + \delta}\boldsymbol{\tau}_e \tag{3.98}$$

is a new interaction vector. Compared with (3.81), it can be seen that the estimated signals $\hat{\boldsymbol{s}}$, $\dot{\hat{\boldsymbol{q}}}_r$, and $\ddot{\hat{\boldsymbol{q}}}_r$ are employed instead of $\boldsymbol{s}$, $\dot{\boldsymbol{q}}_r$, and $\ddot{\boldsymbol{q}}_r$. Therefore, the controller $\hat{\boldsymbol{u}}$ in Eq. (3.97) does not require the acceleration information.

The overall controller is proposed for robotic exoskeletons accordingly:

$$\boldsymbol{u} = \hat{\boldsymbol{u}}_r - \boldsymbol{K}_v(\dot{\boldsymbol{\theta}} - \dot{\boldsymbol{q}}). \tag{3.99}$$

The *quasi-steady-state* system (3.71) can be expressed in terms of $\hat{\boldsymbol{s}}$ as:

$$\begin{aligned}&(\boldsymbol{M}(\boldsymbol{q}) + \boldsymbol{B})\dot{\hat{\boldsymbol{s}}} + \boldsymbol{C}(\dot{\boldsymbol{q}}, \boldsymbol{q})\hat{\boldsymbol{s}} + (\boldsymbol{M}(\boldsymbol{q}) + \boldsymbol{B})\ddot{\hat{\boldsymbol{q}}}_r + \boldsymbol{C}(\dot{\boldsymbol{q}}, \boldsymbol{q})\dot{\hat{\boldsymbol{q}}}_r + \boldsymbol{g}(\boldsymbol{q})\\ &= \hat{\boldsymbol{u}}_r + \boldsymbol{\tau}_e.\end{aligned} \tag{3.100}$$

Substituting Eq. (3.97) into Eq. (3.100) yields:

$$(\boldsymbol{M}(\boldsymbol{q}) + \boldsymbol{B})\dot{\hat{\boldsymbol{s}}} + (\boldsymbol{C}(\dot{\boldsymbol{q}}, \boldsymbol{q}) + \boldsymbol{K}_s)\hat{\boldsymbol{s}} = \hat{\phi}(\boldsymbol{f}_e, \hat{\boldsymbol{s}}) + \boldsymbol{\tau}_e. \tag{3.101}$$

Next, multiplying both sides of Eq. (3.93) with $\boldsymbol{M}(\boldsymbol{q})$ and substituting the right side of Eq. (3.60) into it, we have:

$$\begin{aligned}&\boldsymbol{M}(\boldsymbol{q})\dot{\eta} - \boldsymbol{M}(\boldsymbol{q})\ddot{\boldsymbol{q}} + \boldsymbol{C}(\dot{\boldsymbol{q}}, \boldsymbol{q})\eta - \boldsymbol{C}(\dot{\boldsymbol{q}}, \boldsymbol{q})\dot{\boldsymbol{q}}\\ &= \boldsymbol{K}_y\boldsymbol{y} + \boldsymbol{K}_e\boldsymbol{e}_o,\end{aligned} \tag{3.102}$$

and thus the closed-loop equation of the observer is obtained as:

$$\boldsymbol{M}(\boldsymbol{q})\dot{\boldsymbol{y}} + (\boldsymbol{C}(\dot{\boldsymbol{q}}, \boldsymbol{q}) + \boldsymbol{K}_y)\boldsymbol{y} + \boldsymbol{K}_e\boldsymbol{e}_o = \boldsymbol{0}. \tag{3.103}$$

3.3.4 Stability Analysis

To analyze the stability of the overall robotic exoskeleton system, a Lyapunov-like candidate is developed as:

$$V = \frac{1}{2}\hat{\boldsymbol{s}}^T(\boldsymbol{M}(\boldsymbol{q}) + \boldsymbol{B})\hat{\boldsymbol{s}} + \frac{1}{2}\boldsymbol{y}^T\boldsymbol{M}(\boldsymbol{q})\boldsymbol{y}, \tag{3.104}$$

Differentiating (3.104) with respect to time, we have:

$$\dot{V} = \hat{\boldsymbol{s}}^T(\boldsymbol{M}(\boldsymbol{q}) + \boldsymbol{B})\dot{\hat{\boldsymbol{s}}} + \frac{1}{2}\hat{\boldsymbol{s}}^T\dot{\boldsymbol{M}}(\boldsymbol{q})\hat{\boldsymbol{s}} + \boldsymbol{y}^T\boldsymbol{M}(\boldsymbol{q})\dot{\boldsymbol{y}} + \frac{1}{2}\boldsymbol{y}^T\dot{\boldsymbol{M}}(\boldsymbol{q})\boldsymbol{y}. \tag{3.105}$$

Substituting Eqs. (3.101) and (3.103) into (3.105) and using the properties of the dynamic model, we have:

$$\begin{aligned}\dot{V} =& -\hat{s}^T(\boldsymbol{C}(\dot{\boldsymbol{q}},\boldsymbol{q})+\boldsymbol{K}_s)\hat{s}+\frac{1}{2}\hat{s}^T\dot{\boldsymbol{M}}(\boldsymbol{q})\hat{s}+\hat{s}^T(\hat{\phi}(\boldsymbol{f}_e,\hat{s})+\boldsymbol{\tau}_e)\\ &+\frac{1}{2}\boldsymbol{y}^T\dot{\boldsymbol{M}}(\boldsymbol{q})\boldsymbol{y}-\boldsymbol{y}^T((\boldsymbol{C}(\dot{\boldsymbol{q}},\boldsymbol{q})+\boldsymbol{K}_y)\boldsymbol{y}+\boldsymbol{K}_e\boldsymbol{e}_o)\\ =& -\hat{s}^T\boldsymbol{K}_s\hat{s}+\hat{s}^T(\hat{\phi}(\boldsymbol{f}_e,\hat{s})+\boldsymbol{\tau}_e)-\boldsymbol{y}^T\boldsymbol{K}_y\boldsymbol{y}-\boldsymbol{y}^T\boldsymbol{K}_e\boldsymbol{e}_o.\end{aligned} \tag{3.106}$$

Note that $\hat{s}=\dot{\boldsymbol{q}}-\dot{\hat{\boldsymbol{q}}}_c+\alpha\Delta\hat{\varepsilon}$ and $\boldsymbol{y}=\hat{s}+\beta\boldsymbol{e}_o$, (3.106) is written as:

$$\begin{aligned}\dot{V} =& -\hat{s}^T\boldsymbol{K}_s\hat{s}+\hat{s}^T(\hat{\phi}(\boldsymbol{f}_e,\hat{s})+\boldsymbol{\tau}_e)-\boldsymbol{y}^T\boldsymbol{K}_y\boldsymbol{y}-(\hat{s}+\beta\boldsymbol{e}_o)^T\boldsymbol{K}_e\boldsymbol{e}_o\\ =&\hat{s}^T(\hat{\phi}(\boldsymbol{f}_e,\hat{s})+\boldsymbol{\tau}_e)-\boldsymbol{y}^T\boldsymbol{K}_y\boldsymbol{y}-[\hat{s};\boldsymbol{e}_o]^T\boldsymbol{P}[\hat{s};\boldsymbol{e}_o].\end{aligned} \tag{3.107}$$

where

$$\boldsymbol{P}=\begin{bmatrix}\boldsymbol{K}_s & \frac{1}{2}\boldsymbol{K}_e\\ \frac{1}{2}\boldsymbol{K}_e & \beta\boldsymbol{K}_e\end{bmatrix}. \tag{3.108}$$

Now, we have the following theorem to state the passivity of the rigid-link subsystem (*quasi-steady-state system*).

Theorem 3.3.2 *The proposed controller $\boldsymbol{u}_r$ in Eq. (3.97) guarantees the passivity of robot dynamics (3.101), if the control parameters are chosen, such that:*

$$\beta\lambda_{\min}\{\boldsymbol{K}_s\}\geq\frac{1}{4}\lambda_{\max}\{\boldsymbol{K}_e\}. \tag{3.109}$$

Proof With no interaction-related controller asserted, i.e., $\phi(\boldsymbol{f}_e,\hat{s})=\boldsymbol{0}$, equation (3.107) is written as:

$$\dot{V}+W=\hat{s}^T\boldsymbol{\tau}_e, \tag{3.110}$$

where

$$W=\boldsymbol{y}^T\boldsymbol{K}_y\boldsymbol{y}+[\hat{s};\boldsymbol{e}_o]^T\boldsymbol{P}[\hat{s};\boldsymbol{e}_o]. \tag{3.111}$$

When control parameters are chosen such that (3.109) is satisfied, $\boldsymbol{P}$ is positive definite, and W is positive. Integrating Eq. (3.110) over $[0, t]$ yields:

$$\int_0^t\hat{s}^T(\varsigma)\boldsymbol{\tau}_e(\varsigma)\mathrm{d}\varsigma=\underbrace{V(t)-V(0)}_{\text{stored energy}}+\underbrace{\int_0^t W(\varsigma)\mathrm{d}\varsigma}_{\text{energy consumption}}. \tag{3.112}$$

Since W is positive, we have:

$$\int_0^t \hat{\boldsymbol{s}}^T(\varsigma)\boldsymbol{\tau}_e(\varsigma)\mathrm{d}\varsigma \geq V(t) - V(0), \tag{3.113}$$

which demonstrates the passivity of the dynamics between the input $\boldsymbol{\tau}_e$ and the output $\hat{\boldsymbol{s}}$. The passivity shows that at any time, the energy stored in the system $V(t)-V(0)$ is less than or equal to the energy introduced by the human. □

The passivity of the rigid-link subsystem ensures safe human-robot interaction. Next, we have the following theorem to state the stability of the overall robotic exoskeleton.

Theorem 3.3.3 *The overall multi-odal control scheme described by (3.97) and (3.99) for the compliantly-driven exoskeleton described by (3.60) and (3.61) ensures the stability of the closed-loop system and also the convergence of the region error to zero if the control parameters are chosen such that:*

$$\lambda_{\min}\{\boldsymbol{K}_y\} \geq \frac{1}{2}\lambda_{\max}\{\boldsymbol{M}(\boldsymbol{q})\}, \tag{3.114}$$

$$\lambda_{\min}\{\boldsymbol{P}\} \geq \frac{1}{2}\lambda_{\max}\{\boldsymbol{M}(\boldsymbol{q})+\boldsymbol{B}\}. \tag{3.115}$$

Proof In (3.107), the term $\hat{\boldsymbol{s}}^T(\hat{\phi}(\boldsymbol{f}_e, \hat{\boldsymbol{s}})+\boldsymbol{\tau}_e)$ can be written as:

$$\begin{aligned}
&\hat{\boldsymbol{s}}^T(\hat{\phi}(\boldsymbol{f}_e, \hat{\boldsymbol{s}})+\boldsymbol{\tau}_e)\\
=&\hat{\boldsymbol{s}}^T(\boldsymbol{\tau}_e - w_f(\boldsymbol{f}_e)\boldsymbol{\tau}_e - (1-w_f(\boldsymbol{f}_e))\frac{\hat{\boldsymbol{s}}^T\boldsymbol{\tau}_e+\delta}{|\hat{\boldsymbol{s}}^T\boldsymbol{\tau}_e|+\delta}\boldsymbol{\tau}_e)\\
=&(1-w_f(\boldsymbol{f}_e))(1-\frac{\hat{\boldsymbol{s}}^T\boldsymbol{\tau}_e+\delta}{|\hat{\boldsymbol{s}}^T\boldsymbol{\tau}_e|+\delta})\hat{\boldsymbol{s}}^T\boldsymbol{\tau}_e.
\end{aligned} \tag{3.116}$$

Since $0 \leq w_f(\boldsymbol{f}_e) \leq 1$, we have $1-w_f(\boldsymbol{f}_e) \geq 0$. Then, the sign of $1-\frac{\hat{\boldsymbol{s}}^T\boldsymbol{\tau}_e+\delta}{|\hat{\boldsymbol{s}}^T\boldsymbol{\tau}_e|+\delta}$ can be analysed as follows.

(1) When $\hat{\boldsymbol{s}}^T\boldsymbol{\tau}_e \geq 0$, $1-\frac{\hat{\boldsymbol{s}}^T\boldsymbol{\tau}_e+\delta}{|\hat{\boldsymbol{s}}^T\boldsymbol{\tau}_e|+\delta}=0$, and hence the term $\hat{\boldsymbol{s}}^T(\hat{\phi}(\boldsymbol{f}_e, \hat{\boldsymbol{s}})+\boldsymbol{\tau}_e)=0$.

(2) When $-\delta < \hat{\boldsymbol{s}}^T\boldsymbol{\tau}_e < 0$, it can be derived that $\frac{\hat{\boldsymbol{s}}^T\boldsymbol{\tau}_e+\delta}{|\hat{\boldsymbol{s}}^T\boldsymbol{\tau}_e|+\delta} < 1$, and hence the term $1-\frac{\hat{\boldsymbol{s}}^T\boldsymbol{\tau}_e+\delta}{|\hat{\boldsymbol{s}}^T\boldsymbol{\tau}_e|+\delta} > 0$.

(3) When $\hat{\boldsymbol{s}}^T\boldsymbol{\tau}_e \leq -\delta$, $\frac{\hat{\boldsymbol{s}}^T\boldsymbol{\tau}_e+\delta}{|\hat{\boldsymbol{s}}^T\boldsymbol{\tau}_e|+\delta} \leq 0$, and hence $1-\frac{\hat{\boldsymbol{s}}^T\boldsymbol{\tau}_e+\delta}{|\hat{\boldsymbol{s}}^T\boldsymbol{\tau}_e|+\delta} > 0$.

In summary, $\hat{\boldsymbol{s}}^T(\hat{\phi}(\boldsymbol{f}_e, \hat{\boldsymbol{s}})+\boldsymbol{\tau}_e)$ is always non-positive. Then, (3.107) can be written as:

$$\dot{V} \leq -\boldsymbol{y}^T\boldsymbol{K}_y\boldsymbol{y} - [\hat{\boldsymbol{s}};\, \boldsymbol{e}_o]^T\boldsymbol{P}[\hat{\boldsymbol{s}};\, \boldsymbol{e}_o]. \tag{3.117}$$

If conditions (3.114) and (3.115) are satisfied, we have:

$$\dot{V} \leq -\gamma V \leq 0, \tag{3.118}$$

where γ is a positive constant. The inequality (3.118) indicates that $V \leq V(0)\mathrm{e}^{-\gamma t}$ where $V(0)$ denotes the initial value of the Lyapunov-like candidate. Therefore, the exponential stability of the *quasi-steady-state system* (3.100) is guaranteed. Since both the *boundary-layer system* (3.73) and the *quasi-steady-state system* (3.100) are exponentially stable, from Tikhonov's theorem [44], it can be concluded that the closed-loop system is stable.

From the inequality (3.118), we have $V \to 0$ and thus $\hat{\boldsymbol{s}}$, $\boldsymbol{y} \to \boldsymbol{0}$. Since $\hat{\boldsymbol{s}} = \dot{\boldsymbol{q}} - \dot{\hat{\boldsymbol{q}}}_c + \alpha\Delta\hat{\varepsilon}$ and $\Delta\hat{\varepsilon}_i = w_{qi}(q_i)(q_i - \hat{q}_{ci})$, the convergence of $\hat{\boldsymbol{s}} \to \boldsymbol{0}$ ensures the convergence of $\dot{\boldsymbol{q}} \to \dot{\hat{\boldsymbol{q}}}_c$ and $\boldsymbol{q} \to \hat{\boldsymbol{q}}_c$ if $w_{qi}(q_i) \neq 0$, and $\hat{\boldsymbol{s}} \to \boldsymbol{0}$ ensures the convergence of $\dot{\boldsymbol{q}} \to \dot{\hat{\boldsymbol{q}}}_c$ if $w_{qi}(q_i) = 0$. Since $\boldsymbol{y} = \hat{\boldsymbol{s}} + \beta\boldsymbol{e}_o$, the convergence of $\boldsymbol{y} \to \boldsymbol{0}$ and $\dot{\boldsymbol{q}} \to \dot{\hat{\boldsymbol{q}}}_c$ ensures the convergence of $\boldsymbol{e}_o \to \boldsymbol{0}$. That is, $\boldsymbol{q}_c \to \hat{\boldsymbol{q}}_c$, and thus $\Delta\varepsilon \to \Delta\hat{\varepsilon}$. In addition, $\hat{\boldsymbol{s}} \to \boldsymbol{0}$ and $\dot{\boldsymbol{q}} \to \dot{\hat{\boldsymbol{q}}}_c$ ensure the convergence of $\Delta\hat{\varepsilon} \to \boldsymbol{0}$. Since $\Delta\hat{\varepsilon} \to \boldsymbol{0}$, we have $\Delta\varepsilon \to \boldsymbol{0}$. If $w_f(\boldsymbol{f}_e) = 1$, the *safety-stop* mode is activated, and $\Delta\varepsilon \to \boldsymbol{0}$ implies that $\boldsymbol{q} \to \boldsymbol{q}_s$. If $w_f(\boldsymbol{f}_e) \neq 0$, $\Delta\varepsilon \to \boldsymbol{0}$ is satisfied when $w_i(q_i) = 0$ or $\boldsymbol{q} = \boldsymbol{q}_c$. In either case, $h_{qi}(q_i) \leq 0$ and the robot is controlled to stay inside the joint-space regions. □

Note that condition (3.109) can be satisfied when the control parameters β and $\boldsymbol{K}_s$ are set sufficiently large. Condition (3.114) can also be satisfied when $\boldsymbol{K}_y$ is large, and condition (3.115) is satisfied when $\boldsymbol{K}_s$ is set large.

3.3.5 Results

In this section, experimental results are presented to demonstrate the effectiveness of the proposed rehabilitation control scheme. The purpose of the experiments was to test the functionality of the controller, and it was not a clinical or medical trial to test the biological response of patients. Thus, only healthy subjects were involved. The experimental methods employed were approved by the university, and the subjects provided corresponding informed consent.

Upper-Limb Exoskeleton: First, the proposed control method was implemented in a one-DOF upper-limb exoskeleton as shown in Fig. , which mainly consists of a robot link driven by the SEA and a dSPACE control system. The dSPACE includes an encoder board (DS3002), an ADC board (DS2002), a DAC board (DS2102), and a processor (DS1007). Sensor outputs (encoders, potentiometers, force sensor) are connected to the encoder board or the ADC board, and the processor receives the data and generates the control output based on the proposed algorithm. Then, the control output is transferred to the DAC board to control the motor via the motor driver.

The reference trajectory q_d was specified as a sinusoid trajectory as: $q_d = 0.1\pi\sin(\pi t)$ rad, which has been extensively utilized in upper-limb rehabilitation [47]. The joint-space region was specified to enclose the reference trajectory as:

Table 3.9 Control and observer parameter settings

Parameter	α	K_s	δ	K_v	K_y	β	K_e
Value	20	1	10^{-10}	10^{-3}	1	100	50

$$h_q(q) = \frac{(q - q_d)^2}{0.025^2} - 1 \leq 0, \tag{3.119}$$

where the bound of the region was set as: $b_q = 0.025$ rad. When the robot is near the reference trajectory such that $|q - q_d| \leq b_q$, the region function $h_q(q) \leq 0$, and vice versa.

Then, the force region was specified as:

$$h_f(f_e) = \frac{f_e^2}{15^2} - 1 \leq 0, \tag{3.120}$$

where the bound was set as: $b_f = 15$ N. When the interaction force is negligible such that $|f_e| \leq b_f$, the region function $h_f(f_e) \leq 0$, and vice versa.

The control parameters used in Eqs. (3.97) and (3.99) are summarized in Table 3.9. An observer was implemented to estimate the composite vector, avoiding the need for acceleration information and its derivatives, with observer parameters also listed in the table. The experimental results are shown in Fig. 3.20. As seen from Fig. 3.20a, b, the robot was controlled to follow the moving region around the reference trajectory, for the realization of rehabilitation. When the robot was inside the region, the proposed controller worked in the *robot-assisted* mode such that the free movement of the human was encouraged. When the robot was outside the region, the controller worked in the *robot-dominant* mode and drove the robot back to the region, to correct the motion of the human. During the period of $t \approx 6$ s, 14 s, 23 s, the human cannot follow the movement of robot, resulting in a large interaction force (> 15 N), and the weighting function $w_f(f_e)$ increased to 1 (see Fig. 3.20c). Note that the weighting function was rescaled for the convenience of displaying both the weighting function and the interaction force. Then, the *safety-stop* mode was activated, and the robot was stopped to ensure the safety (see Fig. 3.20a). When the interaction force was reduced such that $w_f(f_e) = 0$, the controller transitioned to the *robot-assisted* mode again, such that the rehabilitation process continued. The observation error is shown in Fig. 3.20d, which is less than 0.03 rad. The sliding vector and the control input are shown in Fig. 3.20e, f, which are bounded. The absolute value of both the sliding vector and the control is small in the *robot-assisted* mode and relatively large in the *robot-dominant* mode, due to the variations of the region error (i.e., nonzero outside the region and zero inside the region), which thus shows the successful transition among different modes.

Next, the frequency of the reference trajectory was increased from 0.5 Hz to 1 Hz as: $q_d = 0.1\pi \sin(2\pi t)$ rad. The joint-space region was enlarged by setting $b_q = 0.05$ rad, to provide more free-moving space for humans. The force region was also

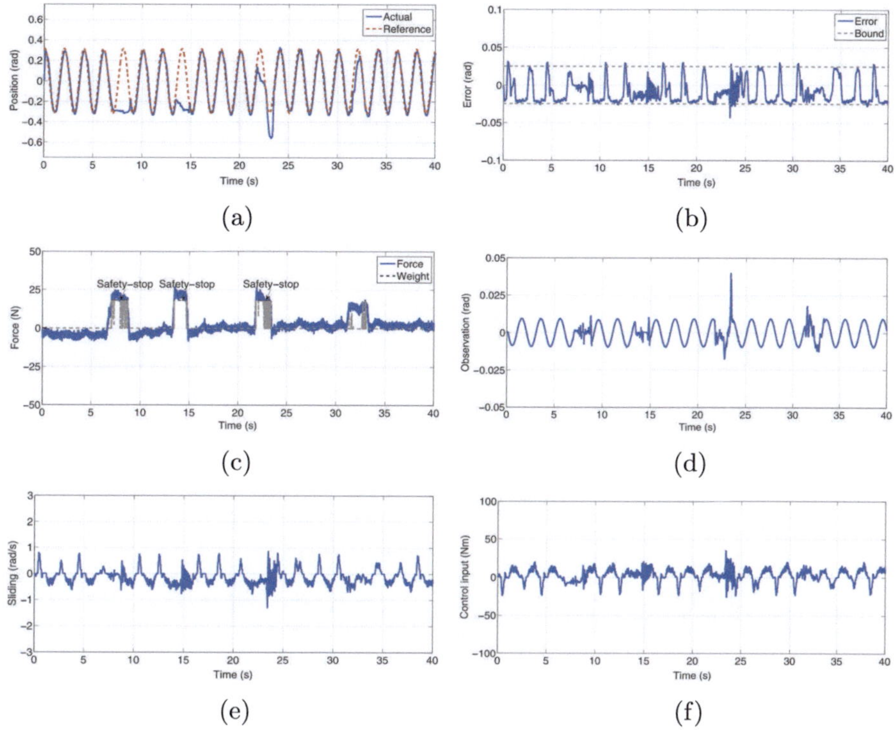

Fig. 3.20 Experiment 1: The reference trajectory was set as: $q_d = 0.1\pi \sin(\pi t)$ rad. **a** Path of robot joint; **b** Tracking error $q - \hat{q}_c$ where $\hat{q}_c$ denotes the estimate of the composite variable q_c; **c** Interaction force f_e; **d** Observation error $e_o = q_c - \hat{q}_c$; **e** Sliding vector $\hat{s}$; **f** Control input u

enlarged as $b_f = 31$ N, to avoid the too-frequent activation of the *safety-stop* mode. The control parameters remained the same. The robot was also controlled to follow the moving region around the reference trajectory, and the experimental results are show in Fig. 3.21. Compared with the first experiment, it was relatively difficult for the human to follow the faster reference trajectory, which then led to the larger interaction force and thus the larger tracking error. Therefore, more robot control efforts were required to correct human's motion outside the joint-space region. At $t \approx 11$ s, 28 s, the *safety-stop* mode was activated to stop the robot, due to large interaction forces. As the interaction force was reduced, the robot was controlled to follow the moving region again. The observation error is shown in Fig. 3.21d, which is around 0.025 rad. The sliding vector and the control input are shown in Fig. 3.21d, f, which are also bounded. The experimental results show the successful implementation of three control modes in the upper-limb robotic exoskeleton.

Lower-Limb Exoskeleton: The exoskeleton robot [48] was shown in Fig. 3.22. The overall system is modular in structure, such that the knee or the ankle module can be easily removed from the entire robot system when the system is used as a one-

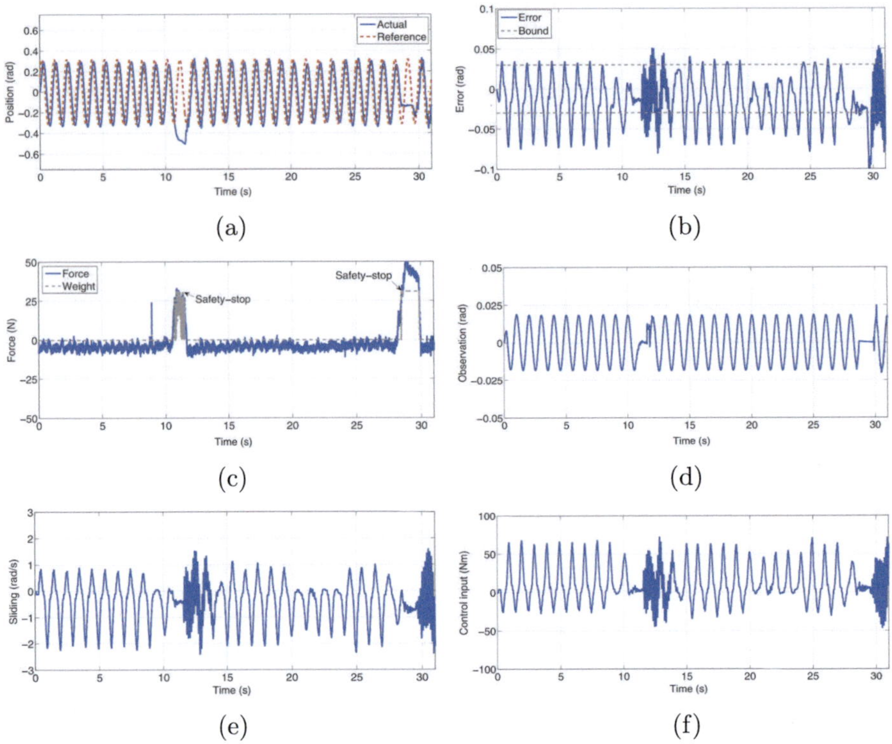

Fig. 3.21 Experiment 2: The reference trajectory was set as: $q_d = 0.1\pi \sin(2\pi t)$ rad. **a** Path of robot joint; **b** Tracking error $q - \hat{q}_c$ where $\hat{q}_c$ denotes the estimate of the composite variable q_c; **c** Interaction force f_e; **d** Observation error $e_o = q_c - \hat{q}_c$; **e** Sliding vector $\hat{s}$; **f** Control input u

DOF robot. Each joint of the robot is driven by the SEA in Sect. 2.1, based on a four-bar linkage shown in Fig. 3.22b. The structure of the robot has been optimized based on the biomechanics of human gait to provide capable and safe assistance during overground walking. The subjects were required to wear the exoskeleton on the right leg, as shown in Fig. 3.22c. Several trials served as practice before the data collection to make the subjects familiar with the exoskeleton and the test scenario. At the beginning of each test, the subjects were asked to stand upright. After the robot is started, the subjects can start walking at any time. The subjects were then asked to perform overground walking in a straight line at a preferred speed for about 30 m and stop at a stand-up posture.

The control algorithm was realized in the NI CompactRIO 9074 embedded control system (National Instruments), and the feedback information was obtained by using the NI data acquisition/output system, which consists of an analog input module (NI 9215), an analog output module (NI 9263), and an encoder module (NI 9516). The reference trajectory is specified as a gait trajectory by analyzing different gait phases of human walking [49]. An example of gait phases and gait events for the knee and

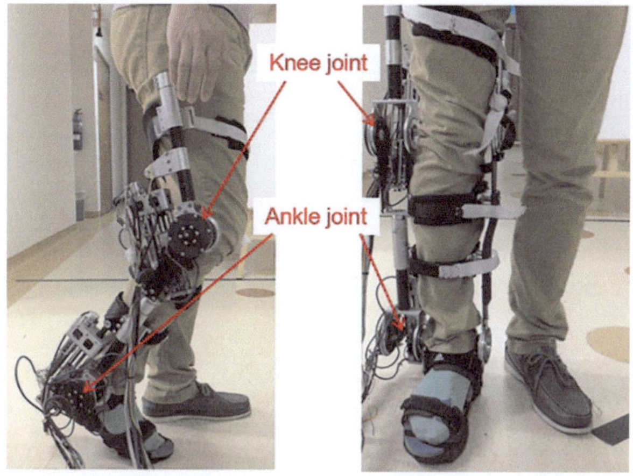

Fig. 3.22 The control method was implemented in a knee-ankle-foot robot: **a** The prototype of the rehabilitation robot; **b** Schematic diagram of the four-bar linkage, where δ_c is the initial angle of the crank, φ_1 and φ_2 are the angles of the four-bar linkage, l_1, l_2 are the length of the bars, and d is the output length of the actuator; **c** A human is wearing the knee-ankle-foot robot

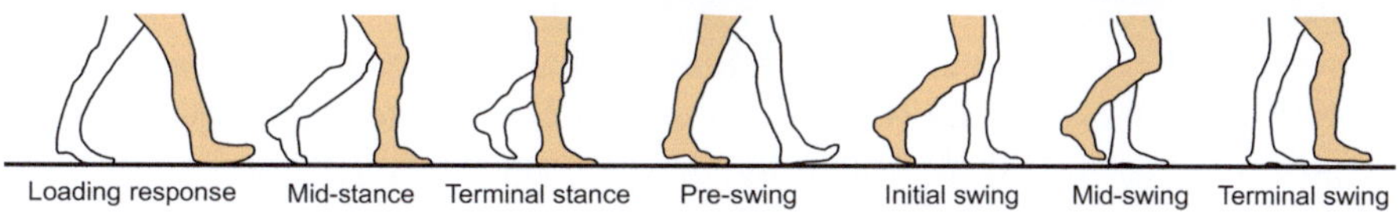

Fig. 3.23 A typical gait cycle consists of seven gait phases: loading response, mid-stance, terminal stance, pre-swing, initial swing, mid-swing, and terminal swing

ankle joints is illustrated in Fig. 3.23. In Fig. 3.23, a typical gait cycle consists of seven gait phases: loading response, mid-stance, terminal stance, pre-swing, initial swing, mid-swing, and terminal swing. The seven gait phases are classified by a sequence of gain event occurrences, including initial contact (IC), opposite toe off (OT), heel rise (HR), opposite initial contact (OI), toe off (TO), feet adjacent (FA), and tibia vertical (TV) [50].

The joint-space regions are specified for the knee joint and the ankle joint, respectively, as

$$h_{q1}(q_1) = \frac{(q_1 - q_{d1})^2}{b_{q1}^2} - 1 \leq 0, \tag{3.121}$$

$$h_{q2}(q_2) = \frac{(q_2 - q_{d2})^2}{b_{q2}^2} - 1 \leq 0, \tag{3.122}$$

where q_1 and q_2 denote the knee angle and the ankle angle respectively, $\boldsymbol{q}_d = [q_{d1}, q_{d2}]^T$ denotes the gait trajectory, and b_{q1} and b_{q2} denote region bounds, which are positive constants. Instead of interaction forces, the output torques of actuators are specified as safety measures in the lower-limb exoskeleton. If the human manages to follow the movement of the robot, the output torque $\boldsymbol{\tau}_o = \boldsymbol{K}(\boldsymbol{\theta} - \boldsymbol{q})$ is small. If the human fails to follow the movement of the robot, the relative position $\boldsymbol{\theta} - \boldsymbol{q}$

Table 3.10 Weighting function and controller settings

Parameter	κ_{qi}	κ_{oi}	$\boldsymbol{K}_v$	$\boldsymbol{K}_s$	α	δ	b_{qi}
Value	0.5	0.5	$200\boldsymbol{I}_2$	$400\boldsymbol{I}_2$	10	10^{-10}	0.087, 0.17, 0.26

increases and thus the output torque $\boldsymbol{\tau}_o$ also increases. Then, two torque regions were formulated to monitor the output torques of the knee joint and the ankle joint as

$$h_{o1}(\tau_{o1}) = \frac{\tau_{o1}^2}{b_{o1}^2} - 1 \leq 0, \tag{3.123}$$

$$h_{o2}(\tau_{o2}) = \frac{\tau_{o2}^2}{b_{o2}^2} - 1 \leq 0, \tag{3.124}$$

where τ_{o1} and τ_{o2} denote the output torques of the knee joint and the ankle joint, respectively, and $b_{o1} = 4.29$ Nm and $b_{o2} = 7.14$ Nm denote region bounds which are positive constants. If the output torques are very large, $h_{oi}(\tau_{oi}) > 0$, and vice versa.

The corresponding weighting functions $w_{qi}(q_i)$ and $w_{oi}(\tau_{oi})$ were formulated based on the region functions $h_{qi}(q_i)$ and $h_{oi}(\tau_{oi})$ by following the similar development, the configuration of these functions, and the controller parameters were set in Table 3.10. The corresponding trajectories are shown in Fig. 3.24, and the tracking errors are shown in Fig. 3.25. Due to the properties of region errors (i.e., $\neq \mathbf{0}$ outside regions and $= \mathbf{0}$ inside regions), it can be seen that when the regions were set small (b_{qi}=0.087 rad), the robot was more dominant during human-robot interaction, which limited the human limb within small bounds around the gait trajectories, and thus the trajectories of robot joints were also close to the gait trajectories with smaller position errors (less than 0.4 rad in the *robot-assisted* or *robot-dominant* mode). When the joint-space regions were set larger, more flexibility was provided such that the human was able to adjust the positions of the human limb within the large regions, and the position errors also increased. Some oscillations can be found in the trajectory of the ankle joint when the regions were set small (see Fig. 3.24b), which resulted from too frequent transitions between the *robot-dominant* mode and the *robot-assisted* mode. The oscillations were attenuated as the regions were scaled up (see Fig. 3.24f).

The output torques of SEA are shown in Fig. 3.26. If the output torques were very large and outside the torque regions such that $h_{oi}(\tau_{oi}) > 0$, the weight factors increased to 1 (e.g., $t \approx 20$ s in Fig. 3.26a, $t \approx 14$ s and $t \approx 25$ s in Fig. 3.26b), and the *safety-stop* mode was activated, such that the robot was stopped to guarantee the safety of the human, until the human was able to follow the gait trajectory again, such that the output torques were reduced. Observers were also constructed to eliminate the requirement of high-order derivatives, and the observer parameters in (3.93) were set as: $\boldsymbol{K}_y = \boldsymbol{I}_2$, $\boldsymbol{K}_e = 50\boldsymbol{I}_2$, and $\beta = 100$. The dynamic parameters in (3.97) and

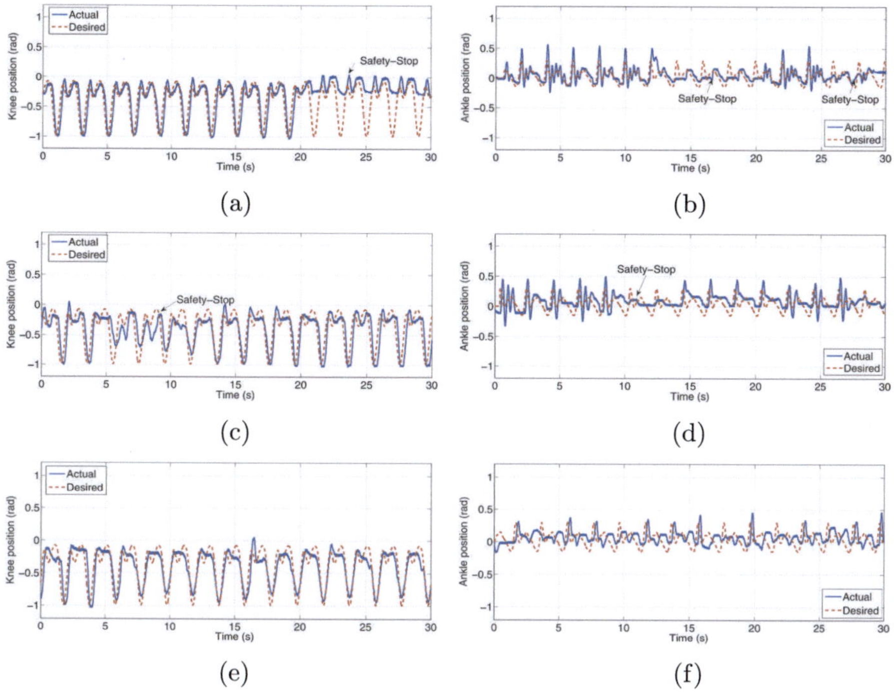

Fig. 3.24 Trajectories of knee joints and ankles for different joint-space regions. **a** Knee joint (b_{qi}=0.087 rad); **b** Ankle joint (b_{qi}=0.087 rad); **c** Knee joint (b_{qi}=0.17 rad); **d** Ankle joint (b_{qi}= 0.17 rad); **e** Knee joint (b_{qi}=0.26 rad); **f** Ankle joint (b_{qi}=0.26 rad)

(3.99) were obtained by referring to [48] with sufficient accuracy. The observation errors are shown in Fig. 3.27, which are less than 0.1 rad. The control inputs for different joint-space regions are shown in Fig. 3.28, which are bounded throughout the experiments. It is found that the control inputs were small when the joint angles were inside the region. This is because that the region errors $\Delta\varepsilon$ remain zero inside the regions and only activated outside the regions to correct the movement of human. The larger joint-space regions lead to fewer non-zero values of the region errors and thus smaller control inputs. The variation of control input also shows the successful transition among different modes.

Experiments with multiple subjects (1) age 27, weight 65 kg; (2) age 32, weight 73 kg; (3) age 34, weight 55 kg were also carried out, where the joint-space regions were specified with two different sizes (b_{qi}=0.087, 0.17 rad) to validate the proposed control scheme. The RMS results for multiple subjects are shown in Fig. 3.29. As the regions were set larger, more free-moving space was provided inside the regions, where the robot can deviate from the reference trajectory, leading to larger tracking errors.

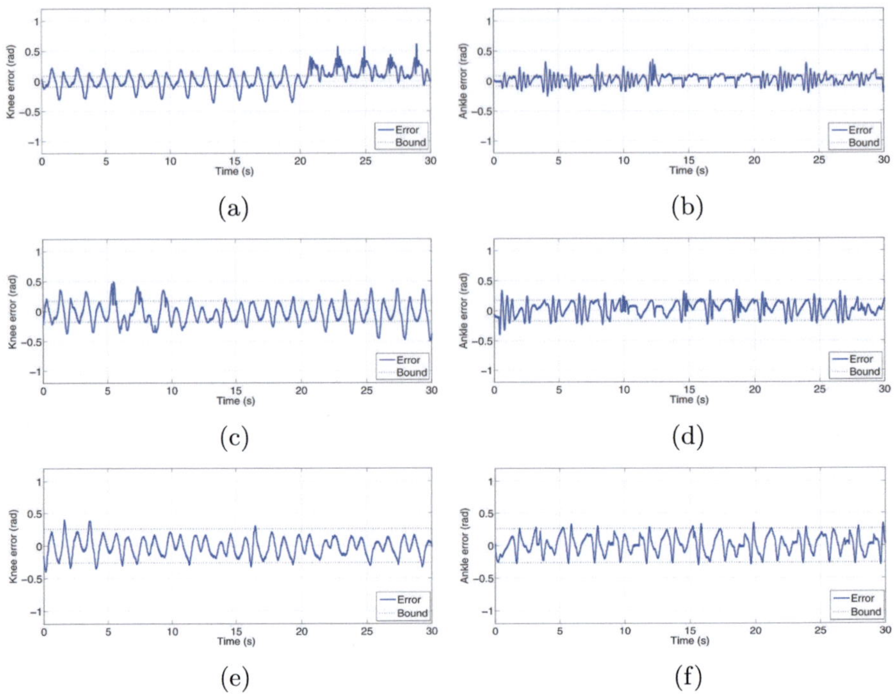

Fig. 3.25 Tracking errors for different joint-space regions. **a** Knee joint (b_{qi}=0.087 rad); **b** Ankle joint (b_{qi}=0.087 rad); **c** Knee joint (b_{qi}=0.17 rad); **d** Ankle joint (b_{qi}=0.17 rad); **e** Knee joint (b_{qi}=0.26 rad); **f** Ankle joint (b_{qi}=0.26 rad)

3.4 Summary and Notes

This chapter first reviews several exoskeleton robots, e.g., the lower-limb exoskeleton can support the weight of wearers, and the upper limb can augment the wearer's manipulation ability. Then, we present two important modules of exoskeleton robotics: intention estimation and interactive control. Intention estimation enables the robot to capture human motion intentions in real time and adjust its trajectories accordingly, ensuring that the assistance provided aligns seamlessly with the user's intent. Interactive control, on the other hand, facilitates effective human-robot interaction between the exoskeleton and the wearer, ensuring precise trajectory tracking and torque control during this process. Together, these components form the foundation for safe and efficient exoskeleton operation.

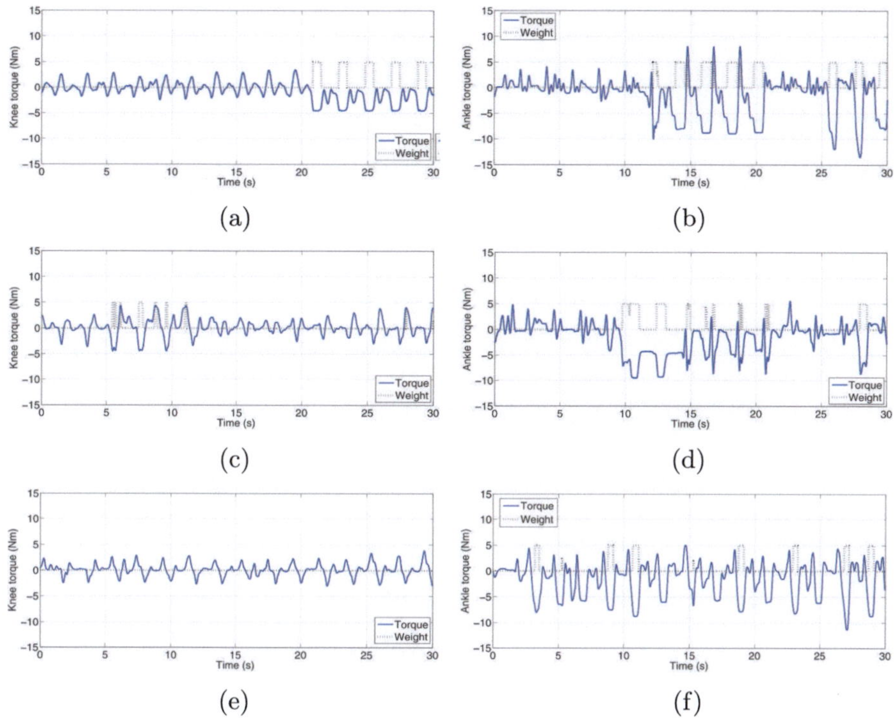

Fig. 3.26 Output torques for different joint-space regions. **a** Knee joint (b_{qi}=0.087 rad); **b** Ankle joint (b_{qi}=0.087 rad); **c** Knee joint (b_{qi}=0.17 rad); **d** Ankle joint (b_{qi}=0.17 rad); **e** Knee joint (b_{qi}=0.26 rad); **f** Ankle joint (b_{qi}=0.26 rad)

Current approaches to intention estimation include methods based on inertial measurement units [51], electromyography sensors [52], or electroencephalogram signals [53]. Our approach leverages the fusion of multiple sensors and employs a dual-channel intention estimation framework. Specifically, one channel outputs the short-term intent based on the physical interaction model; the other channel outputs the long-term intent by exploring the probabilistic model. This enables real-time and highly accurate recognition of the user's motion intent, significantly improving the efficiency of the exoskeleton's assistance while adapting to varying user needs.

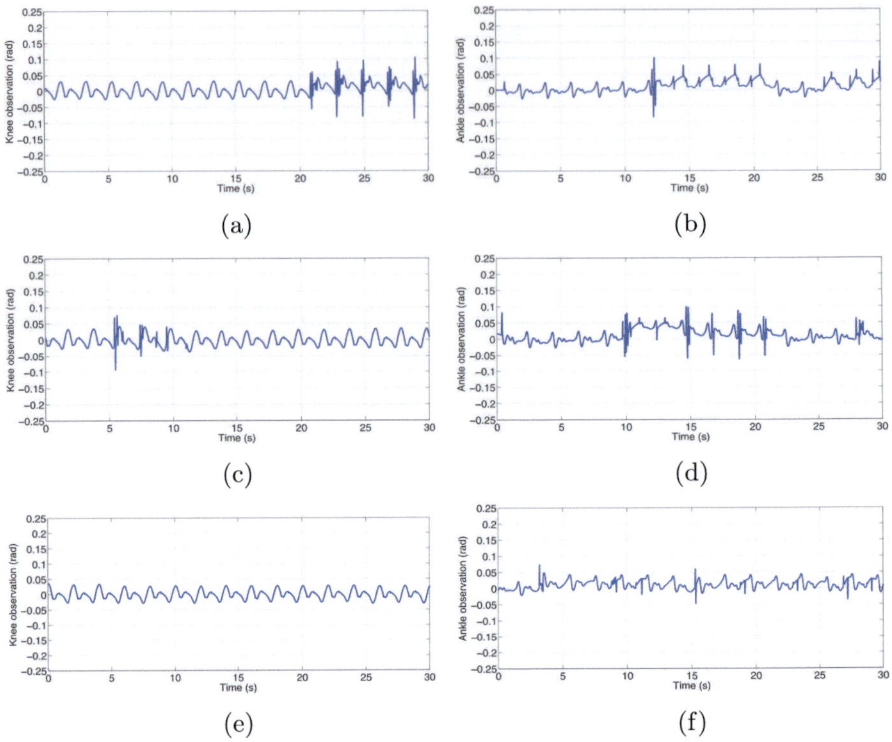

Fig. 3.27 Observation errors for different joint-space regions. **a** Knee joint (b_{qi}=0.087 rad); **b** Ankle joint (b_{qi}=0.087 rad); **c** Knee joint (b_{qi}=0.17 rad); **d** Ankle joint (b_{qi}=0.17 rad); **e** Knee joint (b_{qi}=0.26 rad); **f** Ankle joint (b_{qi}=0.26 rad)

For interactive control, existing methods often rely on PID-based controllers [54] or adaptive control schemes to achieve trajectory tracking for exoskeletons [55]. In contrast, our method incorporates a deeper understanding of human participation during assistance. It adjusts the level of robotic assistance dynamically and ensures a safe stop under special conditions. By utilizing the proposed multi-mode control scheme, we achieve a regulated interaction between the human and the exoskeleton based on a series of regions. The transition between multiple regions is smooth, and hence, multiple interaction modes are stably integrated into a single control input. Such an integration can ensure safe interaction between the robot and the wearer and encourage the voluntary efforts exerted by the wearer to speed up patient recovery.

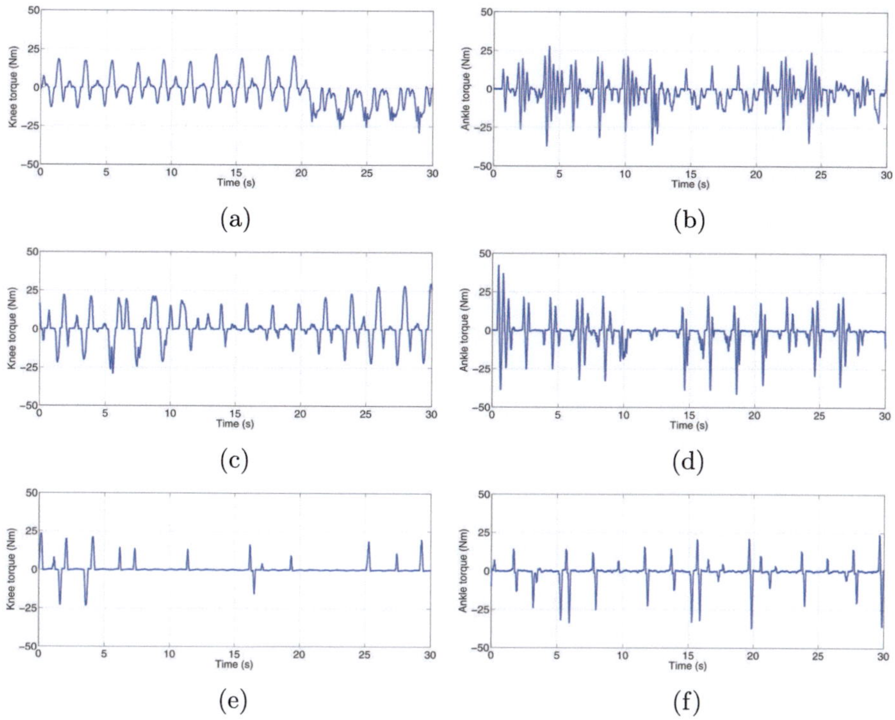

Fig. 3.28 Control inputs for different joint-space regions. **a** Knee joint (b_{qi}=0.087 rad); **b** Ankle joint (b_{qi}=0.087 rad); **c** Knee joint (b_{qi}=0.17 rad); **d** Ankle joint (b_{qi}=0.17 rad); **e** Knee joint (b_{qi}=0.26 rad); **f** Ankle joint (b_{qi}=0.26 rad)

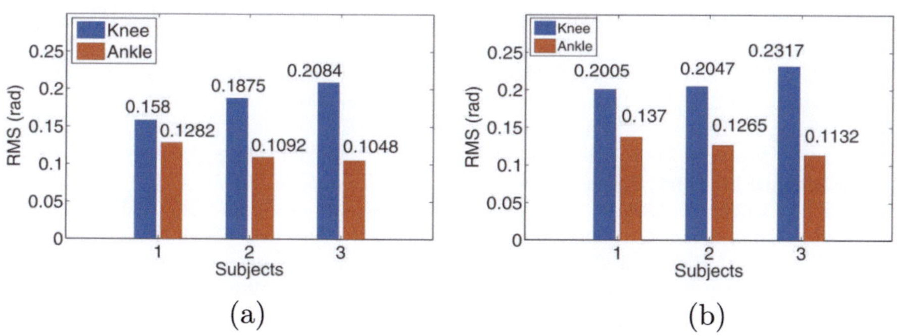

Fig. 3.29 Experimental results for multiple subjects. **a** Smaller joint-space regions (b_{qi}=0.087 rad); **b** Larger joint-space regions (b_{qi}=0.17 rad)

References

1. M. Cestari, D. Sanz-Merodio, J.C. Arevalo, E. Garcia, An adjustable compliant joint for lower-limb exoskeletons. IEEE/ASME Trans. Mechatron. **20**(2), 889–898 (2014)
2. J. Pan, D. Astarita, A. Baldoni, F. Dell'Agnello, S. Crea, N. Vitiello, E. Trigili, Nesm-γ: An upper-limb exoskeleton with compliant actuators for clinical deployment. IEEE Robot. Autom. Lett. **7**(3), 7708–7715 (2022)
3. Z. Li, Z. Huang, W. He, C.-Y. Su, Adaptive impedance control for an upper limb robotic exoskeleton using biological signals. IEEE Trans. Industr. Electron. **64**(2), 1664–1674 (2016)
4. W. Huo, H. Moon, M.A. Alouane, V. Bonnet, J. Huang, Y. Amirat, R. Vaidyanathan, S. Mohammed, Impedance modulation control of a lower-limb exoskeleton to assist sit-to-stand movements. IEEE Trans. Rob. **38**(2), 1230–1249 (2021)
5. K.A. Witte, P. Fiers, A.L. Sheets-Singer, S.H. Collins, Improving the energy economy of human running with powered and unpowered ankle exoskeleton assistance. Sci. Robot. **5**(40), eaay9108 (2020)
6. M.A. Ergin, V. Patoglu, A self-adjusting knee exoskeleton for robot-assisted treatment of knee injuries, in *2011 IEEE/RSJ International Conference on Intelligent Robots and Systems* (IEEE, 2011), pp. 4917–4922
7. K. Seo, J. Lee, Y. Lee, T. Ha, Y. Shim, Fully autonomous hip exoskeleton saves metabolic cost of walking, in *2016 IEEE International Conference on Robotics and Automation (ICRA)* (IEEE, 2016), pp. 4628–4635
8. L. Zhou, W. Chen, W. Chen, S. Bai, J. Zhang, J. Wang, Design of a passive lower limb exoskeleton for walking assistance with gravity compensation. Mech. Mach. Theory **150**, 103840 (2020)
9. A. Nasr, S. Bell, J. McPhee, Optimal design of active-passive shoulder exoskeletons: A computational modeling of human-robot interaction. Multibody Syst. Dyn. **57**(1), 73–106 (2023)
10. T. Chen, R. Casas, P.S. Lum, An elbow exoskeleton for upper limb rehabilitation with series elastic actuator and cable-driven differential. IEEE Trans. Rob. **35**(6), 1464–1474 (2019)
11. J.A. Martinez, P. Ng, S. Lu, M.S. Campagna, O. Celik, Design of wrist gimbal: A forearm and wrist exoskeleton for stroke rehabilitation, in *2013 IEEE 13th International Conference on Rehabilitation Robotics (ICORR)* (IEEE, 2013), pp. 1–6
12. B. Kim, A.D. Deshpande, An upper-body rehabilitation exoskeleton harmony with an anatomical shoulder mechanism: Design, modeling, control, and performance evaluation. Int. J. Robot. Res. **36**(4), 414–435 (2017)
13. S.J. Park, K. Choi, H. Rodrigue, C.H. Park, Soft exosuit based on fabric muscle for upper limb assistance. IEEE/ASME Trans. Mechatron. **28**(1), 26–37 (2022)
14. X. Zhang, E. Tricomi, X. Ma, M. Gomez-Correa, A. Ciaramella, F. Missiroli, L. Mišković, H. Su, L. Masia, A lower limb wearable exosuit for improved sitting, standing, and walking efficiency. IEEE Trans. Robot. (2024)
15. G. Li, L. Cheng, Z. Gao, X. Xia, J. Jiang, Development of an untethered adaptive thumb exoskeleton for delicate rehabilitation assistance. IEEE Trans. Rob. **38**(6), 3514–3529 (2022)
16. H. Kim, L.M. Miller, N. Byl, G.M. Abrams, J. Rosen, Redundancy resolution of the human arm and an upper limb exoskeleton. IEEE Trans. Biomed. Eng. **59**(6), 1770–1779 (2012)
17. M.A.T. Ho, Y. Yamada, Y. Umetani, An adaptive visual attentive tracker for human communicational behaviors using hmm-based td learning with new state distinction capability. IEEE Trans. Rob. **21**(3), 497–504 (2005)
18. L.M. Hiatt, C. Narber, E. Bekele, S.S. Khemlani, J.G. Trafton, Human modeling for human-robot collaboration. Int. J. Robot. Res. **36**(5–7), 580–596 (2017)
19. Q. Wang, W. Jiao, R. Yu, M.T. Johnson, Y. Zhang, Virtual reality robot-assisted welding based on human intention recognition. IEEE Trans. Autom. Sci. Eng. **17**(2), 799–808 (2019)
20. M.C. Khoo, *Physiological Control Systems: Analysis, Simulation, and Estimation* (Wiley, 2018)
21. Y. Li, S.S. Ge, Human-robot collaboration based on motion intention estimation. IEEE/ASME Trans. Mechatron. **19**(3), 1007–1014 (2013)

22. M.M. Rahman, R. Ikeura, K. Mizutani, Investigation of the impedance characteristic of human arm for development of robots to cooperate with humans. JSME Int J., Ser. C **45**(2), 510–518 (2002)
23. A. Kanazawa, J. Kinugawa, K. Kosuge, Adaptive motion planning for a collaborative robot based on prediction uncertainty to enhance human safety and work efficiency. IEEE Trans. Rob. **35**(4), 817–832 (2019)
24. X. Li, Y. Pan, G. Chen, H. Yu, Multi-modal control scheme for rehabilitation robotic exoskeletons. Int. J. Robot. Res. **36**(5–7), 759–777 (2017)
25. C.C. Cheah, X. Li, *Task-Space Sensory Feedback Control of Robot Manipulators*, vol. 73 (Springer, 2015)
26. S. Arimoto, *Control Theory of Non-Linear Mechanical Systems: A Passivity-Based and Circuit-Theoretic Approach* (Oxford University Press, London, U.K., 1996)
27. R. Mortensen, Stochastic stability and control (harold j. kushner). SIAM Rev. **10**(4), 460–462 (1968)
28. X. Li, C.C. Cheah, Stochastic optical trapping and manipulation of a micro object with neural-network adaptation. IEEE/ASME Trans. Mechatron. **22**(6), 2633–2642 (2017)
29. S. Theodoridis, K. Koutroumbas, *Pattern Recognition* (Elsevier, 2006)
30. X. Li, Y.-H. Liu, H. Yu, Iterative learning impedance control for rehabilitation robots driven by series elastic actuators. Automatica **90**, 1–7 (2018)
31. X. Li, Y.-H. Liu, H. Yu, Adaptive impedance control for compliantly actuated robots with a unified safety measure, in *2018 13th World Congress on Intelligent Control and Automation (WCICA)* * (IEEE, 2018), pp. 444–449
32. C.C. Cheah, D.W. Wang, Learning impedance control for robotic manipulators (1998)
33. H.J. Kushner, Stochastic stability, in *Stability of Stochastic Dynamical Systems: Proceedings of the International Symposium Organized by "The Control Theory Centre", University of Warwick, July 10–14, 1972 Sponsored by the "International Union of Theoretical and Applied Mechanics"* (Springer, 2006), pp. 97–124
34. Q.-C. Pham, N. Tabareau, J.-J. Slotine, A contraction theory approach to stochastic incremental stability. IEEE Trans. Autom. Control **54**(4), 816–820 (2009)
35. C.A. Oatis, The use of a mechanical model to describe the stiffness and damping characteristics of the knee joint in healthy adults. Phys. Ther. **73**(11), 740–749 (1993)
36. A. Holobar, M.A. Minetto, A. Botter, F. Negro, D. Farina, Experimental analysis of accuracy in the identification of motor unit spike trains from high-density surface emg. IEEE Trans. Neural Syst. Rehabil. Eng. **18**(3), 221–229 (2010)
37. C.A. Williams, S. Ratel, Definitions of muscle fatigue *Human muscle fatigue* (2009), pp. 3–16
38. B.K. Arya, J. Mohapatra, K. Subramanya, H. Prasad, R. Kumar, M. Mahadevappa, Surface emg analysis and changes in gait following electrical stimulation of quadriceps femoris and tibialis anterior in children with spastic cerebral palsy, in *2012 Annual International Conference of the IEEE Engineering in Medicine and Biology Society* (IEEE, 2012), pp. 5726–5729
39. C.-J. Hsu, J. Kim, E.J. Roth, W.Z. Rymer, M. Wu, Use of pelvic corrective force with visual feedback improves paretic leg muscle activities and gait performance after stroke. IEEE Trans. Neural Syst. Rehabil. Eng. **27**(12), 2353–2360 (2019)
40. J. Zhang, P. Fiers, K.A. Witte, R.W. Jackson, K.L. Poggensee, C.G. Atkeson, S.H. Collins, Human-in-the-loop optimization of exoskeleton assistance during walking. Science **356**(6344), 1280–1284 (2017)
41. H.A. Yousif, A. Zakaria, N.A. Rahim, A.F.B. Salleh, M. Mahmood, K.A. Alfarhan, L.M. Kamarudin, S.M. Mamduh, A.M. Hasan, M.K. Hussain, Assessment of muscles fatigue based on surface emg signals using machine learning and statistical approaches: A review, in *IOP Conference Series: Materials Science and Engineering*, vol. 705, (IOP Publishing, 2019), p. 012010
42. K. Kong, M. Tomizuka, Control of exoskeletons inspired by fictitious gain in human model. IEEE/ASME Trans. Mechatron. **14**(6), 689–698 (2009)
43. M.W. Spong, Adaptive control of flexible joint manipulators: Comments on two papers. Automatica **31**(4), 585–590 (1995)

44. H. Khalil, *Nonlinear Systems* (Macmillan, New York, 1992)
45. J. Zhang, C.C. Cheah, Passivity and stability of human-robot interaction control for upper-limb rehabilitation robots. IEEE Trans. Rob. **31**(2), 233–245 (2015)
46. X. Li, C.C. Cheah, Adaptive neural network control of robot based on a unified objective bound. IEEE Trans. Control Syst. Technol. **22**(3), 1032–1043 (2013)
47. R. Ronsse, N. Vitiello, T. Lenzi, J. van den Kieboom, M.C. Carrozza, A.J. Ijspeert, Human-robot synchrony: Flexible assistance using adaptive oscillators. IEEE Trans. Biomed. Eng. **58**(4), 1001–1012 (2011)
48. H. Yu, S. Huang, G. Chen, N. Thakor, Control design of a novel compliant actuator for rehabilitation robots. Mechatronics **23**, 1072–1083 (2013)
49. G. Chen, P. Qi, Z. Guo, H. Yu, Mechanical design and evaluation of a compact portable knee-ankle-foot robot for gait rehabilitation. Mech. Mach. Theory **103**, 51–64 (2016)
50. M.W. Whittle, *Gait Analysis: An Introduction* (Elsevier, 2003)
51. L. Zhu, Z. Wang, Z. Ning, Y. Zhang, Y. Liu, W. Cao, X. Wu, C. Chen, A novel motion intention recognition approach for soft exoskeleton via imu. Electronics **9**(12), 2176 (2020)
52. L. Bi, C. Guan et al., A review on emg-based motor intention prediction of continuous human upper limb motion for human-robot collaboration. Biomed. Signal Process. Control **51**, 113–127 (2019)
53. A. Buerkle, W. Eaton, N. Lohse, T. Bamber, P. Ferreira, Eeg based arm movement intention recognition towards enhanced safety in symbiotic human-robot collaboration. Robot. Comput.-Integrated Manufact. **70**, 102137 (2021)
54. W. Yu, J. Rosen, Neural pid control of robot manipulators with application to an upper limb exoskeleton. IEEE Trans. Cybernet. **43**(2), 673–684 (2013)
55. B. Brahmi, M. Saad, C. Ochoa-Luna, M.H. Rahman, Adaptive control of an exoskeleton robot with uncertainties on kinematics and dynamics, in *2017 International Conference on Rehabilitation Robotics (ICORR)* (IEEE, 2017), pp. 1369–1374

Chapter 4
Interactive Learning

The primary objective of an exoskeleton robot varies depending on its application. For assistive exoskeletons, the goal is often to enhance their physical capabilities based on a given task and human motion intention. In contrast, rehabilitation-based exoskeletons may focus on providing specific resistance or assistance to promote effort, motor learning, and recovery. In both applications, considerable data is generated during the physical interaction between the human and the robot. This chapter discusses the assistive application and presents the results of how the robot shapes and improves its assistance via interaction learning (that is, obtaining and utilizing data via physical interaction, dealing with potential conflict during the interaction, and improving the wearer's comfort and the robot's assistance). The content covers both the lower limb and the upper limb exoskeleton robots, where the former supports the body weight, and the latter augments the manipulation ability.

The interactive learning method for the lower-limb exoskeleton is proposed under the framework of Human-In-the-Loop, which refers to an optimization paradigm where human physiological or behavioral responses are actively incorporated into the control and adaptation process of the exoskeleton. In this framework, real-time user feedback, such as metabolic cost, muscle activation, or interaction forces, is utilized to iteratively adjust the exoskeleton's assistance parameters, ensuring personalized and adaptive support tailored to the wearer's needs. It begins by delving into anomaly detection and constructing an end-to-end network to detect conflicts between the wearer and the robot with proprioceptive sensors. Then, the constructed network outputs the anomaly score and guides the generation of an individualized trajectory, aiming to optimize the reference trajectory of the robotic exoskeleton based on the wearer's specific needs and movements. The advantage of the proposed method is the generalization to multiple subjects across various tasks, exploring how these trajectory generation methods can be applied universally, considering diverse individuals and tasks within human-robot interaction.

The interactive learning method for the upper-limb exoskeleton is proposed by referring to its repetitive nature in many tasks. Specifically, an iterative learning

X. Li and H. Yu, *Compliantly-Driven Robots with Application to Exoskeletons*,
https://doi.org/10.1007/978-981-96-9544-7_4

scheme is applied to approximate the unmodeled disturbance in the absence of wearers. Then, the estimated disturbance is compensated for, and a trajectory-tracking controller is implemented to drive the robot to track the desired trajectory for assistance in the presence of wearers. The advantage of the proposed method is that it is able to isolate the disturbance torques from the interaction ones without external forces, hence rejecting the former and utilizing the latter.

4.1 Lower-Limb Exoskeletons

Most of the current lower-limb exoskeletons are designed for a specific task or personalized for individual users. When they are deployed for different tasks or users, manual programming or even fine-tuning is typically necessary to set the assistance strategy or reference trajectory. However, the ability to learn online and adapt to multiple tasks and wearers is important to ensure that the robot provides effective assistance in practical applications, which has relatively less progress. The proposed interactive learning method in this section aims to overcome the aforementioned limitations.

4.1.1 Anomaly Score

A robotic exoskeleton provides assistive forces to the human subject via the close and intensive coupling between the wearer and the robot. However, a mismatch between the wearer and the robot may result in physical conflict. Such discordant scenarios are prevalent not only in rehabilitation contexts but also in other applications involving compliance-driven robots, such as human-machine collaboration. Consequently, to enhance human-machine interaction, it is imperative to detect anomalies with high accuracy. The challenge lies in promptly and accurately identifying these conflicts, given the variable dynamic model of human limbs and the inherent uncertainty in human motion intention. Failure to detect such conflicts could affect assistance efficiency or even compromise safety. Given that conflicts can be successfully detected, the exoskeleton robot should also properly adjust its behavior to address or relax said conflicts. As depicted in Fig. 4.1, a conflict emerges when the movement of the exoskeleton robot is either excessively rapid or too slow in comparison to the wearer's pace. Such discrepancies can potentially harm the wearer. Another example is that the robot's assistance does not suit the intention of human motion.

In general, physical conflicts during human-robot interaction are generally treated as anomalies. Anomaly detection methods can detect abnormal values in the monitoring data so as to find abnormal conditions during operation, such as machine failure, operation error, and so on.

Compared to the model-based method, deep anomaly detection (DAD) technology is able to capture the relationship between multi-modal data and the execut-

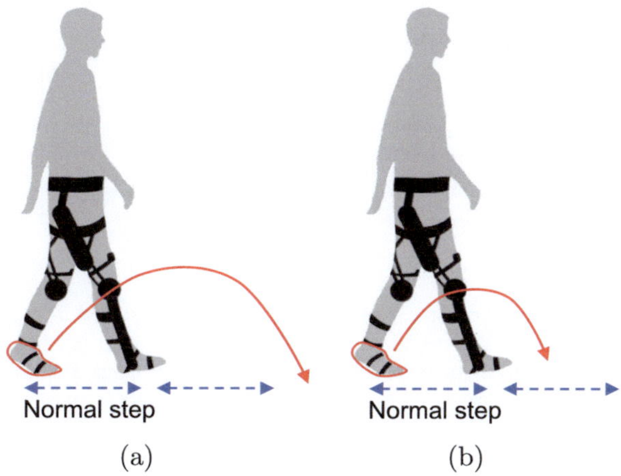

Fig. 4.1 The illustration of conflict in asynchronous footstep scenarios: **a** the subsequent step being disproportionately large; **b** the subsequent step being disproportionately small

ing task, thereby learning complex features to deal with large-scale data sets [1]. For example, k-mean clustering algorithms [2] and the Local Outlier Factor (LOF) algorithm [3, 4] have been employed to solve the anomaly detection problem. Additionally, deep residual autoencoder technology has been used to detect anomalies in an unsupervised form [5] accurately. In the realm of time-series data, methods like LSTM-based anomaly detection [6] and LSTM-based variational autoencoders [7] have been introduced to model the probability distribution of observations. Others include GRU-based multivariate time series anomaly detectors [8].

This chapter studies a new learning scheme to detect physical conflicts between humans and robots. The constructed learning network receives multi-modal information from proprioceptive sensors and then outputs the anomaly score to specify the physical conflict, which score is further used to continuously adjust the robot impedance to ensure a safe and efficient interaction. Such a formulation allows the robot to explore the semantic information during the interaction (e.g., gait phases, imbalance, human fatigue) and hence react properly to the physical conflict.

The key component of the detection network is the variable auto-encoder (VAE), and its principle is illustrated in Fig. 4.2.

Variational Autoencoders (VAEs) are one of the most popular deep-learning-based generative models besides Generative Adversarial Networks (GANs). VAE inherits a similar network frame to plain Autoencoder (AE), which is comprised of two parts: an encoder and a decoder. The encoder maps the high-dimensional input data $x \in \Re^d$ into a low-dimensional variable $l \in \Re^h$ in latent space for feature extraction. The decoder maps the latent variable back to high-dimensional space $\hat{x} \in \Re^d$, attempting to reconstruct the original input. Therefore, VAE intrinsically captures the main feature of input data only.

VAE presumes that the latent variables obey a specific prior distribution and computes the posterior as

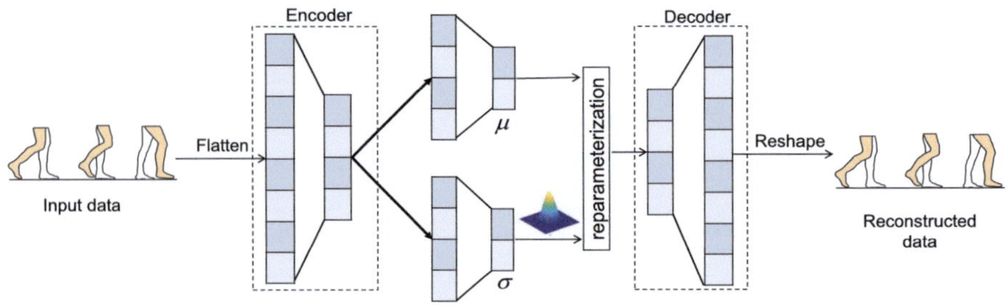

Fig. 4.2 In the context of a VAE, the encoder is used to map the input data into a latent space, defined by the distribution parameters μ (mean) and σ (standard deviation). This is followed by the reparameterization step, which facilitates backpropagation through stochastic nodes. Subsequently, a decoder is employed to reconstruct the input data from this latent representation

$$p_\theta(l|x) = \frac{p_\theta(x|l)p_\theta(l)}{p_\theta(x)}, \tag{4.1}$$

However, the computation of marginal likelihood

$$p_\theta(x) = \int p_\theta(l)p_\theta(x|l)\mathrm{d}l \tag{4.2}$$

is expensive and intractable in most cases, especially when the latent variable l is high-dimensional, which leads the aforementioned posterior $p_\theta(l|x)$ to be intractable.

Variational inference can be applied to address this issue by introducing a function to approximate the posterior as

$$q_\phi(l|x) \approx p_\theta(l|x), \tag{4.3}$$

where $q_\phi(l|x)$ is tractable, such as KL-divergence, which is introduced to describe the similarity between true posterior $p_\theta(l|x)$ and its approximate $q_\phi(l|x)$. The logarithmic marginal likelihood can be given as

$$log p_\theta(x) = KL(q_\phi(l|x)||p_\theta(l|x)) + \mathcal{L}(\theta, \phi; x). \tag{4.4}$$

Since KL-divergence is non-negative, $\mathcal{L}(\theta, \phi; x)$ is referred to as evidence lower bound loss function (ELBO), hence the problem of minimizing KL-divergence is transformed into a problem of maximizing the ELBO.

This chapter refers to the use of proprioceptive sensors for anomaly detection by feeding the information from proprioceptive sensors into the above VAE network. Proprioception, which is also known as kinesthesia, describes the perception of movement, torque, and limb position. For many creatures, proprioception plays an important role in body balance and coordinating limb movement. Physiological studies have shown that limb movement, position, and velocity are encoded by a cluster of sensory neurons. The proprioceptive sense is widely accepted to be composed of

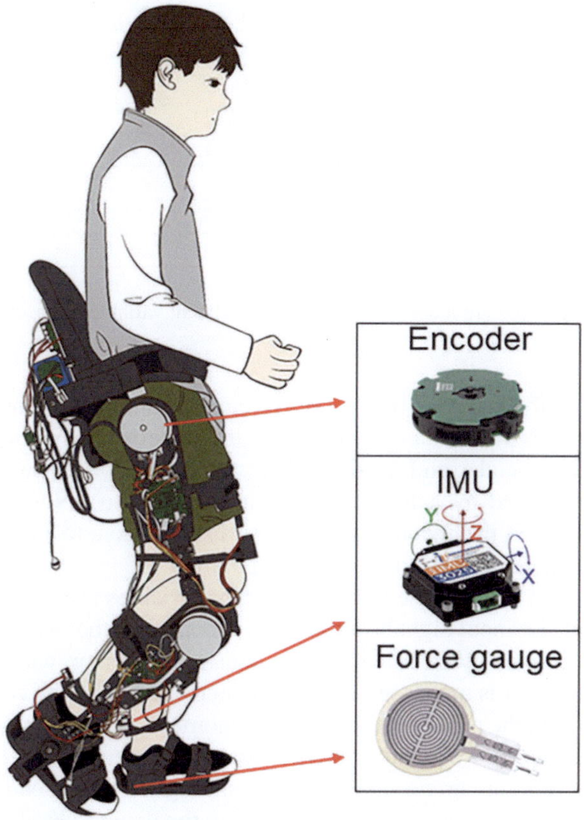

Fig. 4.3 The exoskeleton system is equipped with three types of proprioceptive sensors: IMUs, encoders, and force gauges. These sensors play a crucial role in assessing the walking state by providing vital information about the wearer's movement dynamics and interaction forces

signals from stretch mechanoreceptors in muscles and ligaments and sensory neurons in the inner ear.

Three typical proprioceptive sensors and their purposes for a robotic exoskeleton include (see Fig. 4.3):

(1) *IMUs*: to identify the gait phases;
(2) *Torques*: to measure the physical interaction between human and robot;
(3) *Encoders*: to assess the synchronization between human and robot.

The use of multi-modal information helps the robot effectively understand and detect physical conflicts at a semantic level. The incorporation of proprioceptive sensors, without external ones, also ensures that the approach is both easy to implement and cost-effective.

The aforementioned three sensory inputs are fused in the VAE network to learn the normal patterns of those signals during movement. So, if any anomaly occurs, which means collected movement signals are beyond normal patterns, the detection network can perceive it and its intensity; subsequently, measures are taken to ensure the safety of wearers.

Table 4.1 The architecture of detection network

Layer	Input size	Output size
Flatten	(100, 10)	(1000, 1)
FC[1]	(1000, 1)	(400, 1)
FC (μ, σ)[2]	(400, 1)	(50, 1)
FC	(50, 1)	(400,1)
FC	(400, 1)	(1000, 1)
Reshape	(1000, 1)	(100, 10)

[1] FC denotes fully-connected layer
[2] μ and σ are mean and variance of the latent variables respectively

To handle the time-series signals, the sliding window is introduced to represent the segment of data and then fed into the network. In addition, the architecture of both the encoder and decoder in this setup comprises two fully connected layers, each activated by a Rectified Linear Unit (ReLu). Notably, the initial fully-connected layer of the encoder functions as an equivalent to a flattened layer. The details of this architecture are summarized in Table 4.1.

In the proposed network, latent variables are utilized to capture normal patterns of the input signals and then to reconstruct the original data. The anomaly score is obtained by calculating the reconstruction error, which increases when the physical conflict is more significant and decreases when the conflict is relaxed.

In the process of training an anomaly detection model, the collected exoskeleton equipment data are standardized and converted first, and the input data are divided into multivariate time series by sliding window. Send the divided training data into the model training module, capture the data mode of the exoskeleton during normal operation, learn the model that outputs the abnormal score of exoskeleton operation, and then use the pot method to automatically select the abnormal threshold through the threshold selection module.

GRU is used in the network to capture the complex time dependence of the observed values of various devices in the multivariate time series. The input sensor observations and other data are mapped to the potential space, and the series vector of random variables and GRU hidden variables is obtained. In addition, the linear Gaussian state space model (SSM) is used to model the time correlation between random variables and series vectors in potential space. To train the proposed network, the latent variable l is reparameterized and given as

$$l = \mu + \sigma \odot \epsilon, \tag{4.5}$$

where μ and σ are parameters obtained from the encoder, which represent mean and standard deviation, respectively. Sample noise is set to obey a standard normal distribution, i.e., $\epsilon \sim \mathcal{N}(0, \boldsymbol{I})$.

Next, the loss function is formulated as

$$\mathcal{L}_{\text{train}}(\mathbf{x}^t) = ||\mathbf{x}^t - \hat{\mathbf{x}}^t||^2 + KL[\mathcal{N}(\mu, \sigma), \mathcal{N}(0, \boldsymbol{I})], \tag{4.6}$$

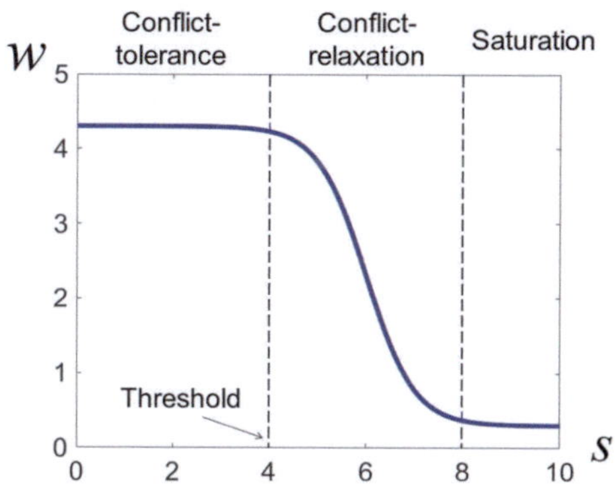

Fig. 4.4 An example of the weighting function, whose variation is to modulate the robot's impedance. Specifically, $\lambda_1 = 2$, $\lambda_2 = 2.3$, $m = 1$, $h = 6$. Note that the length of the "conflict-relaxation" zone is given as $4 < s < 8$, where the robot is to lower its output impedance to become passive. When $s > 8$, the weighting function is saturated to maintain the minimal assistance

which comprises both the reconstruction error and the KL divergence, where $\hat{\mathbf{x}}^t$ denotes the output of the decoder.

Then, the anomaly score can be defined as

$$s = MSE(\mathbf{x}^t, \hat{\mathbf{x}}^t) = ||\mathbf{x}^t - \hat{\mathbf{x}}^t||^2 \tag{4.7}$$

The exoskeleton robot refers to the anomaly score to adjust its assistance and then relaxes physical conflicts with the wearer. The adjustment should fulfill the following requirements.

(1) When the anomaly score oscillates around a small value, the robot should ignore such minor conflicts to avoid overreaction;
(2) When the anomaly score exceeds a certain threshold, significant conflicts may arise, such that the robot should quickly react to address them.

To achieve it, the anomaly score is mapped into a weighting function as

$$w(s) = \lambda_1 \tanh(-\frac{s}{m} + h) + \lambda_2 \tag{4.8}$$

where λ_1 and λ_2 are two positive constants, determine the range and middle of the weighting function, m is a constant to normalize the anomaly score into a specified small range, and h represents the offset of the weighting function from the origin of coordinates along the horizontal axis's positive direction. An example of the weighting function is shown in Fig. 4.4.

The dataset for the training and evaluation of the detection network was collected during ground-walking trials; that is, the human subject wore the robot and walked by following a predefined trajectory under the fixed impedance model. To standardize

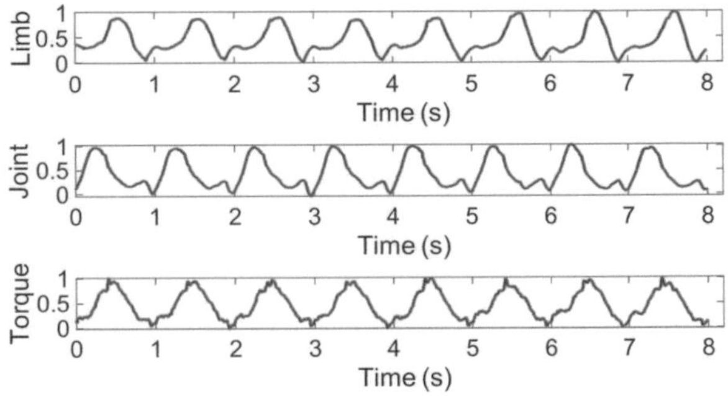

Fig. 4.5 Processed data from proprioceptive sensors, including the angle of human limb (top, measured with IMUs), the angle of robot joint (middle, measured with encoders), and the interaction torque (bottom, measured with encoders and known stiffness)

the sensory data for consistency, the min-max normalization technique was applied, ensuring all data were on the same scale, as illustrated in Fig. 4.5. Data collected when the human-robot interaction proceeded smoothly, allowing the human (wearing the exoskeleton) to follow the set trajectory accurately, was categorized as normal. Conversely, instances where physical conflicts disrupted this synergy between the human and the robot were labeled as abnormal.

During the experiment, the human subject intentionally held the robot, which resulted in a simulated conflict, and the output of the detection network was shown in Fig. 4.6a. As can be seen, the reconstruction error increased significantly when the conflict arose, proving the anomalous issue could be recognized and successfully separated. The performance of the detection network was also validated in another simulated conflict, where the human subject lost balance due to fatigue. The results are shown in Fig. 4.6b, also proving the successful detection of such a conflict.

In addition, ablation studies were carried out to show that using the multi-modal information achieves more accurate detection compared to single sensory input; the comparison results are shown in Fig. 4.7. Note that the curve of receiver operating characteristic (ROC) was introduced to describe the detection performance, which implies how well a classification model can distinguish between classes. In general, a larger area under the curve (AUC) indicates a better performance, and vice versa. It can be seen from Fig. 4.7 that the multi-modal detection has a higher AUC than single sensory input in both scenarios (i.e., asynchronization and imbalance), that is, the proposed scheme with multi-modal information ensured more accurate detection. This was because the multi-modal information was able to handle different anomaly scenarios with specific manifestations. For example, in the presence of asynchronization between the human and robot, the interaction torque quickly increases. When an imbalance issue arises due to human fatigue, the angular information about the human limb, such as IMUs signals, would change more obviously.

According to the experimental results, the proposed anomaly detection network based on neural networks can accurately identify various anomalies and provides quantified evaluation metrics for these anomalies, namely, anomaly scores.

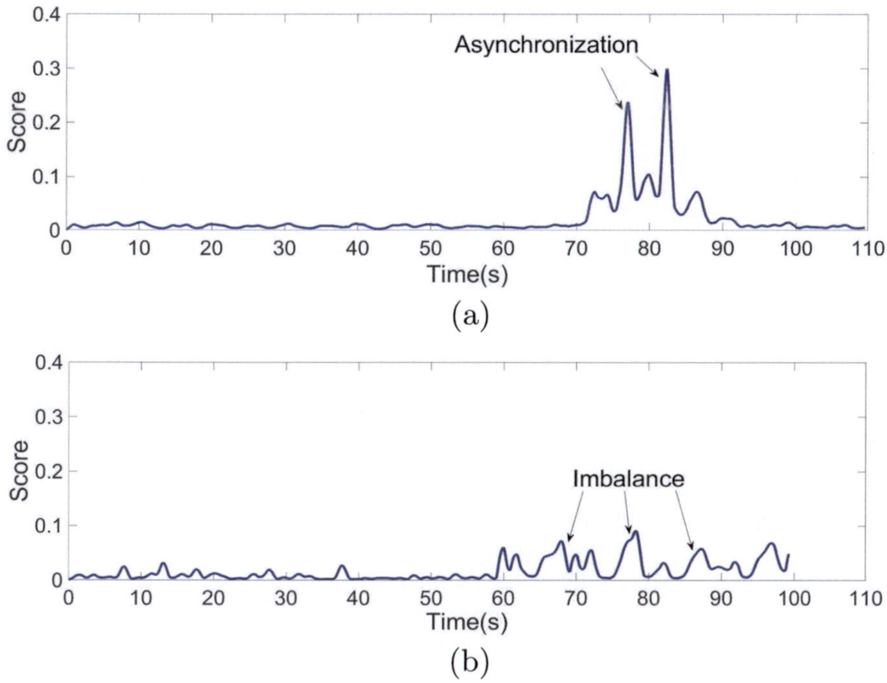

Fig. 4.6 Anomaly scores generated by the physical-detection network: **a** The human intentionally held the robot to simulate the synchronization; **b** The imbalance issue arose due to human fatigue

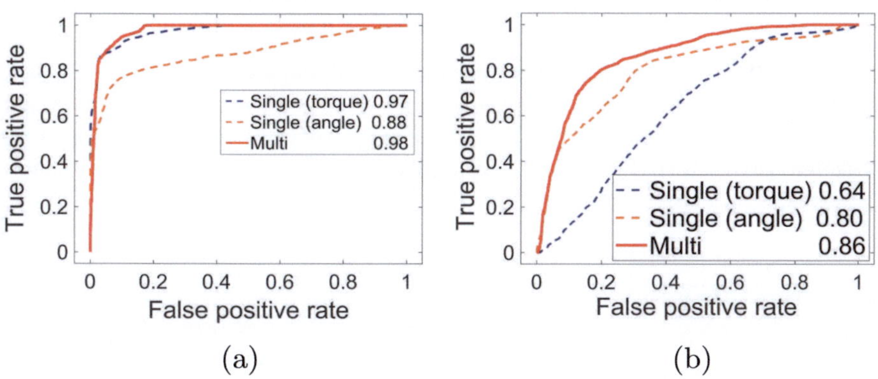

Fig. 4.7 The AUC value (shown in the legend) and the ROC curve of multi-modal detection with respect to the single one (using the torque or the IMU information only): **a** the scenario of asynchronization; **b** the scenario of imbalance

4.1.2 Human-in-the-Loop Optimization

This section details the human-in-the-loop (HIL) learning and adaptation framework for exoskeleton robots to improve their performance in various tasks and for different wearers. To suit different wearers, an individualized walking trajectory is generated online using dynamic movement primitives and Bayesian optimization. To accommodate various tasks, a task translator is constructed using a neural network to generalize a trajectory to more complex scenarios. In addition, the aforementioned anomaly detection network is developed to quantitatively evaluate the wearer's comfort, which is considered in the trajectory learning procedure. The proposed framework is easy to implement because it requires proprioceptive sensors only to perform and deploy data-efficient learning schemes. This makes the exoskeleton practical for deployment in complex scenarios, accommodating different walking patterns, habits, tasks, and conflicts.

The procedure of this HIL optimization is summarized as follows:

(1) An unsupervised detection network is constructed to address anomalies during the task. This network prevents the robot from learning or providing assistance based on incorrect patterns without affecting its generalizability, while optimizing trajectory generation to minimize conflicts.
(2) An online individualizing approach is developed, using dynamic movement primitives (DMP) to parameterize the assistance trajectory and Bayesian optimization to minimize the cost function, enabling the personalized assistance to multiple wearers.
(3) A neural network (NN) is trained to serve as the task translator. This allows optimized assistance for ground walking to be transferred to a variety of tasks that may not be suitable for online learning, such as stair climbing.

The workflow of the proposed learning and control framework is shown in Fig. 4.8. The individualized trajectory is generated through the offline learning phase, and HIL optimization is performed. In the offline stage, multiple human users wear the exoskeleton to perform various tasks (e.g., walking on the ground or ascending stairs).

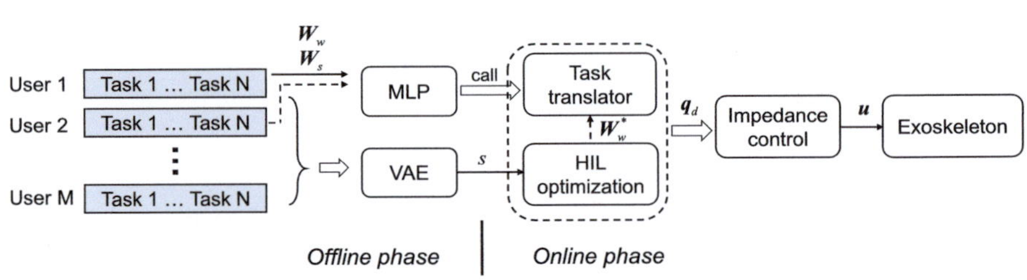

Fig. 4.8 Workflow of the proposed learning and control method, where M is the number of human users, N denotes the number of tasks, $\boldsymbol{W}_w$ and $\boldsymbol{W}_s$ denote the learned parameters from different tasks, s is the anomaly score, $\boldsymbol{W}_w^*$ represents the optimized parameters for a specific human user, $\boldsymbol{q}_d$ is the generated trajectory, and $\boldsymbol{u}$ is the control input

The joint trajectories of the exoskeleton robot are recorded and represented as the joint motion of human subjects. During the collection of datasets, the robot passively follows the subject under a zero-impedance model [9] to simulate the user's natural movements without the exoskeleton.

The collected datasets are used to train anomaly detection and task translation modules. The anomaly detection module shapes the impedance parameter (with the weighting function) and guides HIL optimization, while task translation establishes the mappings among multiple tasks.

In the online phase, HIL optimization is performed to customize the gait trajectory for a human user, considering the body parameters and walking habits among other factors. This trajectory can be adapted for new tasks (e.g., ascending stairs), using the trained task translator. The trajectory is integrated into a variable impedance model, and the robot tracks this model to provide safe and effective assistance. Both HIL optimization and impedance control operate simultaneously in the online phase.

This framework offers several advantages, including ease of implementation and adaptability to multiple tasks and subjects:

(1) Only proprioceptive sensors, which are typically integrated into the exoskeleton, are required for trajectory learning and interaction control, resulting in cost-effectiveness and reduced complexity;
(2) Using common sensors, the robot can effectively adapt to new users and tasks that are not included in the offline training;
(3) This adaptation process requires the data of only a few users to fit the nonlinear relationships between different patterns.

Trajectory Parameterization: Next, we will introduce how to use DMP to encode trajectories, thereby obtaining a dimensionality-reduced representation, along with the generalization methods for different wearers and tasks. The premise of optimizing assistance and transferring tasks relies on dimensionally reducing the assistance trajectories to minimize the exploration space significantly. Here, we use the chosen DMP method as an example to demonstrate how to encode and reduce the dimensions of the trajectories, resulting in a low-dimensional weighting matrix that can encompass most of the trajectory information. Basically, the trajectory is encoded via an *output system*:

$$\dot{\omega}_i = -2\pi \left[\alpha_o(\beta_o \cdot q_{\mathrm{DMP},i} + \omega_i) + \frac{\sum_{j=1}^{J} \Psi_j w_{ij}}{\sum_{j=1}^{J} \Psi_j} \right], \tag{4.9}$$

$$\dot{q}_{\mathrm{DMP},i} = 2\pi \cdot \omega_i, \tag{4.10}$$

$$\Psi_j = \exp[h(\cos(2\pi \cdot t - c_j) - 1)], \tag{4.11}$$

where $q_{\mathrm{DMP},i}$ and ω_i correspond to the normalized joint angle and velocity, respectively, for the i-th joint. The parameters α_o and β_o are positive constants. The term w_{ij} represents the weights of the i-th joint to be updated, which can be treated as the trajectory parameters, and Ψ_j represents Gaussian-like kernel functions with a width h and center c_j. In practice, after determining the number of kernels J, h is typically set as $2.5J$, and c_j is evenly spaced in $[0, 2\pi]$. Under such formulation, the trajectory parameters are first initiated in the offline phase with the dataset, and then the parameters are updated in the online phase to generate the desired trajectory.

Given that the dataset records a series of joint trajectories for multiple wearers in multiple tasks, the mean trajectory is selected for demonstration and denoted as the vector $\boldsymbol{q}_{\text{demo}}$. In the offline phase, the trajectory parameters are initialized by setting them as $\boldsymbol{q}_{\text{DMP}}=\boldsymbol{q}_{\text{demo}}$, where $\boldsymbol{q}_{\text{DMP}}=[q_{\text{DMP},1}, \cdots, q_{\text{DMP},i}, \cdots, q_{\text{DMP},n}] \in \Re^n$, and then the parameters (i.e., the weighting matrix) are learned through Incremental Locally Weighted Regression (ILWR) [10] as

$$w_{ij}(t+1) = w_{ij}(t) + \Psi_j Q_{ij}(t+1) e_{r,ij}(t), \tag{4.12}$$

$$Q_{ij}(t+1) = \frac{1}{\Lambda}\Big(Q_{ij}(t) - \frac{Q_{ij}(t)^2}{\frac{\Lambda}{\Psi_j} + Q_{ij}(t)}\Big), \tag{4.13}$$

$$\begin{aligned} e_{r,ij}(t) &= \frac{1}{(2\pi)^2}\ddot{q}_{\text{demo},i} \\ &+ \alpha_o\big(\beta_o \cdot q_{\text{demo},i} + \frac{1}{2\pi}\dot{q}_{\text{demo},i}\big) - w_{ij}(t), \end{aligned} \tag{4.14}$$

where Λ denotes a forgetting factor, and Q_{ij} and $e_{r,ij}$ denote the auxiliary variable and regression error, respectively, with $q_{\text{demo},i}$ being the demonstration on joint i (i.e., the i-th element of the vector $\boldsymbol{q}_{\text{demo}}$). ILWR learning weights offer advantages of high computational efficiency, avoiding overfitting, and mitigating the impact of noisy data.

In the online phase, the trajectory parameters are updated with the proposed HIL method (which is to be detailed later) and now the desired trajectory can be computed in a forward manner as

$$\boldsymbol{q}_{\text{DMP}} = f_p(t, \boldsymbol{W}, \Omega_c), \tag{4.15}$$

where Ω_c is the gait frequency, $\boldsymbol{W}$ is the weight matrix with elements w_{ij}, where each row of $\boldsymbol{W}$ corresponds to the weights associated with a single joint, and f_p is the function that integrates the *output system* with a subsequent integral process. To determine the frequency Ω_c, another *canonical system* is implemented as

$$\dot{\phi}_{ij} = j\Omega_i - k_{c1} e_i \sin(\phi_{ij}), \tag{4.16}$$

$$\dot{\Omega}_i = -k_{c1} e_i \sin(\phi_{i1}), \tag{4.17}$$

$$\tilde{q}_i = \sum_j A_{ij} \cos(\phi_{ij}), \tag{4.18}$$

$$\dot{A}_{ij} = k_{c2} e_i \cos(\phi_{ij}), \tag{4.19}$$

$$e_i = q_i - \tilde{q}_i, \tag{4.20}$$

where q_i is the current state of the i-th joint, obtained using encoders installed on the joints of the robot. The terms $\tilde{q}_i$ and e_i are defined as the auxiliary state and the learning error for joint i, respectively, ϕ_{ij} and A_{ij} represent the phase and amplitude of the j-th component in the i-th joint, respectively, and k_{c1} and k_{c2} are positive constants. Then, we have

$$\Omega_c = \frac{1}{N_a} \sum_{i}^{N_a} \Omega_i, \tag{4.21}$$

where N_a is the number of active joints. Note that the gait frequency is computed by detecting the discrete events of heel strikes with the strain gauge at the footplate.

Now, the desired trajectory can be specified as $\boldsymbol{q}_d = \boldsymbol{q}_{\text{DMP}}$. Note that discrete movements [11], such as sitting and standing, can also be represented using DMP within a unified framework by encoding them as a weight matrix $\boldsymbol{W}$. In addition, as the time required to perform the movement does not vary significantly across individuals, speed modulation is not required. Therefore, the generated trajectory can be expressed as

$$\boldsymbol{q}_d = \begin{cases} f_p(t, \boldsymbol{W}, \Omega_c), & \textit{periodic movement} \\ f_d(t, \boldsymbol{W}), & \textit{discrete movement} \end{cases} \tag{4.22}$$

where $f_d(t, \boldsymbol{W})$ is the second-order dynamic system under discrete movement conditions.

Trajectory Generation: As shown in Fig. 4.8, the dataset of multiple human users in multiple tasks is used to train the previously mentioned anomaly detector, and it is also employed to train a task translator to enable the transfer of assistance trajectories between different tasks.

We are now introducing a task translator based on a Multi-Layer Perceptron (MLP). Unlike VAE-based training, each entry fed into the task translator represents parameterized trajectories (i.e., weight matrices) for various tasks performed by a single user. The identified relationships allow the robot to transfer learned tasks to new ones in the subsequent online HIL phase. These tasks are typically not repeatable (e.g., ascending stairs and standing up) and thus impractical for online learning.

Figure 4.9 illustrates the workflow of task translation, where $\boldsymbol{w}_{wm}$ and $\boldsymbol{w}_{sm}$ represent the encoded weight vectors for user m during the tasks of walking and ascending stairs, respectively. These vectors are processed through DMP. To train the task translator, gait trajectories in the different tasks are collected in weight matrix form as a training set $(\boldsymbol{W}_{wm}, \boldsymbol{W}_{sm})$, where $\boldsymbol{W}_{wm}$ and $\boldsymbol{W}_{sm}$ are the weight matrices of user m during the tasks of walking and ascending stairs, respectively. Then, the MLP is used to determine the mapping relationship of the weight matrices under different tasks. The predictive weight matrix in the new task is obtained as

$$\hat{\boldsymbol{W}}_{sm} = \boldsymbol{F}_r(\boldsymbol{W}_{wm}), \tag{4.23}$$

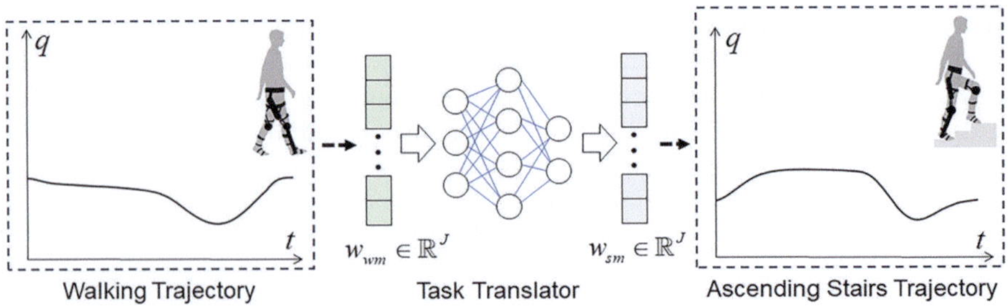

Fig. 4.9 The task translator can generate the trajectory for a new task that is impractical to learn online (e.g., ascending stairs). Here, $\boldsymbol{w}_{wm}$ and $\boldsymbol{w}_{sm}$ represent the encoded weight vectors for user m during walking and stair climbing tasks, respectively

where $\boldsymbol{F}_r$ is a nonlinear regression function yielded by a neural network. The following loss function is minimized

$$\mathcal{L} = \sum_{m}^{M} l(\boldsymbol{W}_{sm}, \hat{\boldsymbol{W}}_{sm}), \tag{4.24}$$

where l is a differentiable convex loss function that represents the distance between the prediction $\hat{\boldsymbol{W}}_{sm}$ and target $\boldsymbol{W}_{sm}$, and M is the number of users. Function l is written in the quadratic form as:

$$l(\boldsymbol{W}, \hat{\boldsymbol{W}}) = \|\boldsymbol{W} - \hat{\boldsymbol{W}}\|_F^2, \tag{4.25}$$

where $\|\cdot\|_F$ indicates the Frobenius norm.

Compared with techniques commonly used for prediction, such as K-Nearest Neighbors (KNN), Gaussian Process Regression (GPR), Extreme Gradient Boosting (XGBoost), Gradient Boosting (GBoost), and Support Vector Machines (SVM), the proposed NN-based task translator exhibits better generalization performance over a limited dataset, as validated experimentally.

Notably, the weight matrix is an intrinsic property of the user, which implicitly includes parameters affecting the gait trajectory, such as the height and arm length [12]. For trajectory recording, the proposed task translator requires only proprioceptive sensors.

Trajectory Individualization: Next, the generated trajectory is optimized under the HIL framework. Specifically, a comprehensive cost function is formulated using both the parameterized trajectory and anomaly score as

$$\mathcal{J}_{HIL}(s, \boldsymbol{q}, \boldsymbol{W}) = \frac{\lambda}{T} \sum_{i}^{N_a} \sum_{j}^{T} (q_{di}^{(j)} - q_i^{(j)})^2 + \frac{1-\lambda}{T} \sum_{j}^{T} s^{(j)}, \tag{4.26}$$

where the superscript and subscript indicate the time index and joint index in a gait cycle, respectively. Furthermore, T represents the duration of one gait cycle. The first term ensures that the generated trajectory matches the specific user, and the second term helps ignore the abnormal patterns of the user (likely attributable to physical conflicts) during the learning process. The weight of these two terms is governed by the hyperparameter $\lambda \in (0, 1)$. Specifically, when $\lambda = 0$, optimization is based on only the collected dataset. Conversely, when $\lambda = 1$, the emphasis shifts to optimizing the trajectory alignment.

Bayesian optimization is applied to determine the global minimum of $\mathcal{J}_{HIL}(s, \boldsymbol{q}, \boldsymbol{W})$ under a bounded feasible set, which has proven effective in previous HIL protocols [13]. To this end, the selected matrices and their associated costs are collated into $\mathcal{W}_{past}$ and $\boldsymbol{\mathcal{J}}_{past}$ respectively. Note that $\boldsymbol{\mathcal{J}}_{past}$ is formulated as a vector. Subsequently, we model the target function using Gaussian processes as

$$\begin{bmatrix} \mathcal{J}_{new} \\ \boldsymbol{\mathcal{J}}_{past} \end{bmatrix} \sim \mathcal{N}\left(\mathbf{0}, \begin{bmatrix} K_{nn} & \boldsymbol{K}_{pn}^T \\ \boldsymbol{K}_{pn} & \boldsymbol{K}_{pp} \end{bmatrix}\right), \tag{4.27}$$

where $\mathcal{J}_{new}$ represents the cost corresponding to the weight matrix $\mathbf{W}^{new}$ to be identified. Notably, setting the mean as zero simplifies the calculation, which has proven effective in practice [13] as most of the information is contained in the covariance. Using the radial basis function as the kernel function, $k(\boldsymbol{x}, \boldsymbol{x}')$, we introduce kernel matrices K_{nn}, $\boldsymbol{K}_{pn}$,, and $\boldsymbol{K}_{pp}$ with the following elements:

$$K_{nn} = k(Vec(\mathbf{W}^{new}), Vec(\mathbf{W}^{new})), \tag{4.28}$$

$$K_{in} = k(Vec(\boldsymbol{W}_i), Vec(\mathbf{W}^{new})), \quad \boldsymbol{W}_i \in \mathcal{W}_{past} \tag{4.29}$$

$$K_{ij} = k(Vec(\boldsymbol{W}_i), Vec(\boldsymbol{W}_j)). \quad \boldsymbol{W}_i, \boldsymbol{W}_j \in \mathcal{W}_{past} \tag{4.30}$$

Here, $Vec(\cdot)$ denotes a vectorization function. Following the construction of the posterior distribution, we use the Lower Confidence Bound (LCB) as the acquisition function for guiding the selection of the subsequent point for exploration:

$$\mathbf{W}^{new} = \underset{\mathbf{W}^{new}}{\operatorname{argmin}} \mu_{new} - \upsilon\sigma_{new}, \tag{4.31}$$

$$\mu_{new} = \boldsymbol{K}_{pn}^T \boldsymbol{K}_{pp}^{-1} \boldsymbol{\mathcal{J}}_{past}, \tag{4.32}$$

$$\sigma_{new} = K_{nn} - \boldsymbol{K}_{pn}^T \boldsymbol{K}_{pp}^{-1} \boldsymbol{K}_{pn}, \tag{4.33}$$

where υ is a positive hyperparameter. The minimization of (4.31) is implemented using the L-BFGS algorithm. This process enables iterative updates of the posterior distribution based on historical data, and the appropriate weight matrix is then derived from the acquisition function. To accelerate the convergence over the extended dimensions compared with previous research, the initial parameters are specified based on the sampled gait. Subsequently, the optimal weight matrix $\boldsymbol{W}^*$ for the walking task can be learned based on the proposed cost function.

Minimizing (4.26) can yield an individualized and anomaly-robust trajectory (i.e., $\boldsymbol{q}_d$), allowing the robot to learn the correct and appropriate gait trajectory under all conditions to improve the effectiveness of assistance and the wearer's comfort.

After the walking optimization $\boldsymbol{W}^*_{wm}$ for user m is obtained, the optimal weight matrix for a new task can be predicted using the proposed task translator as

$$\hat{\boldsymbol{W}}^*_{sm} = \boldsymbol{F}_r(\boldsymbol{W}^*_{wm}), \tag{4.34}$$

which enables the provision of predictive assistance $\hat{\boldsymbol{W}}^*_{sm}$ without the demonstration of a new task based on the known optimization result. Thus, the proposed HIL optimization framework can effectively adapt to multiple wearers and tasks.

4.1.3 Results

The proposed scheme was implemented in a bilateral lower-limb exoskeleton robot to assess its performance, as shown in Fig. 4.10. All four joints of the robot (two hip joints and two knee joints) were driven by SEAs. The spring stiffness of each SEA was 635 Nm, and the output torque was determined based on the spring deflection (measured using encoders) and known stiffness. The robot was equipped with two types of proprioceptive sensors: strain gauges and encoders (with 2048 lines). These sensors provided comprehensive task-specific information, including gait phases (measured using strain gauges), interaction torque, and movement trajectory (measured using encoders). During the experiment, four wireless surface EMG sensors (Ws450, Biometrics Ltd.) were mounted on human leg muscles, specifically, the quadriceps femoris (QF) and tibialis anterior muscle (TA) of the left and right legs, to collect the feedback of human limb and hence assess the performance of the proposed controller.

An embedded main board facilitated communication with the PC, as shown in Fig. 4.10. The detection network was implemented on the PC, and the control algorithm was programmed and realized by the main board. The main board transmitted the control commands through the pulse width modulation (PWM) signals to the four slave boards to drive the motion of the four robot joints. Furthermore, to minimize the influence of the motors on the wearer's movements during data collection and reduce variable dimensions in optimization, only the two motors located at the knee joint were activated in this experiment.

Ten able-bodied participants with no prior experience with exoskeletons performed multiple tasks while wearing the exoskeleton. Table 4.2 presents an overview of the participants' statistical attributes, namely age (25.6 ± 1.9 years), weight (66.3 ± 14.8 kg), and height (173.4 ± 9.1 cm). All participants signed a written informed consent form prior to the experimental sessions.

As shown in Fig. 4.11, the participants were instructed to perform five tasks at their own pace: walking, ascending stairs, descending stairs, squatting, and standing up, as summarized in Table 4.3. Throughout data collection, the exoskeleton operated in a transparent mode to allow the wearer to perform multiple tasks in a comfortable manner, and the motion frequency was recorded using (4.21). During squatting and

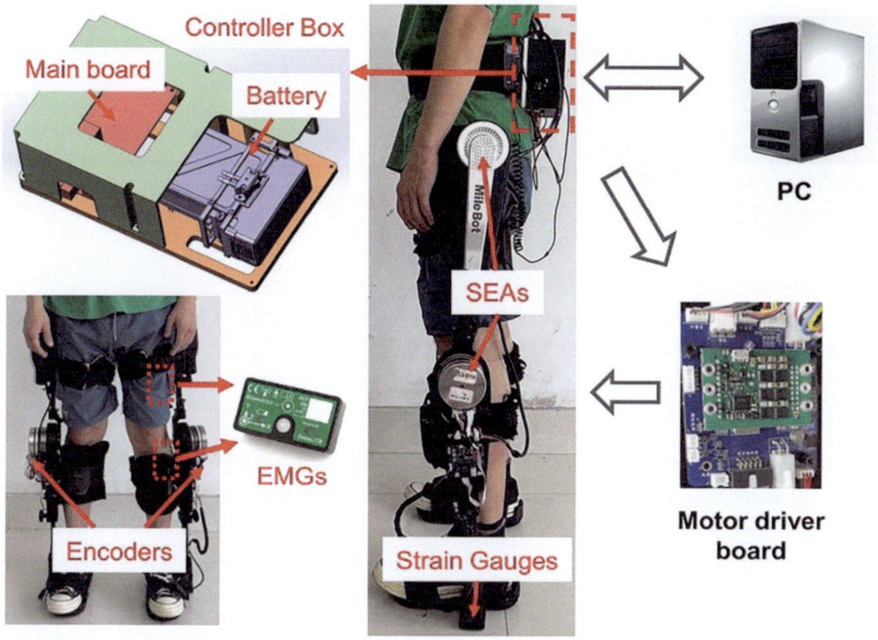

Fig. 4.10 Experimental setup of the exoskeleton robot, where all four joints of both legs are driven by the SEAs. The controller box is composed of the main board and battery. The commands are transmitted to four slave boards to drive the robot joints

Table 4.2 Statistical information of subjects

Subject	Gender	Age (y)	Weight (kg)	Height (cm)
1	Male	25	81	185
2	Male	25	70	178
3	Male	26	78	177
4	Male	24	57	172
5	Male	25	74	183
6	Male	30	87	181
7	Male	27	72	172
8	Female	23	46	159
9	Female	26	48	160
10	Female	25	50	167

standing tasks, the exoskeleton applied a constant torque to support the wearer, with the magnitude of the torque being meticulously adjusted for each individual to ensure comfortable movement. Additionally, each participant underwent a six-minute pre-training session to become familiar with each task. Figure 4.12 shows the angle profiles of the left knee joint under different tasks. The constructed dataset was used to train the anomaly detection and task translation modules.

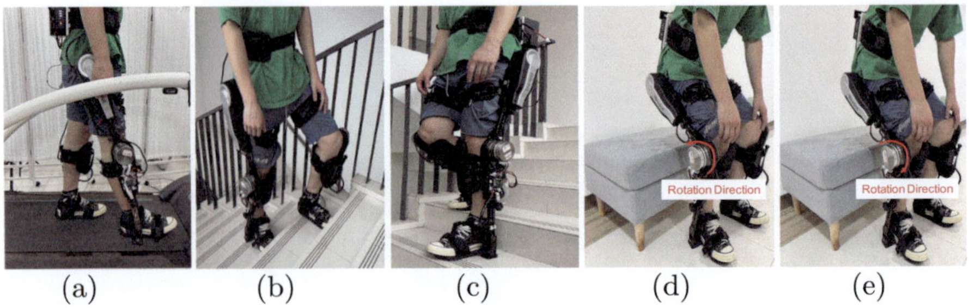

Fig. 4.11 Participants wearing the exoskeleton perform various tasks: **a** Walking; **b** Ascending stairs; **c** Descending stairs; **d** Squatting; **e** Standing up

Table 4.3 Tasks

Task	Description
Walking	To walk at a comfortable pace
Ascending stairs	To ascend a flight of stairs at a comfortable pace.
Descending stairs	To descend a flight of stairs at a comfortable pace
Squatting	To squat and hold the position for a brief period
Standing up	To stand up from a seated position

Anomaly Detector: For anomaly detection across different tasks, the anomaly detector can be retrained with the collected data to improve its generalization to various anomalies. During the experiment, a participant intentionally held the robot to simulate conflict. The output of the detection network in this condition is shown in Fig. 4.13. The reconstruction error significantly increased when the conflict arose, indicating the wearer's discomfort. This finding highlighted that the anomaly could be successfully detected. Furthermore, the performance of the detection network was validated in another simulated conflict, where the participant lost balance due to fatigue. The results are shown in Fig. 4.14, further reinforcing the network's capacity to successfully detect such conflicts.

This finding highlights that the anomaly detection network enables precise detection for different tasks while incorporating the normal trajectories of new tasks. For instance, in the presence of asynchronization between the human and the robot, the interaction torque rapidly increased. Conversely, in cases of imbalance due to human fatigue, angular information on the human limb, such as encoder signals, changed more significantly.

Task Translator: To accommodate trajectories across different tasks, the collected data were used to train a task translator. This process was modeled as a regression problem to promote generalization across diverse tasks. Owing to the limited sample size, we used a leave-one-out cross-validation approach and applied various techniques to address the regression problem, namely ridge regression, KNN, GPR, XGBoost, GBoost, SVM, and NN.

Trajectories were encoded into weight matrices corresponding to distinct tasks to allow these techniques to predict the transformation of the weight matrix from one task to another, for example, the transition from walking to ascending stairs. The root

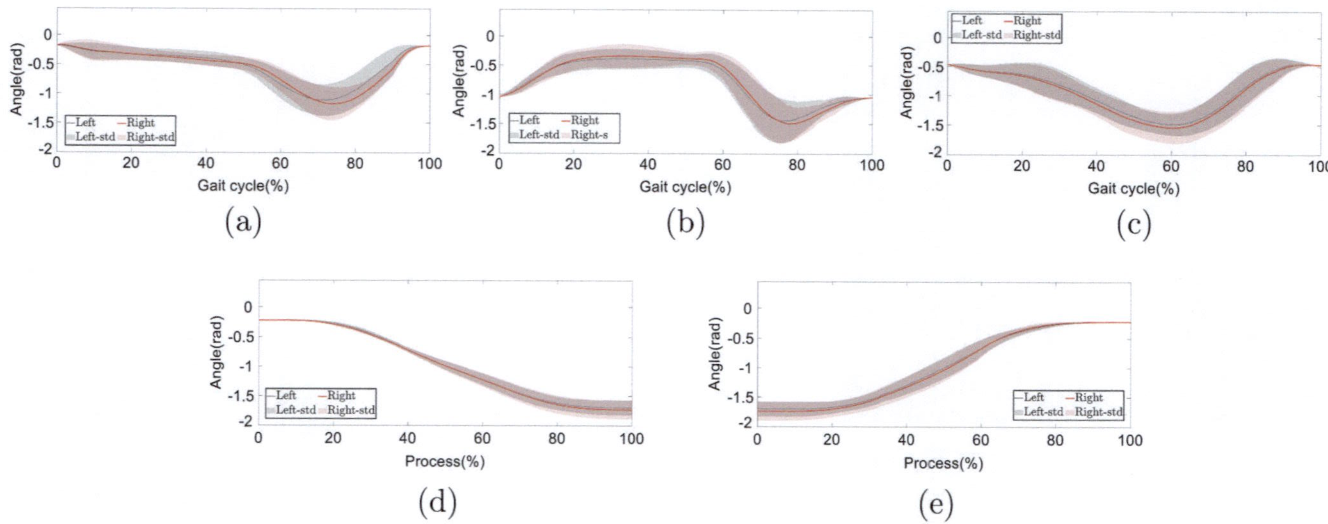

Fig. 4.12 Gait profiles of the knees over different tasks: **a** Walking; **b** Ascending stairs; **c** Descending stairs; **d** Squatting; **e** Standing up. The mean trajectory of the right knee is depicted by a solid red line, accompanied by a red-shaded area that indicates its standard deviation. Similarly, the mean trajectory of the left knee is illustrated by a solid black line, with the corresponding standard deviation represented by a black-shaded area. The RMSEs (rad) for the trajectories of the two legs across different tasks for all ten subjects are as follows: **a** 0.082; **b** 0.083; **c** 0.143; **d** 0.031; **e** 0.034

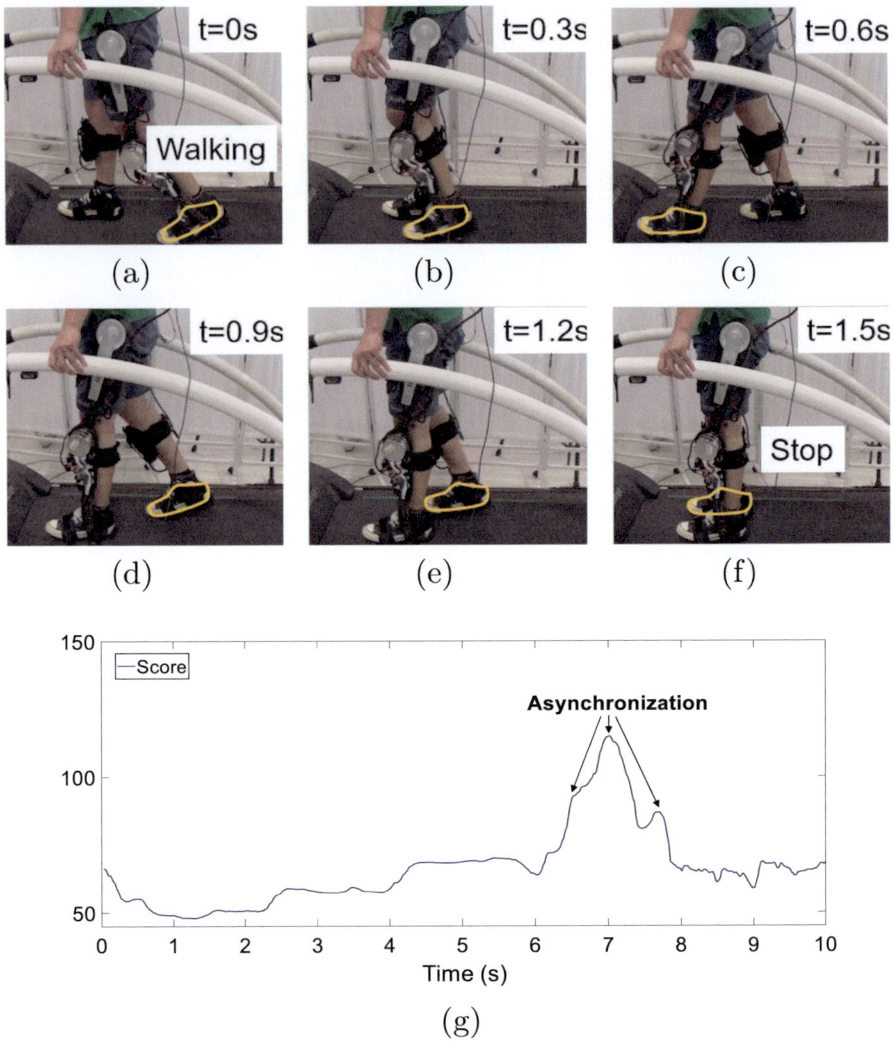

Fig. 4.13 **a–f** Snapshots of conflict due to asynchronization; **g** anomaly score during asynchronization. The red contour represents the foot during the stepping motion

mean square error (RMSE) within the joint space was used as the evaluation metric for this regression task. The RMSE values are expressed in degree units. A comparison of the translation from walking to ascending stairs, using the aforementioned techniques, is presented in Table 4.4.

Note that the complexity of the regression task increases along with the number of kernels. In addition, among the compared models, the NN-based translator demonstrated superior performance. Considering both the fitting error and expression of DMP, the task translator included the NN, and the optimal number of kernels, in this situation, was determined to be $J = 20$.

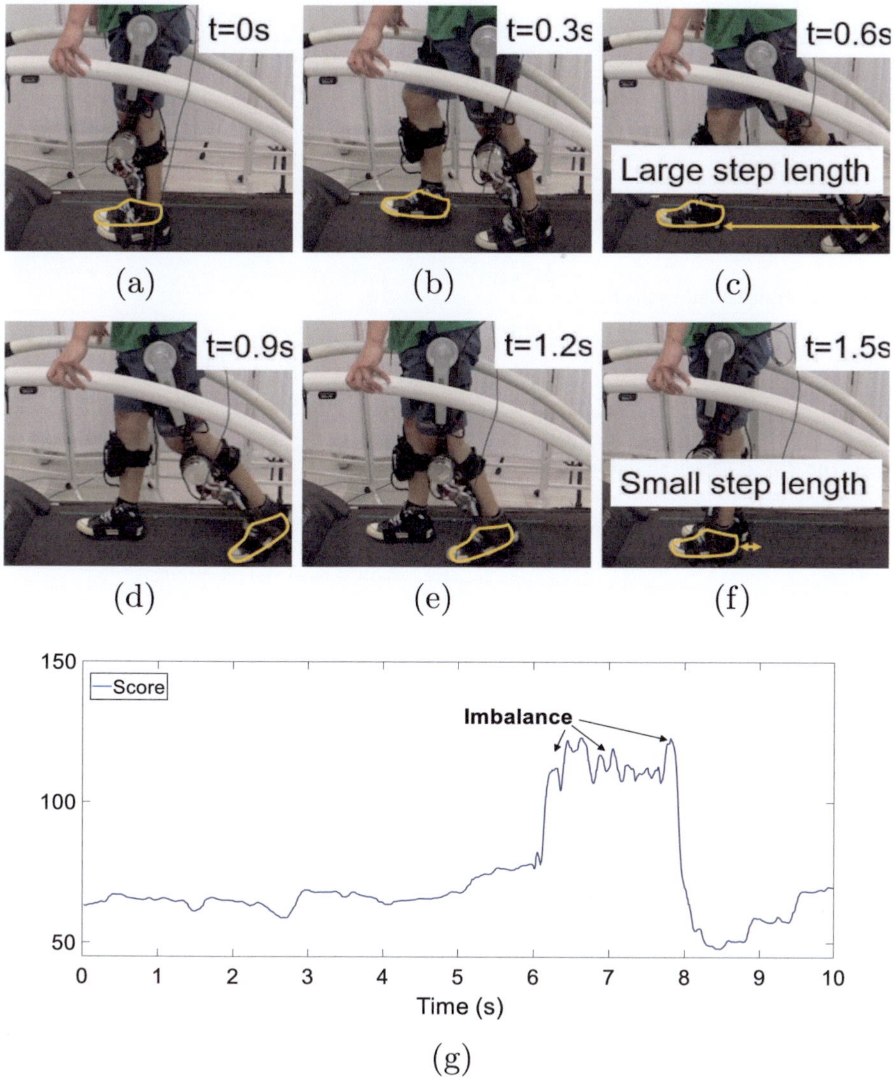

Fig. 4.14 **a–f** Snapshots of conflict due to imbalance; **g** anomaly score during imbalance. The red contour represents the foot during the stepping motion

After determining the kernel number for the rhythmic motion, the same approaches were used to execute the translation into discrete movements, specifically, the transitions between the squatting and standing postures. The translation outcomes for standing are summarized in Table 4.5. Among the evaluated methods, the NN outperformed the other techniques and was thus selected for the translation of these discrete movements. Considering both the fitting error and DMP expression, the final kernel number was selected as $J_d = 15$.

Table 4.4 Translation results for ascending stairs

J	Ridge	KNN	GPR	XGBoost	GBoost	SVM	NN
15	4.74	5.12	5.34	6.57	5.61	4.89	**2.94**
20	4.81	5.19	5.31	6.44	5.63	5.06	**3.33**
25	4.87	5.23	5.39	6.74	5.91	5.08	**3.97**

Note The RMSE values are presented in degree units
J is the kernel number for rhythmic movement

Table 4.5 Translation results for standing

J_d	Ridge	KNN	GPR	XGBoost	GBoost	SVM	NN
10	4.54	4.60	4.69	4.35	4.55	4.43	**2.71**
15	4.81	4.87	4.86	4.66	4.86	4.69	**2.89**
20	4.95	5.01	4.97	4.83	4.75	4.87	**2.95**

Note: The RMSE values are presented in degree units
J_d is the kernel number for discrete movement
The number of kernels for rhythmic movement is set as $J = 20$

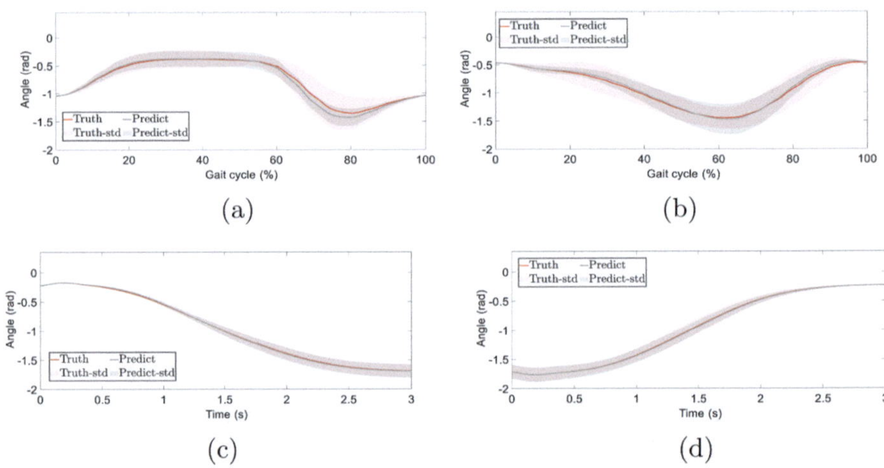

Fig. 4.15 Translation results across different tasks: **a** Ascending stairs; **b** Descending stairs; **c** Squatting; **d** Standing up. The red solid line represents the mean of the true profile, with the red shade indicating its standard variation. Conversely, the black solid line indicates the prediction from the task translator, with the surrounding black shade denoting its standard variation

Hence, within the framework of the task translator, we selected the NN for translation. The translation results, derived from the application of the leave-one-out cross-validation, are visualized in Fig. 4.15. Note that the reproduced profile aligns satisfactorily with the collected gait profile across different tasks, despite the limitations of a small dataset. This is evident in the metric of the average RMSE, which yields values of 3.33 for ascending stairs, 2.64 for descending stairs, 2.93 for squatting, and 2.89 for standing up.

Table 4.6 Characteristics of subjects in HIL

Subject	Gender	Age (y)	Weight (kg)	Height (cm)
1	Male	23	65	172
2	Male	30	85	183

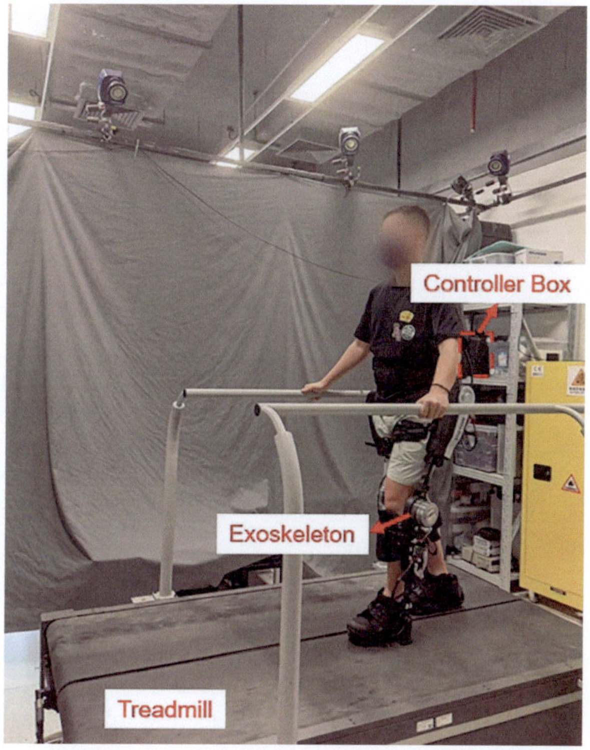

Fig. 4.16 Experimental setup for HIL optimization, where the participant walks on a treadmill with the exoskeleton

To optimize the comfort of exoskeleton assistance, we recruited two participants who were not included in the pre-collected dataset for the application of the proposed HIL optimization. The attributes of these participants are summarized in Table 4.6. The participants were instructed to walk at a comfortable pace on a treadmill (FIT, Bertic, USA), and the frequency of reference trajectory was set to a constant according to (4.21). This speed remained constant during the optimization process, as depicted in Fig. 4.16.

To evaluate the performance of various participants in the walking task, we used the metric defined in (4.26) to guide trajectory optimization. Only the motors in the knees were activated to aid the wearers, such that the weight matrix $\boldsymbol{W}$ was reduced to a simple weight vector. During the initial phase of standing, the first 10 elements of the weight vectors, which represent the encoded trajectory, were similar across all walking tasks. To alleviate the computational complexity, these elements were fixed to the mean of the data. The assistive profile was updated upon every seventh heel strike. The initial six gaits were reserved for recording sensor signals to

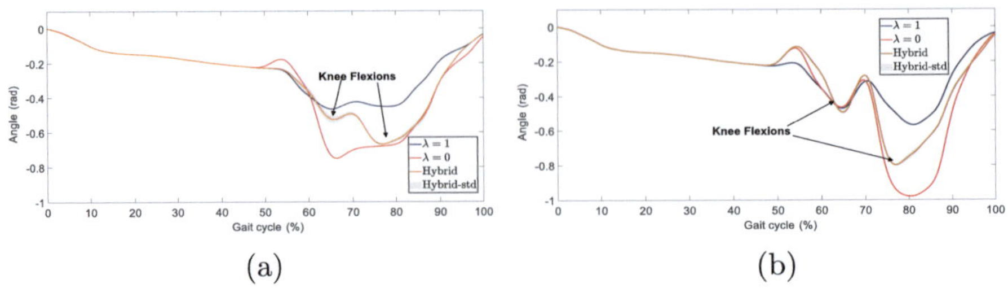

Fig. 4.17 HIL optimization results of different hyperparameters λ for different participants: **a** Participant 1; **b** Participant 2. The blue and red solid lines represent the optimized trajectories in the context of $\lambda = 1$ and $\lambda = 0$, respectively. The black solid line is the mean of the trajectories, with $\lambda \in (0, 1)$, and the surrounding black shade indicates its standard variation

evaluate the performance with human interaction, and the final gait was input to the Bayesian optimization algorithm. This approach prevented any sudden changes in the assistive profile that may have affected the balance of the wearers during consistent walking.

To accelerate optimization, we confined the parameter boundaries to 30% around the identified optimal vector. Subsequently, we halved the boundaries whenever the optimal vector remained static over the last 10 steps. This method allowed the optimization process to refine the optimal weights, thereby enhancing the search efficiency without compromising the convergence certainty. Optimization for each walking task required approximately 14 min, and 10 random restarts were implemented to bypass local minima.

HIL Optimization: Fig. 4.17 illustrates the HIL optimization outcomes. The blue and black solid lines correspond to the optimal trajectories when $\lambda = 1$ and 0, respectively; the red trajectory depicts the mean and variance of the optimization outcomes for λ values between (0, 1), i.e., $\lambda = 0.25, 0.5, 0.75$. The results revealed that an optimal trajectory with $\lambda = 1$ tended towards lazy assistance because the optimization term overly emphasized trajectory alignment, thereby causing the trajectory to converge to a curve with a smaller stretch amplitude. In contrast, for $\lambda = 0$, where the cost function consisted solely of the anomaly scores output by the NN, the final optimized trajectory exhibited heightened and frequent fluctuations. When λ ranged between 0 and 1, its effect on the trajectory optimization was significantly similar. This may be attributable to λ eliminating the gap between the dataset and participant actions with anomaly-based optimization. Essentially, the first term in (4.26) was optimized to a lower value, and then the trajectory details were improved according to the human-machine interaction results, thereby minimizing conflict and enhancing comfort.

Additional tests were performed to evaluate the effectiveness of the optimization results. Specifically, an indirect, quantitative evaluation of user comfort and assistance effectiveness was performed using EMG signals from the TA and QF, along with the average power transferred by the exoskeleton per step. Figure 4.18 shows the experimental outcomes in the transparent mode, with the optimized trajectory and without exoskeleton wear. Considering the minimal influence of λ on the trajec-

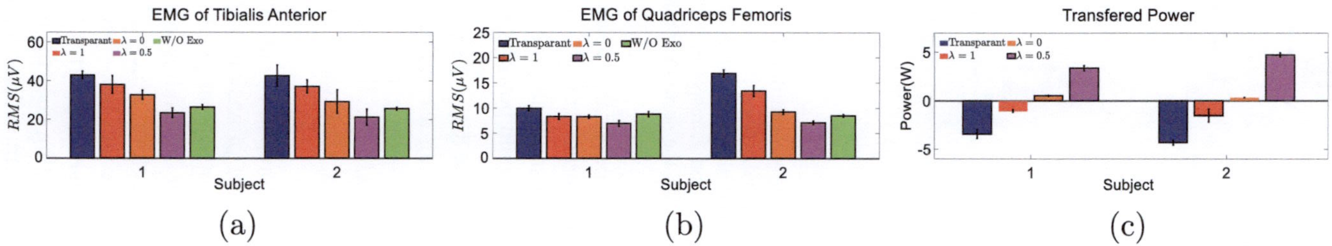

Fig. 4.18 Comparative evaluation of different modes across participants: **a** RMS of EMG signals from the tibialis anterior, indicative of user comfort; **b** RMS of EMG signals from the quadriceps femoris; **c** Average power transferred by the exoskeleton per step, with negative and positive values indicating resistance and assistance, respectively

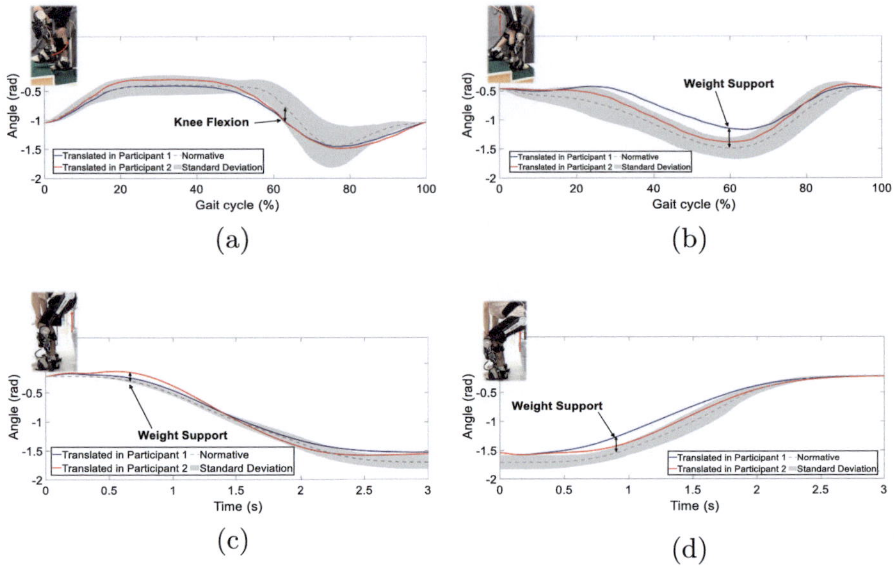

Fig. 4.19 Translation of optimized assistance across different tasks: **a** Ascending stairs; **b** Descending stairs; **c** Squatting; **d** Standing up. The red arrows denote the corresponding interactive forces exerted by the exoskeleton assistance. The blue and red solid lines represent the translation of optimized trajectories from Participants 1 and 2, respectively. The black dashed line represents the prediction from the task translator, with the surrounding black shade denoting its standard variation

tory optimization results, we included only cases where λ equaled 0, 0.5, and 1. The outcomes suggested that a trajectory produced with $\lambda = 0.5$ could not only ensure optimal comfort, as evidenced by the small amplitude of the EMG signal, but also provide effective assistance, i.e., maximum power transfer under all conditions.

Translated Assistance: The trajectories derived from HIL optimization were used to accommodate various tasks, including ascending stairs, descending stairs, and the combined tasks of squatting and standing up. Figure 4.19 offers a visualization of these translated trajectories along with the trajectory acquired in the transparent mode. Within our dataset, the average trajectory is referred to as the "normative trajectory".

The findings demonstrated that, compared with the normative trajectories, the translated trajectories could enable the provision of appropriate assistance. When ascending stairs, the elevated stepping leg received assistance in knee flexion to facilitate foot landing. Conversely, for descending stairs, the translated trajectories were consistent with the desired trajectory, which allows the robot to support the body weight through the impedance controller. Similarly, tasks such as squatting and standing up were assisted by weight support achieved through trajectory optimization. At the beginning of the squatting task, the translated trajectories exhibited an upward curve that deviated from the immediate decline observed in the normative trajectory, suggestive of preparatory actions for weight support.

Table 4.7 Modes

Mode	Character
Transparent	Unrestricted mobility without apparent resistance
Normative	$\boldsymbol{q}_d$ is the mean trajectory from the database
Optimized	$\boldsymbol{q}_d$ is the translation result of the optimized trajectory
W/O Exo	Unrestricted mobility without assistance

To validate the effectiveness of the translated assistance, we compared the EMG signals of the TA and QF and the average power transfer per step under different conditions: transparent mode, normative mode, optimized mode, and a scenario without the exoskeleton, as summarized in Table 4.7.

The findings from these experiments are illustrated in Fig. 4.20. Notably, the tasks of squatting and standing up were treated as one combined task, with a one-second separation used to distinguish the two. In terms of minimizing the amplitude of the EMG signal and augmenting power transfer, the translated assistance demonstrated a modest improvement over the normative mode. This outcome suggested that the optimized and translated trajectories converged in the same latent space. Specifically, the translator, which was trained using the collected dataset, could be directly applied to the results of the HIL optimization. This task translator could thus enhance comfort and assistance across various tasks based on the HIL results from the walking task. Therefore, this HIL optimization framework not only optimized the task trajectories that posed challenges for online optimization but also reduced the time required for task transfer.

This method is evaluated using the training samples of 10 users and the validation samples of two new users. Experiments and comparative studies are performed over four tasks (i.e., walking, stair ascent, stair descent, and sit-to-stand) to demonstrate the effectiveness of the exoskeleton robot.

4.2 Upper-Limb Exoskeletons

This section considers the upper-limb exoskeletons, where the joint is actuated by both the cable-driven mechanism and the SEA. The advantages of cable-driven exoskeleton robots with series elastic actuators can be summarized in twofold: (1) the inertia of the robot joint is relatively low, which is more friendly for human-robot interaction; (2) the elastic element is tolerant to impacts and hence provides structural safety. As trade-offs, the overall dynamic model of such a system is of high order and subject to both unmodelled disturbances and external torques, opening up challenges for controller development.

This section utilizes interactive learning to approximate and compensate for the unmodeled disturbances to obtain more accurate dynamic models. Then, a suboptimal model predictive controller is proposed to drive the robot to track the desired trajectory. While controlling such a robot is not trivial, the proposed control scheme exhibits the advantages of force-senselessness, high accuracy, and low complexity compared with other methods in experiments.

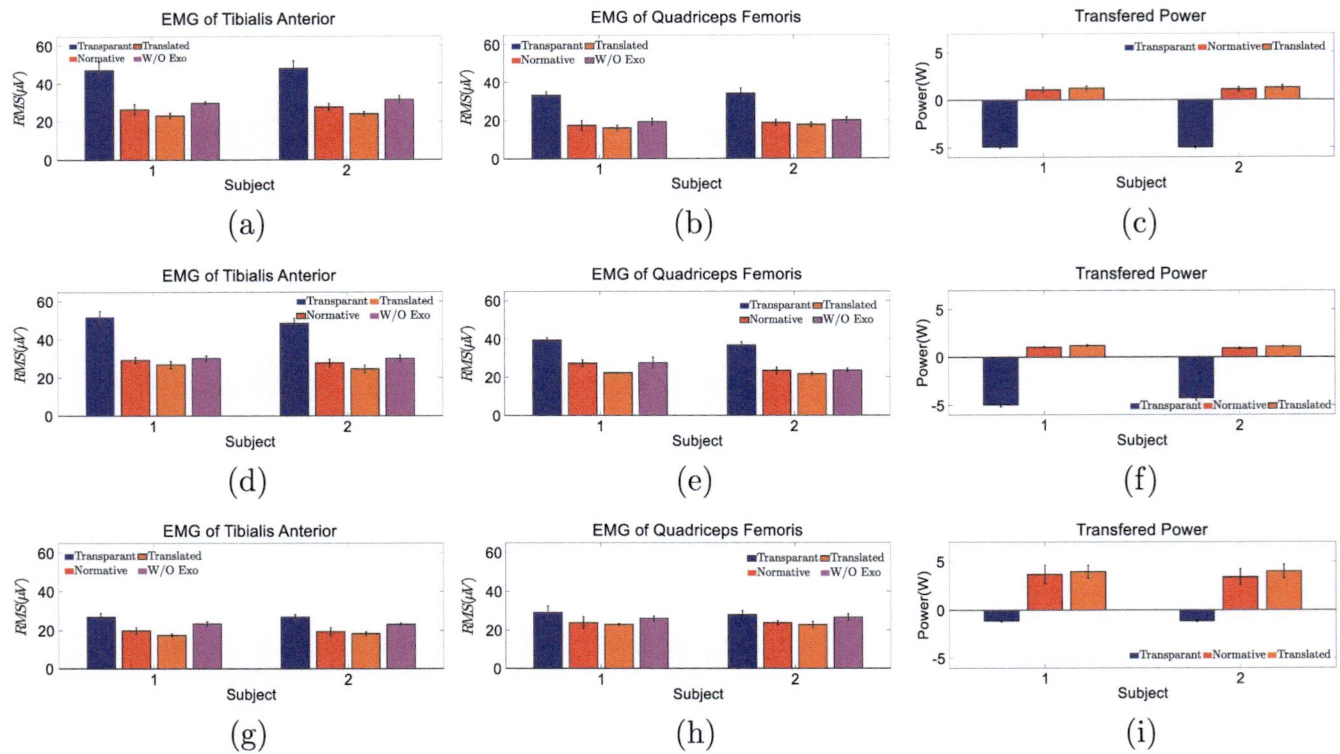

Fig. 4.20 Comparative evaluation of different modes across participants. Translation results for **a–c** stair ascent; **d–f** stair descent; **g–i** squatting and standing up; Each column depicts different metrics: RMS of EMG signals from Tibialis Anterior and RMS of EMG signals from Quadriceps Femoris, indicating user comfort; and average power transferred by the exoskeleton per step, with negative values indicating resistance and positive values denoting assistance

4.2.1 System Structure

This section considers a cable-driven upper-limb exoskeleton developed by Tsinghua University and MileBot Robotics, as shown in Fig. 4.21, which consists of five active joints and one passive joint; Note that Joint 1–5 are active, and the corresponding motions are given as: Joint 1—forearm internal and external rotation; Joint 2—elbow flexion/extension; Joint 3—upper-arm internal and external rotation; Joint 4—shoulder flexion/extension; Joint 5—shoulder abduction/adduction; and Joint 6 is passive and used to coordinate the eccentric movement of the shoulder joints to break through the limitation of the upper limb movement range and increase the human subject's movement training space.

The block diagram of the robot is shown in Fig. 4.22, where each joint is equipped with an encoder (QY2204-SSI for Joint 1 and Joint 3, 3590S-2-104L for Joint 2, Joint 4, and Joint 5) and a harmonic deceleration servo motor (AK80-64 for Joint 1–3; LSG-32-142-80 for Joint 4 and Joint 5). In addition, the embedded main board of the robot communicates with slave boards to drive five active joints through the controller area network (CAN), and PC communicates with the equipment through a universal asynchronous receiver/transmitter (UART).

All the active joints are driven by our self-developed series elastic actuators (SEA), and its CAD model and mechanical structure are shown in Fig. 4.23. The tight end

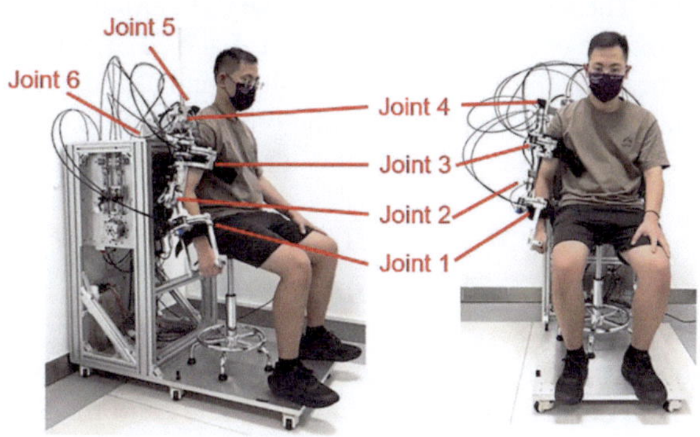

Fig. 4.21 The developed cable-driven upper-limb exoskeleton (left-side view, right-front view), where the robot consists of five active joints (Joint 1–5) and one pass joint (i.e., Joint 6)

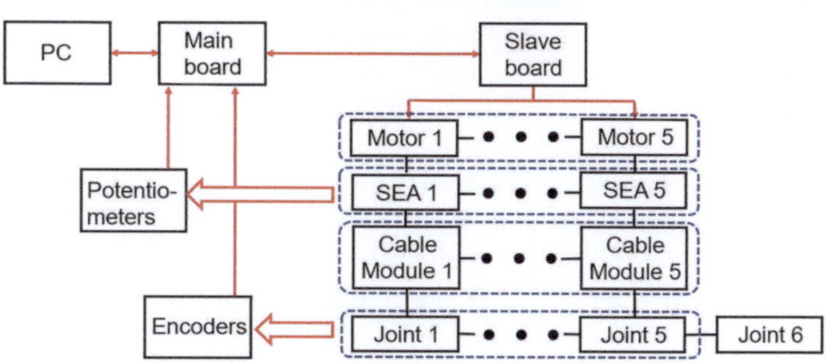

Fig. 4.22 The block diagram of the developed exoskeleton robot, where the red arrow represents the electrical transmission, and the black line represents the mechanical connection

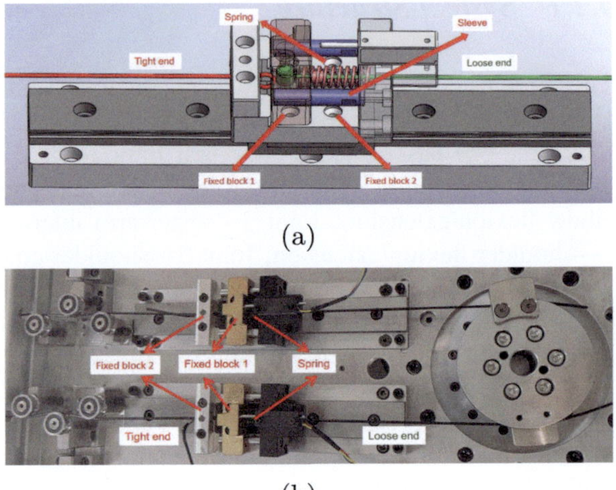

Fig. 4.23 All the robot joints are driven by the self-developed SEAs: **a** CAD model; **b** mechanical structure

of the wire rope is connected with the pulley at the motor end, the fixed block 1 can move on the sleeve, the sleeve is fixed on the fixed block 2, and the loose end of the wire rope is connected with the moving joint parts. When the tight end of the wire rope moves to the right, the fixed block 1 moves to the right along with the rope head to compress the spring. After the spring is compressed, fixed block 2 gives the right force. The fixed block 2 moves to the right along with the loose end of the wire rope to realize the overall right movement. Each SEA is equipped with two potentiometers to measure the spring compression, and the difference between the value of the potentiometers is divided by the pulley radius to obtain the motor shaft angle.

The dynamic model of such a robot can be given as [14]:

$$\boldsymbol{M}(\boldsymbol{q})\ddot{\boldsymbol{q}} + \boldsymbol{C}(\dot{\boldsymbol{q}}, \boldsymbol{q})\dot{\boldsymbol{q}} + \boldsymbol{g}(\boldsymbol{q}) = \boldsymbol{K}(\boldsymbol{\theta} - \boldsymbol{q}) + \boldsymbol{\tau}_e + \boldsymbol{\tau}_f, \tag{4.35}$$

$$\boldsymbol{B}\ddot{\boldsymbol{\theta}} + \boldsymbol{K}(\boldsymbol{\theta} - \boldsymbol{q}) = \boldsymbol{u}, \tag{4.36}$$

Note that the overall dynamic model described by (4.35) and (4.36) is a high-order system consisting of the joint-side (4.35) and the actuator-side (4.36), and stabilizing and controlling such a system is not trivial. In addition, both $\boldsymbol{\tau}_e$ and $\boldsymbol{\tau}_f$ are coupled on the joint side and cannot be isolated with a single force/torque sensor. However, distinguishing each other is important to achieve the control objective, that is, rejecting the unmodelled disturbances $\boldsymbol{\tau}_f$ and utilizing the interaction torque $\boldsymbol{\tau}_e$.

To ensure a safe interaction, the robot is controlled to track a time-varying trajectory under a desired impedance model:

$$\boldsymbol{M}_d(\ddot{\boldsymbol{q}} - \ddot{\boldsymbol{q}}_d) + \boldsymbol{C}_d(\dot{\boldsymbol{q}} - \dot{\boldsymbol{q}}_d) + \boldsymbol{K}_d(\boldsymbol{q} - \boldsymbol{q}_d) = \boldsymbol{\tau}_e, \tag{4.37}$$

where $\boldsymbol{q}_d \in \Re^n$ is the vector of the reference trajectory, $\boldsymbol{M}_d, \boldsymbol{C}_d, \boldsymbol{K}_d \in \Re^{n\times n}$ denote the desired inertia, the desired damping, the desired stiffness matrices respectively, which are diagonal and positive definite. Tracking the reference trajectory under the impedance model allows the patient to deviate from the trajectory and hence provide a certain level of compliance to improve safety.

4.2.2 Disturbance Compensation

The proposed two stages are compared and summarized in Table 4.8. Specifically, *Stage I* (offline phase) is to approximate and then compensate unmodelled disturbances, that is, $\boldsymbol{\tau}_f$ in (4.35); *Stage II* (online phase) is to track the desired trajectory under the desired impedance model (4.37); The controllers used in two stages do not require dedicated force sensors and both are computationally efficient for the high-order system.

First, at *Stage I*, no human subject is involved at this stage such that (4.35) becomes

$$\boldsymbol{M}(\boldsymbol{q})\ddot{\boldsymbol{q}} + \boldsymbol{C}(\dot{\boldsymbol{q}}, \boldsymbol{q})\dot{\boldsymbol{q}} + \boldsymbol{g}(\boldsymbol{q}) = \boldsymbol{K}(\boldsymbol{\theta} - \boldsymbol{q}) + \boldsymbol{\tau}_f. \tag{4.38}$$

Then, an iterative learning impedance controller is designed for (4.36) and (4.38) by referring to [15]. The development of such a controller follows the backstepping approach [16, 17]. That is, by rewriting (4.38) as

$$\boldsymbol{M}(\boldsymbol{q})\ddot{\boldsymbol{q}} + \boldsymbol{C}(\dot{\boldsymbol{q}}, \boldsymbol{q})\dot{\boldsymbol{q}} + \boldsymbol{g}(\boldsymbol{q}) + \boldsymbol{K}\boldsymbol{q} = \boldsymbol{K}\boldsymbol{\theta}_d + \boldsymbol{K}\Delta\boldsymbol{\theta} + \boldsymbol{\tau}_f \tag{4.39}$$

where $\Delta\boldsymbol{\theta} = \boldsymbol{\theta} - \boldsymbol{\theta}_d$, a desired input vector $\boldsymbol{\theta}_d \in \Re^n$ is proposed as

$$\begin{aligned}\boldsymbol{\theta}_d =& \boldsymbol{q} + \boldsymbol{K}^{-1}(-\boldsymbol{K}_s\boldsymbol{s} \\ &+ \boldsymbol{M}(\boldsymbol{q})\ddot{\boldsymbol{q}}_r + \boldsymbol{C}(\dot{\boldsymbol{q}}, \boldsymbol{q})\dot{\boldsymbol{q}}_r + \boldsymbol{g}(\boldsymbol{q}) + \boldsymbol{m}),\end{aligned} \tag{4.40}$$

where $\boldsymbol{K}_s \in \Re^{n\times n}$ is a positive-definite and diagonal matrix, and $\boldsymbol{s} \in \Re^n$ is a sliding vector defined as

$$\begin{aligned}\boldsymbol{s} =& \dot{\boldsymbol{q}} - \dot{\boldsymbol{q}}_r \\ =& \dot{\boldsymbol{q}} - \dot{\boldsymbol{q}}_d + \alpha(\boldsymbol{q} - \boldsymbol{q}_d),\end{aligned} \tag{4.41}$$

where α is a positive constant, and $\dot{\boldsymbol{q}}_r \triangleq \dot{\boldsymbol{q}}_d - \alpha(\boldsymbol{q} - \boldsymbol{q}_d)$ is defined as a reference vector, and $\boldsymbol{m} \in \Re^n$ denotes a feed-forward control vector which is updated by the following law

$$\boldsymbol{m}_{k+1} = \boldsymbol{m}_k - \beta\boldsymbol{s}_k, \tag{4.42}$$

where β is a positive constant, and the subscript of k denotes the kth iteration.

Table 4.8 Comparison between two stages

	Stage I	Stage II
Purpose	Compensation	Rehab. training
Interaction	$\tau_e = \mathbf{0}$	$\tau_e \neq \mathbf{0}$
Controller	ILC	Suboptimal MPC
Force sensor	Nil	Nil

Next, to ensure that the actual state of the actuator converges to the desired input (i.e., $\boldsymbol{\theta} \to \boldsymbol{\theta}_d$), the actual control input exerted on the actuator side is proposed as

$$\boldsymbol{u}_I = \boldsymbol{K}(\boldsymbol{\theta} - \boldsymbol{q}) + \boldsymbol{B}\ddot{\boldsymbol{\theta}}_r - \boldsymbol{K}_\theta \Delta\boldsymbol{\theta}, \tag{4.43}$$

where $\boldsymbol{K}_\theta \in \Re^{n\times n}$ is diagonal and positive-definite, $\ddot{\boldsymbol{\theta}}_r$ is the time derivative of $\dot{\boldsymbol{\theta}}_r$, and $\dot{\boldsymbol{\theta}}_r$ is another reference vector defined as $\dot{\boldsymbol{\theta}}_r = \dot{\boldsymbol{\theta}}_d - \gamma\Delta\boldsymbol{\theta}$, where γ is a positive constant.

The proposed controller described by (4.40) and (4.43) leads to the closed-loop system as

$$\boldsymbol{M}(\boldsymbol{q})\dot{\boldsymbol{s}} + (\boldsymbol{C}(\dot{\boldsymbol{q}}, \boldsymbol{q}) + \boldsymbol{K}_s)\boldsymbol{s} = \boldsymbol{K}\Delta\boldsymbol{\theta} + \boldsymbol{\tau}_f + \boldsymbol{m}, \tag{4.44}$$

$$\boldsymbol{B}\Delta\ddot{\boldsymbol{\theta}} + \gamma\Delta\dot{\boldsymbol{\theta}} + \boldsymbol{K}_\theta\Delta\boldsymbol{\theta} = \boldsymbol{0}. \tag{4.45}$$

Next, it can be proven by following the development in [15] that $\boldsymbol{m}+\boldsymbol{\tau}_f \to \boldsymbol{0}$ at steady state. That is, the unmodelled disturbances are well compensated.

Using the proposed iterative learning controller (4.43) to compensate for disturbances has the following advantages:

(1) It isolates and approximates the disturbances from other dynamic parameters without any force sensors;
(2) It explores the repetitive nature of rehabilitation training to realize efficient computing; such compensation is feedforward and updated during the cycle intervals only.

Note that the desired time-varying trajectory $\boldsymbol{q}_d$ in (4.41) is set as the same as the one used in the formal rehabilitation training (i.e., *Stage II*) such that $\boldsymbol{\tau}_f$ does not vary significantly at both stages.

4.2.3 Trajectory Tracking

After unmodelled disturbances are well compensated, another control scheme is introduced to enable the robot to interact with the human subject to perform the training task.

The development of the controller at *Stage II* follows the approach of singular perturbation [18, 19]. This is because (1) the overall system can be separated into a slower joint subsystem (4.35) and a faster actuator subsystem (4.36), and (2) all the dynamic parameters, including the disturbances, have been obtained.

Therefore, the overall control input at *Stage II* is given as

$$\boldsymbol{u}_{II} = \boldsymbol{u}_f + \boldsymbol{u}_s, \tag{4.46}$$

where $\boldsymbol{u}_f$ and $\boldsymbol{u}_s$ represent the fast and the slow time-scale control terms, stabilizing both subsystems of (4.35) and (4.36), respectively.

First, $\boldsymbol{u}_f$ is designed as [19]:

$$\boldsymbol{u}_f = -\boldsymbol{K}_v(\dot{\boldsymbol{\theta}} - \dot{\boldsymbol{q}}), \tag{4.47}$$

where $\boldsymbol{K}_v \in \Re^{n\times n}$ is a diagonal and positive-definite matrix.

Next, $\boldsymbol{u}_s$ is designed under the framework of MPC to realize the impedance control task. Specifically, an extended state vector is formulated based on (4.35) as $\boldsymbol{x} = [\boldsymbol{q}, \dot{\boldsymbol{q}}, z, \dot{z}, \hat{\boldsymbol{\tau}}_e, \hat{\boldsymbol{\tau}}_f]^T$, where $z \in \Re^n$ is formulated as an impedance vector as

$$z = \dot{\boldsymbol{q}} - \dot{\boldsymbol{q}}_d + \lambda(\boldsymbol{q} - \boldsymbol{q}_d) - \boldsymbol{\tau}_l, \tag{4.48}$$

where $\boldsymbol{\tau}_l$ denotes a low-pass-filtered signal of $\boldsymbol{\tau}_e$, and λ is a positive constant that is set according to the desired impedance parameters of $\boldsymbol{M}_d, \boldsymbol{C}_d, \boldsymbol{K}_d$ in (4.37); It has been shown in [15] that the convergence of $z \to \boldsymbol{0}$ implies the realization of the desired impedance mode in the low-frequency range. In addition, $\hat{\boldsymbol{\tau}}_e$ and $\hat{\boldsymbol{\tau}}_f$ denote the estimates of $\boldsymbol{\tau}_e$ and $\boldsymbol{\tau}_f$ respectively. Note that $\hat{\boldsymbol{\tau}}_e$ is obtained with the nonlinear disturbance observer [20], and $\hat{\boldsymbol{\tau}}_f$ has already been obtained at *Stage 1* (i.e., $\hat{\boldsymbol{\tau}}_f = -\boldsymbol{m}$).

Now, the dynamic model of the slower joint subsystem can be described as

$$\dot{\boldsymbol{x}} = \boldsymbol{f}(\boldsymbol{x}(t), \boldsymbol{u}_s(t)), \tag{4.49}$$

where the control input $\boldsymbol{u}_s$ is subject to the following constraint

$$\boldsymbol{u}_s(t) \in \boldsymbol{U}, \forall t \geq 0, \tag{4.50}$$

where $\boldsymbol{U}$ is a set which is introduced as

$$\boldsymbol{U}_k = [\boldsymbol{u}_s^T(k|k), \boldsymbol{u}_s^T(k+1|k), \ldots, \boldsymbol{u}_s^T(k+N_p-1|k)]^T,$$

and N_p is the prediction horizon.

The control objective is to ensure that the state vector in the prediction time domain is as close as possible to the reference value; That is, the cumulative error between the prediction state vector and the reference value should be minimized. In addition, the control action should not be too large and hence is subject to constraints. Therefore, the cost function of MPC is defined as [21]:

$$J(\boldsymbol{x}(t), \boldsymbol{u}_s(t)) = \int_t^{t+N_p} (\|\boldsymbol{x}(\zeta)\|_Q^2 + \|\boldsymbol{u}_s(\zeta)\|_R^2) d\zeta, \tag{4.51}$$

subject to

$$\boldsymbol{u}_s(\zeta) \in \boldsymbol{U}, \zeta \in [t, t+N_p] \tag{4.52}$$

where $\boldsymbol{Q} \in \Re^{n\times n}$, $\boldsymbol{R} \in \Re^{n\times n}$ are symmetric positive-definite weighting matrices.

Therefore, the control input is generated by solving the quadratic optimization problem

$$\boldsymbol{u}_s^{\star} = arg \min_{\boldsymbol{u}_s} J(\boldsymbol{x}, \boldsymbol{u}_s), \tag{4.53}$$

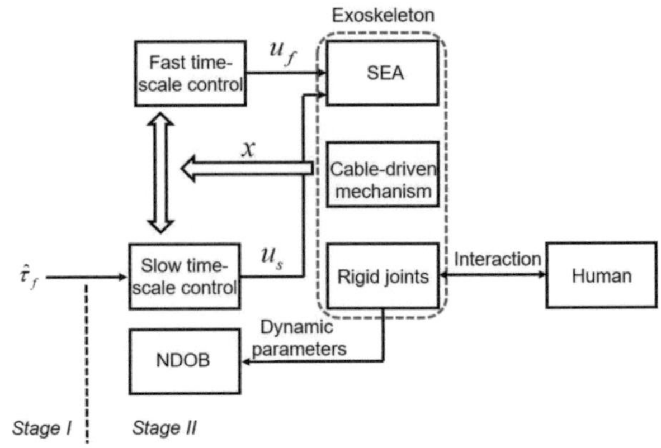

Fig. 4.24 The development of the controller at *Stage II* follows the approach of singular perturbation, where u_f and u_s denote the fast and slow time-scale control terms

specifically, extracting the first element of $\boldsymbol{U}_k$ as the control quantity of this control cycle.

To guarantee the real-time performance of the controller, the maximum number of iterations can be set; When the controller reaches the maximum number of iterations, the solving procedure stops and generates a suboptimal solution. The structure of the proposed control scheme (4.46) at *Stage II* is illustrated in Fig. 4.24, which has the following advantages:

(1) While the overall dynamic model is a high-order system, the control input does not require high-order time derivatives, and hence it is not computationally involved;
(2) While the track control task is achieved under a desired impedance model, the controller development does not require the force sensor either.

It will be shown in the subsequent experiments that the proposed controller works smoothly even on a microprocessor with very limited resources.

4.2.4 *Results*

Simulation studies are carried out to validate the performance of the proposed two-stage control scheme. Specifically, the simulation model of a single-DOF cable-driven robot is established by setting the dynamic parameters and the desired impedance parameters as in Table 4.9. In addition, the desired trajectory is set as

$$q_d = 0.1 \sin(0.2\pi \cdot t - \frac{\pi}{2}). \tag{4.54}$$

First, the proposed ILC controller (4.43) at *Stage I* was implemented by following the Algorithm 4.1, where the disturbances were predefined and set as the ground truth. Specifically, a constant disturbance (i.e., $\tau_f = 0.02$ Nm) and a cyclic disturbance (i.e., $\tau_f = 0.02 \sin(0.2\pi \cdot t - \pi/2)$ Nm) were applied to the robot successively, which is the disturbances from the cable-driven mechanism.

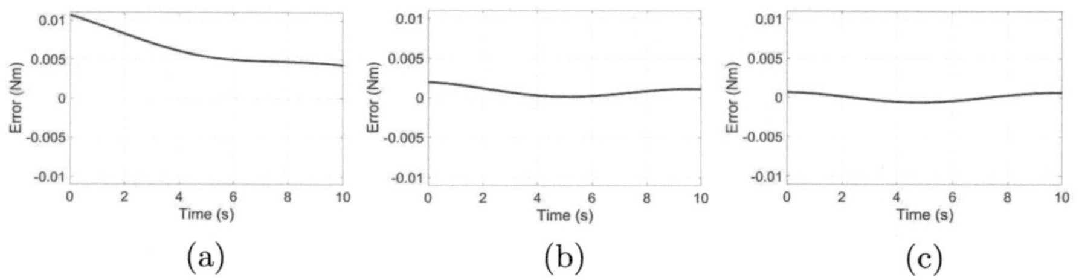

Fig. 4.25 Compensation of constant disturbances: **a** after 3 iterations; **b** after 10 iterations; **c** after 30 iterations

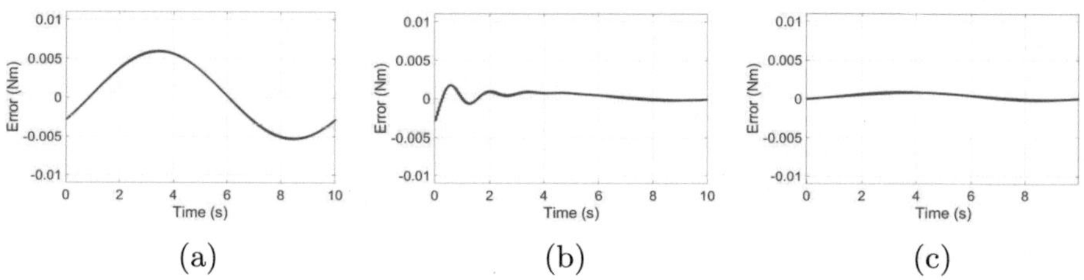

Fig. 4.26 Compensation of varying disturbances: **a** after 3 iterations; **b** after 10 iterations; **c** after 30 iterations

During each cycle of the trajectory, a total of $N = 100$ sampling points were considered, and the value of the sliding vector s at each sampling point was recorded; At the end of each cycle, the feedforward term m for the whole cycle was updated according to (4.42), where the update gains were set as 0.3 and 1.0 for constant disturbance and time-varying disturbance respectively; During the next cycle, the updated m at each sampling point was then added into the input θ_d accordingly.

The results are shown in Fig. 4.25 (constant disturbance) and Fig. 4.26 (time-varying disturbance). It is seen that convergence of $m + \tau_f \to 0$ was achieved (error less than 0.001 Nm) after 30 iterations.

Next, the proposed MPC controller at *Stage II* was implemented by following the Algorithm 4.2. Comparison has also been carried out to show the effectiveness of disturbance compensation at *Stage I*. As seen in Fig. 4.27, with such compensation, the impedance error (i.e., the vector z in (4.48)) is smaller at the steady state, implying a better realization of the desired impedance model (4.37).

Table 4.9 Parameters of simulation model

	Parameter	Value
Impedance model	M_d	$1.0\ \mathrm{Nm} \cdot \mathrm{s}^2$
	C_d	$13.0\ \mathrm{Nm} \cdot \mathrm{s}$
	K_d	4.0 Nm
Dynamic model	M	$1.0\ \mathrm{Nm} \cdot \mathrm{s}^2$
	B	$0.01\ \mathrm{Nm} \cdot \mathrm{s}^2$
	C	0
	K	2 Nm/rad
	g	$9.80\ \mathrm{m/s}^2$

Algorithm 4.1 Iterative Learning Control

Require: N: Number of sampling points in a cycle P: Maximum Iterations $\beta, M, C, g, B, K_s, K_\theta, K, \alpha, \gamma$ m_0: Value to initialize the unknown disturbance.
Ensure: Estimated values of m and control input u.
for $i = 0$ **to** $P - 1$ **do**
  for $j = 0$ **to** $N - 1$ **do**
    Compute and record s_j using Eq. (4.41).
    Compute θ_{dj} using Eq. (4.40).
    Compute u_j with updated m using Eq. (4.43).
  end for
  Update m:
  for $j = 0$ **to** $N - 1$ **do**
    $m_{j+1} = m_j - \beta s_j$.
  end for
end for

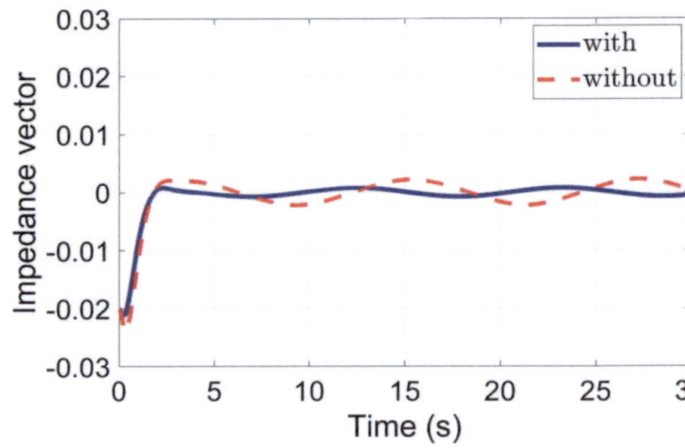

Fig. 4.27 The impedance errors with (blue) and without (red) the compensation obtained at *Stage I*

Experiments have been performed to validate the performance of the proposed control scheme by implementing it on Joint 1, 2, and 5 of the robot. The upper-limb rehabilitation device consists of the upper-limb assembly, the aluminum alloy bracket, and the lifting stool. The main function of the aluminum alloy bracket is to support the mass of all the components; The lifting stool can adjust the height of the subject's sitting posture to fit the exoskeleton structure; The output torque of the motor is transmitted to the joints through the stretching movement of the cables to control its movement.

In the experiment, the desired trajectory was set as

$$\boldsymbol{q}_d = \begin{bmatrix} \frac{\pi}{12} \cdot \sin(0.4\pi \cdot t) \\ \frac{\pi}{12} \cdot \sin(0.4\pi \cdot t) \\ -\frac{\pi}{10} + \frac{\pi}{12} \cdot \sin(0.2\pi \cdot t + \frac{\pi}{2}) \end{bmatrix}. \tag{4.55}$$

First, *Stage I* was to estimate and compensate for the unmodelled disturbances; Hence, the exoskeleton robot was controlled to follow the predefined trajectory without human subjects involved (i.e., $\boldsymbol{\tau}_e = \mathbf{0}$), and the snapshots at different time instants can be found at Fig. 4.28a–d.

Algorithm 4.2 Model Predictive Control

Require: The state space equation of the slow system:
$\dot{x} = f(x, u_s)$ N_u, N_p : the control horizon and the prediction horizon $u_{init} = [u_s(0), u_s(1), \ldots, u_s(N_p - 1)]$: Value to initialize control vector.
Ensure: Reference value sequence D and maximum number of iterations *MaxIterations*.
Discretize the prediction model: $x(k+1) = \overline{A}x(k) + \overline{B}u_s(k)$
while task not completed **do**
for $i = 0$ **to** $N_p - 1$ **do**
Compute:$x(k+i|k) = \overline{A}^i x(k) + \sum_{n=0}^{i-1} \overline{A}^{i-1-n}\overline{B}u_s x(k+n|k)$
end for
Compute reference sequence:

$$D = [d(k+1)^T, d(k+2)^T, \ldots, d(k+N_p)^T]$$

Calculate cost function using Eq. (17).
for $i = 0$ **to** *MaxIterations*-1 **do**
Iteratively optimize:

$$\min_{U_k} J(U_k)$$

to solve U_k.
end for
end while

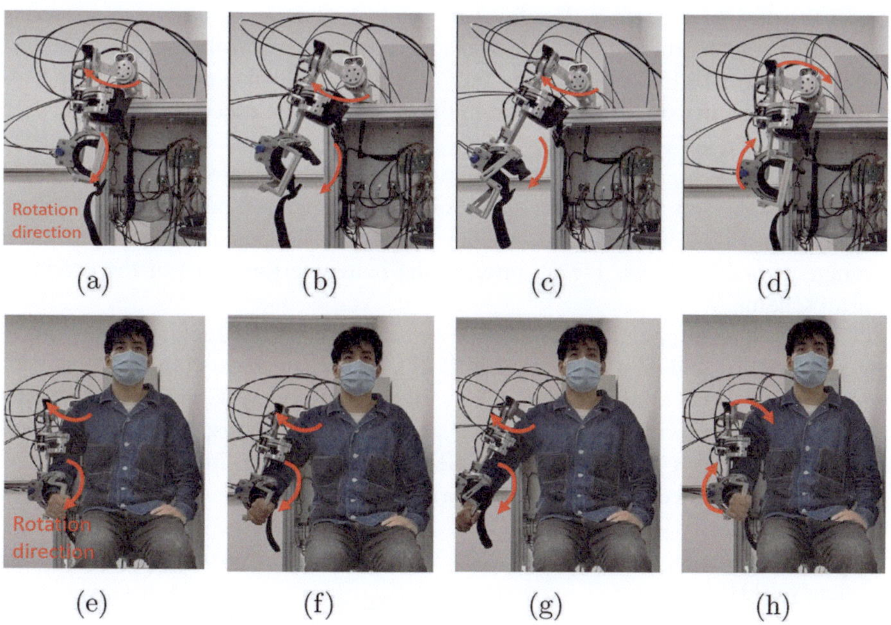

Fig. 4.28 **a–d** Snapshots of *Stage I* at different time instants: $t = 0, 3.0s, 5.0s, 10.0s$; **e–f** Snapshots of *Stage II* at different time instants: $t = 0, 3.0s, 5.0s, 10.0s$

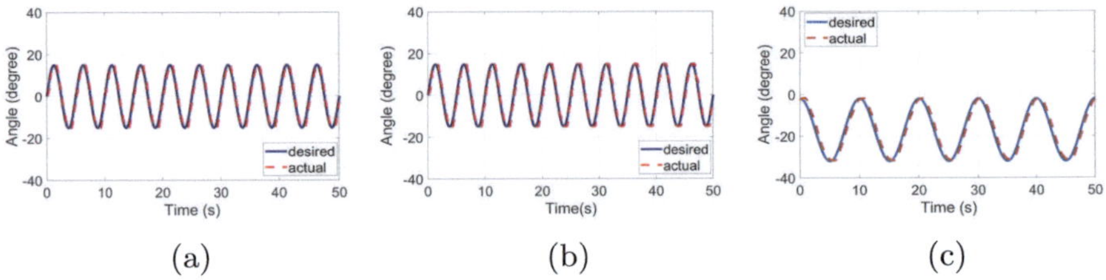

Fig. 4.29 Joint-space trajectory: **a** Joint 1; **b** Joint 2; **c** Joint 5

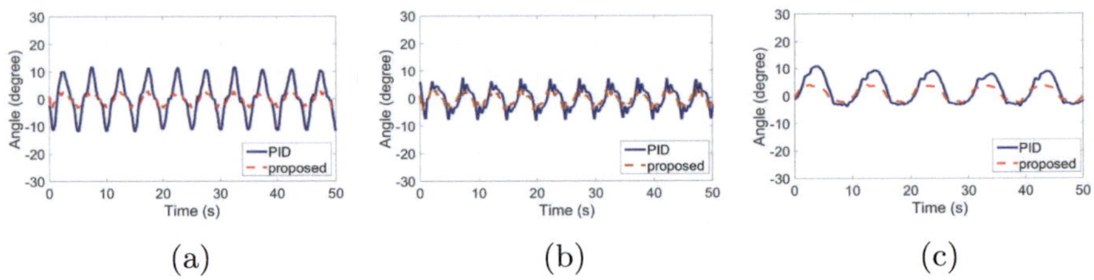

Fig. 4.30 Tracking errors: **a** Joint 1; **b** Joint 2; **c** Joint 5

Then, *Stage II* rejected the disturbances and employed the NDOB to design the impedance control input, so that the robot was controlled to lead the human to follow the desired trajectory. The fast time-scale control input $\boldsymbol{u}_f$ was proposed to stabilize the compliant actuator, and the slow time-scale control input $\boldsymbol{u}_s$ was based on the suboptimal MPC algorithm to minimize the impedance error. The snapshots at different time instants can be found in Fig. 4.28e–h. The trajectories of Joint 1, 2, and 5 are shown in Fig. 4.29.

As a comparison, the PID control algorithm was also implemented to track the same trajectory, and the tracking errors of PID control and our proposed method are shown in Fig. 4.30. It is seen that our method ensures smaller errors for joints (average errors: $3°42'$ vs. $1°73'$), proving the better performance of the proposed control scheme.

The PID algorithm and our proposed two-stage trajectory-tracking control strategy are implemented on a 168 MHz MCU (STM32F4). The execution time of each control loop is set at approximately 10 ms using a time delay. Compared with other control algorithms with similar tracking accuracy, our proposed strategy requires lower hardware cost and sampling frequency. For example, in Ref. [22], the xPC system is used to execute the control algorithm at the sampling frequency of 1 kHz, and the MAE of two joints was 1.30/1.26° , respectively. At the same time, the average MAE of our proposed strategy was 1.92°, which was executed at the sampling frequency 100 Hz.

4.3 Summary and Notes

Exoskeletons serve as essential tools for human augmentation [23] and rehabilitation [12], enhancing mobility, strength, and endurance in both healthy individuals and those with physical impairments. This chapter focuses on augmentation in healthy individuals, emphasizing the significance of interactive learning in optimizing human-exoskeleton interaction. Regardless of whether the exoskeleton is used for augmentation or rehabilitation, a vast amount of interaction data is generated, including joint trajectories and interaction torques during movement. These data are crucial as they implicitly capture the degree of comfort between the wearer and the exoskeleton, as well as potential conflicts arising during movement [24]. Existing works on interactive learning primarily utilize these data to fit dynamical models [14] and adaptively adjust the level of exoskeleton assistance [25]. In this chapter, we apply interactive learning techniques to both lower-limb and upper-limb exoskeletons.

For lower-limb exoskeletons, existing HIL optimization approaches suffer from several limitations: some rely on metabolic rate measurements, which are time-consuming [26], while others focus on single-joint optimization [27], limiting their effectiveness in multi-joint movements. Moreover, most existing works concentrate solely on walking-based HIL optimization, without considering its potential for tasks such as stair climbing or squatting. In our approach, we introduce an anomaly score-based HIL optimization method and further generalize the optimized results to different lower-limb tasks using a neural network model. This enables the seamless transfer of personalized assistance across multiple locomotion tasks, including stair ascent, stair descent, and squatting. The ability to adapt exoskeleton assistance across multiple users and diverse tasks significantly enhances its practicality and broadens its applicability, paving the way for exoskeleton deployment in real-world scenarios.

For upper-limb exoskeletons, the mechanical design structure is a critical factor influencing usability. Existing exoskeletons often suffer from design trade-offs: fully motorized joints lead to excessive weight and noticeable inertial effects [28], while cable-driven mechanisms have typically been deployed in a limited number of joints, failing to provide a sufficient range of motion for daily upper-limb activities [29]. The use of interactive learning to separate disturbance forces from actual interaction torques is crucial in achieving precise control. By effectively decoupling these forces, the introduced approach enables accurate force control and impedance regulation, ensuring a more natural and intuitive human-exoskeleton interaction.

References

1. R. Chalapathy, S. Chawla, Deep learning for anomaly detection: A survey (2019). arXiv:1901.03407
2. G. Münz, S. Li, G. Carle, Traffic anomaly detection using k-means clustering, in *Gi/itg workshop mmbnet*, vol. 7 (2007)
3. X. Xu, Y. Lei, X. Zhou, A lof-based method for abnormal segment detection in machinery condition monitoring, in *2018 Prognostics and System Health Management Conference (PHM-Chongqing)* (IEEE, 2018), pp. 125–128
4. Z. Gan, X. Zhou, Abnormal network traffic detection based on improved lof algorithm, in *2018 10th International Conference on Intelligent Human-Machine Systems and Cybernetics (IHMSC)*, vol. 1 (IEEE, 2018), pp. 142–145
5. D.J. Samuel, F. Cuzzolin, Unsupervised anomaly detection for a smart autonomous robotic assistant surgeon (saras) using a deep residual autoencoder. IEEE Robot. Autom. Lett. **6**(4), 7256–7261 (2021)
6. P. Malhotra, L. Vig, G. Shroff, P. Agarwal, et al., Long short term memory networks for anomaly detection in time series, in *Esann*, vol. 2015 (2015), p. 89

7. D. Park, Y. Hoshi, C.C. Kemp, A multimodal anomaly detector for robot-assisted feeding using an lstm-based variational autoencoder. IEEE Robot. Autom. Lett. **3**(3), 1544–1551 (2018)
8. Y. Su, Y. Zhao, C. Niu, R. Liu, W. Sun, D. Pei, Robust anomaly detection for multivariate time series through stochastic recurrent neural network, in *Proceedings of the 25th ACM SIGKDD International Conference on Knowledge Discovery & Data Mining* (2019), pp. 2828–2837
9. Y. Zimmermann, E.B. Küçüktabak, F. Farshidian, R. Riener, M. Hutter, Towards dynamic transparency: Robust interaction force tracking using multi-sensory control on an arm exoskeleton, in *2020 IEEE/RSJ International Conference on Intelligent Robots and Systems (IROS)* (IEEE, 2020), pp. 7417–7424
10. S. Schaal, C.G. Atkeson, Constructive incremental learning from only local information. Neural Comput. **10**(8), 2047–2084 (1998)
11. A.J. Ijspeert, J. Nakanishi, H. Hoffmann, P. Pastor, S. Schaal, Dynamical movement primitives: learning attractor models for motor behaviors. Neural Comput. **25**(2), 328–373 (2013)
12. X. Li, X. Zhang, X. Li, J. Long, J. Li, L. Xu, G. Chen, J. Ye, Bear-h: an intelligent bilateral exoskeletal assistive robot for smart rehabilitation. IEEE Robot. Autom. Mag. **29**(3), 34–46 (2021)
13. M. Kim, C. Liu, J. Kim, S. Lee, A. Meguid, C.J. Walsh, S. Kuindersma, Bayesian optimization of soft exosuits using a metabolic estimator stopping process, in *2019 International Conference on Robotics and Automation (ICRA)* (IEEE, 2019), pp. 9173–9179
14. Z. Li, Z. Huang, W. He, C.-Y. Su, Adaptive impedance control for an upper limb robotic exoskeleton using biological signals. IEEE Trans. Industr. Electron. **64**(2), 1664–1674 (2016)
15. X. Li, Y.-H. Liu, H. Yu, Iterative learning impedance control for rehabilitation robots driven by series elastic actuators. Automatica **90**, 1–7 (2018)
16. A. Saberi, P.V. Kokotovic, H.J. Sussmann, Global stabilization of partially linear composite systems. SIAM J. Control. Optim. **28**(6), 1491–1503 (1990)
17. Y. Pan, H. Wang, X. Li, H. Yu, Adaptive command-filtered backstepping control of robot arms with compliant actuators. IEEE Trans. Control Syst. Technol. **26**(3), 1149–1156 (2017)
18. P. Kokotović, H.K. Khalil, J. O'reilly, *Singular Perturbation Methods in Control: Analysis And Design* (SIAM, 1999)
19. X. Li, Y. Pan, G. Chen, H. Yu, Multi-modal control scheme for rehabilitation robotic exoskeletons. Int. J. Robot. Res. **36**(5–7), 759–777 (2017)
20. A. Mohammadi, H.J. Marquez, M. Tavakoli, Nonlinear disturbance observers: Design and applications to euler? lagrange systems. IEEE Control Syst. Mag. **37**(4), 50–72 (2017)
21. P.O. Scokaert, D.Q. Mayne, J.B. Rawlings, Suboptimal model predictive control (feasibility implies stability). IEEE Trans. Autom. Control **44**(3), 648–654 (1999)
22. Y. Wang, S. Li, D. Wang, F. Ju, B. Chen, H. Wu, Adaptive time-delay control for cable-driven manipulators with enhanced nonsingular fast terminal sliding mode. IEEE Trans. Industr. Electron. **68**(3), 2356–2367 (2020)
23. L. Brenner, Exploring the psychosocial impact of ekso bionics technology. Arch. Phys. Med. Rehabil. **97**(10), e113 (2016)
24. Y. Chen, S. Miao, G. Chen, J. Ye, C. Fu, B. Liang, S. Song, X. Li, Learning to assist different wearers in multitasks: efficient and individualized human-in-the-loop adaptation framework for lower-limb exoskeleton. IEEE Trans. Robot. (2024)
25. S. Han, H. Wang, H. Yu, Human–robot interaction evaluation-based aan control for upper limb rehabilitation robots driven by series elastic actuators. IEEE Trans. Robot. (2023)
26. J. Zhang, P. Fiers, K.A. Witte, R.W. Jackson, K.L. Poggensee, C.G. Atkeson, S.H. Collins, Human-in-the-loop optimization of exoskeleton assistance during walking. Science **356**(6344), 1280–1284 (2017)
27. Y. Ding, M. Kim, S. Kuindersma, C.J. Walsh, Human-in-the-loop optimization of hip assistance with a soft exosuit during walking. Sci. Robot. **3**(15), eaar5438 (2018)
28. Q. Wu, X. Wang, F. Du, Development and analysis of a gravity-balanced exoskeleton for active rehabilitation training of upper limb. Proc. Inst. Mech. Eng. C J. Mech. Eng. Sci. **230**(20), 3777–3790 (2016)
29. Y. Mao, X. Jin, G.G. Dutta, J.P. Scholz, S.K. Agrawal, Human movement training with a cable driven arm exoskeleton (carex). IEEE Trans. Neural Syst. Rehabil. Eng. **23**(1), 84–92 (2014)

Chapter 5
Rehabilitation Robots

While the previous chapter discusses the application of exoskeleton in strength augmentation for healthy subjects, this chapter presents another application, that is, rehabilitation training, where the robot is worn by a stroke patient and provides assistance to help perform repetitive motions and regain motor functions. The deployment of exoskeletons can alleviate the shortage of experienced therapists and play a vital role in countries with aging populations.

Actually, the demand for robotic rehabilitation can be reflected in many developed countries, where aging populations and stroke patient numbers have been increasing rapidly in recent years. There is also a large population of stroke patients in China, which is now in the tens of millions, and the number is growing at a rate of 10% each year. Hence, there is a lack of around 400, 000 therapists in China, which is far more than the total number of the nation's active therapists [1]. The nature of manual rehabilitation is highly labor-intensive, in which a therapist must hold a patient's limb to assist him/her in carrying out repetitive movements. Moreover, the quality of this rehabilitation is dependent on the therapist's experience, overall treatment time, and consistency of the procedure. In this regard, deploying rehabilitation robots may help to overcome the limitations of manual rehabilitation by exploring their ability to produce high repetition and consistency.

5.1 Lower-Limb Exoskeleton

While the potential impact of exoskeleton robots on rehabilitating training is significant, the intelligence level of existing exoskeletons is relatively low, where a robot cannot adapt to either the online change of a subject's motion (e.g., the gait pattern) or the variation of his/her body parameters (i.e., a new subject who is going to wear the robot), potentially resulting in conflict between human and robot, and possibly even leading to physical damage. As such, the application of a robotic exoskeleton

X. Li and H. Yu, *Compliantly-Driven Robots with Application to Exoskeletons*,
https://doi.org/10.1007/978-981-96-9544-7_5

in clinical studies is limited. This section introduces a new lower-limb exoskeleton robot for rehabilitation (i.e., BEAR-H), where the main novelty is the integration of multiple intelligent features, such as gait recognition and synchronization, cloud-computing diagnosis, and individualized gait generation. Such an integration helps the robot to understand the patient's condition better and hence provide effective assistance, which, in the end, achieves smart rehabilitation.

5.1.1 Overall Structure

As mentioned, the performance of existing prototypes/products is commonly affected by several open issues. First, the robot joints are commonly driven with rigid actuators, which cannot provide structural safety. Second, the robot joints may not fully cover the entire lower limb, in the sense that some of the patient's motions cannot be achieved. Third, most robots are in passive training mode, such that the natural functionality after the treatment is not guaranteed. Those issues actually limit the applications of robotic exoskeletons in clinical scenarios.

BEAR-H is developed to address the above issues, which is a new bilateral exoskeletal assistive robot developed by Shenzhen MileBot Robotics Co., Ltd, China [2]. A comparison between BEAR-H and other robotic exoskeletons for rehabilitation is shown in Table 5.1. Compared with other products, BEAR-H integrates multiple intelligent features in the training mode: (1) it can automatically recognize the patient's gait event to synchronize the robot's angular trajectory with respect to the patient's actual gait; (2) it can inspect the body parameters of a patient and then customize an individualized gait, corresponding to the gait profile of a healthy subject with similar body parameters; (3) it carries a cloud-based artificial intelligence (AI) diagnosis platform, which allows doctors and family members to monitor the progress when the patient is under home rehabilitation. The integration of the aforementioned features helps BEAR-H better understand the patient's condition and then generate a more "intelligent" gait for rehabilitation. Additionally, an adaptive impedance control scheme is proposed to drive BEAR-H to track such gait, which guarantees a safe interaction and also encourages the patient to execute voluntary

Table 5.1 Comparison of exoskeletons

Feature	Objective	Actuation	DOFs	Certificate	Training
HAL [3]	Paraplegia, Hemiplegia	Rigid	6	FDA	Passive
Ekso [4]	Paraplegia	Rigid	4	FDA	Passive
ReWalk [5]	Paraplegia	Rigid	4	CE, FDA	Passive
UGO210 [6]	Paraplegia	Rigid	4	NMPA	Passive
BEAR-H [2]	Hemiplegia	Compliant	6	NMPA	Active

efforts and hence speed up the recovery. Moreover, compliant actuators are fully deployed in the hip, knee, and ankle joints of BEAR-H to further ensure safety at the structural level. BEAR-H has been successfully implemented in clinical applications, and comparative studies have been carried out for 30 hemiplegic patients. Experimental results from different aspects are presented to validate the effectiveness of BEAR-H.

Specifically, BEAR-H is a lower-limb exoskeleton robot designed to facilitate smart rehabilitation. Specifically, the robot fully assesses the body parameters, online status, and history data of the patient and then customizes the rehabilitation strategy to better suit his/her condition, and hence improve the rehabilitation effect. The overall system consists of a backpack (housing the control module), segments of waist and both legs, and a touch pad as the interface. Leg length can be adjusted to suit users with different heights, from 150 m (i.e., children) to 190 cm (i.e., adults). The dimension of the waist joint can also be adjusted to suit users who weigh up to 85 kg. Each leg has three active DoFs and two passive DoFs. Hence, the whole robot has six active DOFs in total. The active DoFs are designed as the flexion/extension around the hip, knee, and ankle joints, while the passive DoFs are designed as the inversion/eversion of the ankle joint and the adduction and abduction of the hip joint. Therefore, the whole robot has ten DoFs in total, and such a high-DOF structure supports full rehabilitation for lower limbs. All the active joints are driven with compliant actuators, guaranteeing safety from the structural aspect. A module in the shape of a shoe is designed at the end of both legs to support the patient's feet. The aforementioned can be found in Fig. 5.1.

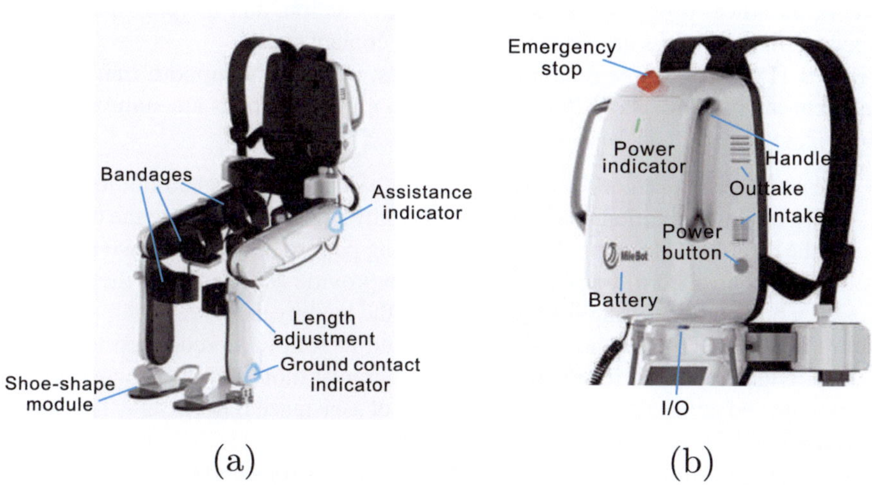

Fig. 5.1 Structure of BEAR-H: **a** each leg consists of five DOFs (i.e., three active DOFs and two passive DOFs), and the length can be adjusted for patients with different heights; **b** the backpack carries the battery and the embedded control system and also has a power indicator, emergency stop, and multiple I/Os for extension

Table 5.2 Interaction modes

MODE	Degree of Severity	Level of Assistance
Weight support	*Serious* - No full gait cycle	Passive following
Training	*Moderate* - Full gait cycle but abnormal patterns	Active assist. w/ predefined gait traj.
Intell. Interact.	*Mild* - Several full cycles but abnormal patterns	Active assist. w/ online sync. + opt. indiv. traj.

The central control system is embedded into the backpack to realize multiple interaction modes and the algorithm of intelligent rehabilitation. The backpack also integrates the modules of an emergency stop, I/Os, and indicators to show the level of assistance and the ground contact status. The battery inside the backpack can support BEAR-H's continuous operation for around eight hours. There are also two handles on the backpack, allowing therapists to hold the handles to help patients put on the robot (see Fig. 5.1b). The touch screen conveys information during rehabilitation, where therapists specify the training mode and parameters according to the progress of rehabilitation. During the development, such an interface is useful for debugging as well.

A range of sensors are installed in BEAR-H to capture the change of human limbs and to monitor the robot's posture. The robot refers to the multi-modal sensory data and then intelligently recognizes the gait of the patient to output proper assistance at different joints. Specifically, the encoders are used to measure the flexion/extension of the robot's joints, the inertial measurement units (IMUs) are used to measure the angle of the lower limb and the acceleration information, and the pressure sensors embedded in both shoes are used to inspect the contact status.

BEAR-H provides three rehabilitation modes, i.e., weight support, training, and intelligent interaction. The differences between multiple modes are summarized in Table 5.2, i.e.,

(1) The *weight support* mode is designed for the early stages of rehabilitation. Under such a mode, the motion on the patient's healthy side is converted to the trajectory for the disabled side, according to the walking pattern between the two legs. The robot is controlled to track the trajectory to provide assistance for the disabled side while compensating for the weight of the healthy side.
(2) In the *training* mode, the robot is controlled to follow a predefined trajectory to provide assistance for the patient. The therapist monitors the progress of the training and adjusts the frequency and level of assistance if necessary. This mode is designed for patients whose motor functionalities are partially restored.
(3) In the *intelligent interaction* mode, the level of assistance and frequency of the robot's trajectory are automatically adjusted online by monitoring the patient's walking pattern. This mode is designed for rehabilitation at the recovery stage, where the patient is able to walk independently with several abnormal patterns.

These multiple modes can meet the demands of patients with different backgrounds, and a criterion is also proposed to identify the patient's degree of severity (see Table 5.2). Both modes of *Training* and *Intelligent Interaction* belong to active training, which is important for pushing the patient to master the normal and regular gait pattern. For patients with dyskinesias, active training is even more significant in the sense that a closed loop is established in the central neural system to repair the disordered functionality and lead to a positive long-term training effect.

The following development mainly focuses on the *intelligent interaction* mode to specify the generation of the *intelligent* gait pattern. Note that the initial frequency of the robot's trajectory in the *intelligent interaction* mode is set as 0.3 Hz, and it is progressively scaled up or scaled down to suit the actual walking frequency of the patient. When the rate of synchronization is over 75%, the corresponding frequency and level of assistance are displayed in the touch panel. In addition to the aforementioned modes, BEAR-H also has the ability to customize the highly individualized gait according to specific body parameters, which can better suit patients with different conditions.

5.1.2 Compliant Actuation

Wearing a robotic exoskeleton involves substantial physical interactions, and hence, the safety of patients is always the main concern. Most robotic exoskeletons in the market are driven with rigid actuators (see Table 5.1). While the structure of rigid actuators is relatively simple, it is unable to provide additional flexibility for the movement of the human body. More importantly, as the output force of the rigid actuator is directly transferred to the robot joint, forces that are too large might cause physical damage to the human body.

BEAR-H is designed with the deployment of compliant actuators, which makes it safe from a structural point of view. The compliant actuator is developed by referring to the concept of the series elastic actuator (SEA) [7], where an elastic element is installed between the driving motor and the robot joint. Such a design has several attractive features:

(1) The elastic property is explored to store excessive energy and hence counterbalance an impact that is too large.
(2) The output torque/force can be obtained by simply measuring the deflection of the elastic element (usually with the encoders' reading on both ends of the element).
(3) The actuator is backdrivable such that a certain level of free motion is allowed for the patient.

While various designs of compliant actuators have been reported in the literature, BEAR-H is the first commercialized exoskeleton robot (i.e., a medical product with official registration) that converts the prototype of compliant actuators into an actual implementation. The structure of compliant actuators used in BEAR-H is illustrated

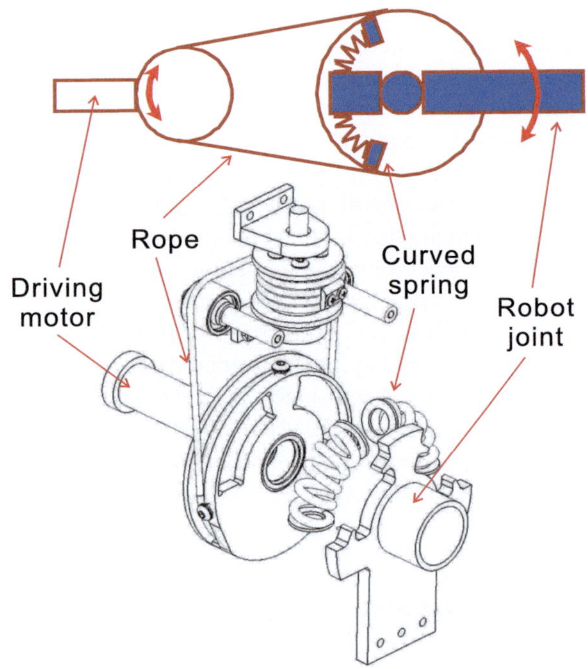

Fig. 5.2 The joints of BEAR-H are driven by compliant actuators, where a curved spring is installed between the driving motor and the robot joint, whose motion is pulled by the rope via a rotary table

in Fig. 5.2. The motor drives the motion of the rope coiled around a spool, and the curved spring is then pulled by the rope via a rotary table. Then, the motion is transmitted to the output rod so as to realize the rotation of the robot joint at the end. Such a design achieves a balance between the small size, lightweight, high-density output, and relatively low cost, thus guaranteeing the actual implementation. Table 5.3 summarizes the specifications of the developed compliant actuators.

Table 5.3 Specification of compliant actuators

SPEC.	VALUE	SPEC.	VALUE
Spring diameter	70 mm	Weight	2.5 kg
Spring thickness	9 mm	Length	130 mm
Maximum rotation	±8 deg	Encoder resolution	17 *bit*
Maximum torque	60 Nm	Reduction rate	100 : 1

Table 5.4 Gait phases and gait events

CODE	PHASE	EVENT
LR	loading response	initial contact (IC)
MSt	mid-stance	opposite toe off (OT)
TSt	terminal stance	heel rise (HR)
PSw	pre-swing	opposite initial contact (OI)
ISw	initial swing	toe off (TO)
MSw	mid-swing	feet adjacent (FA)
TSw	terminal swing	tibia vertical (TV)

5.1.3 Gait Synchronization

BEAR-H considers seven phases of the human gait as shown in Table 5.4, and each phase is initiated with a gait event. The seven phases of healthy subjects almost follow the same sequence, and the duration of each phase is also similar. For patients, abnormal patterns might be detected in the gait, e.g., one or several missed phases, too long or too short durations, or asymmetry between both legs. Hence, the phases of the gait can be treated as a measure to assess the quality of rehabilitation. The gait events are illustrated in Fig. 5.3.

In BEAR-H, a Hidden Markov Model (HMM) with seven states is initialized to describe seven gait phases as

$$[LR, MSt, TSt, PSw, ISw, MSw, TSw], \tag{5.1}$$

where the notation from LR to TSw represents different gait events, as shown in Table 5.4. The HMM can be described with the tuple of parameters as $\lambda = (\pi, \boldsymbol{A}, \boldsymbol{B})$, where $\pi, \boldsymbol{A}, \boldsymbol{B}$ denote the prior probability, the state transition probability, and the observation probability, respectively. Then, the observation vector is defined as

$$z_t = [\omega_R, \alpha_R, \omega_L, \alpha_L, \theta_{R_{knee}}, \theta_{R_{shank}}], \tag{5.2}$$

where ω_R and ω_L denote the angular speed of both feet, and α_R and α_L are their first-order derivatives (i.e., acceleration), and $\theta_{R_{knee}}$ and $\theta_{R_{shank}}$ are the angles of the right knee and the right shank, respectively. Note that the value of those variables can be obtained with sensors mounted on the human limb.

Given the observation vector (5.2), the Baum-Welch learning algorithm [8] is to find the optimal value of model parameters by solving $\lambda^\star = \arg\max_\lambda P(z_t; \lambda)$, where $P(\cdot)$ is the probability. This is basically achieved by following the EM steps, that is, (E-step): To calculate the expectations of the states; (M-step): To estimate the model parameters by maximizing the expectation. Then, with the well-trained HMM, the most likely gait phase during overground walking can be recognized.

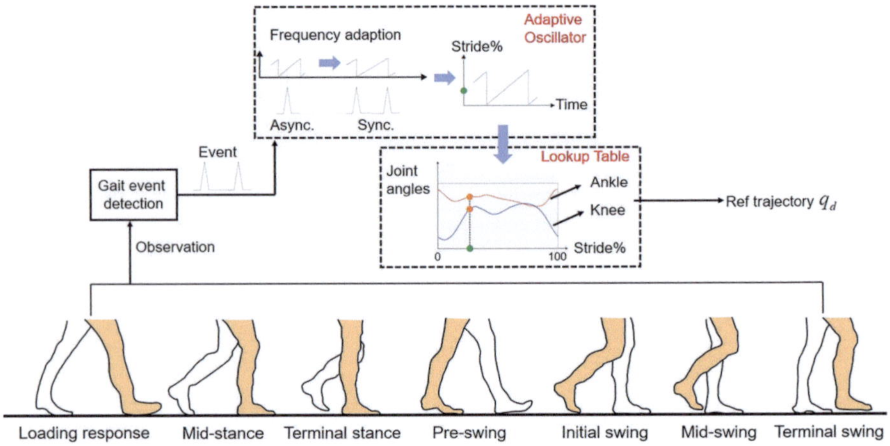

Fig. 5.3 In general, the gait cycle consists of seven phases, which are represented by the events of *initial contact (IC)*, *opposite toe off (OT)*, *heel rise (HR)*, *opposite initial contact (OI)*, *toe off (TO)*, *feet adjacent (FA)*, and *tibia vertical (TV)*. The adaptive oscillator updates the frequency according to gait events and then computes the stride percentage. The lookup table receives the stride percentage to normalize the reference trajectory

To quickly detect the specific gait events, the adaptive thresholding scheme was proposed in [9], where a linear regression model was introduced to approximate the relationship between three signal inputs (i.e., thigh angle, thigh angular velocity, and forward axial acceleration) and the output value of the threshold. The proposed scheme has been validated in the event detection of HR and TO [9].

The detected events can be used to adjust the robot's reference trajectory online for synchronization between the robot and the patient. To achieve this, an adaptive oscillator was developed for BEAR-H in [8]. The adaptive oscillator refers to the gait events to update its frequency as

$$\dot{\omega} = \varepsilon \cdot (\Omega - \phi) + \sum_{k=0}^{\infty} \delta(t - kT) \cdot Q \cdot G(\omega, \Delta\phi), \tag{5.3}$$

where ω denotes the frequency, ε denotes the adaptation rate, $\sum_{k=0}^{\infty} \delta(\cdot)$ is a periodic Dirac delta function describing the occurrence of gait event, Ω is the corresponding phase of actual gait, T is the period of the gait cycle, Q is a positive value related to the amount of the frequency change, $\Delta\phi = \frac{\phi-\Omega}{2} + 2\pi$, representing the phase difference between the adaptive oscillator and the actual gait, and $G(\cdot)$ is a function curving the phase response defined as

$$\begin{aligned} G(\omega, \Delta\phi) = &\max(\frac{\sin(\Delta\phi)}{2\pi}, 0)(\omega_{\min} - \omega) \\ &- \min(\frac{\sin(\Delta\phi)}{2\pi}, 0)(\omega_{\max} - \omega), \end{aligned} \tag{5.4}$$

where $\omega_{\max}$ and $\omega_{\min}$ represent the upper and lower bounds of frequency respectively. According to (5.4), when $0 < \phi - \Omega \leq 2\pi$, the oscillator leads the gait event, $\sin(\Delta\phi) \geq 0$, $G(\cdot) < 0$, its frequency is lowered; When $-2\pi \leq \phi - \Omega < 0$, the oscillator lags behind the gait event, $\sin(\Delta\phi) \leq 0$, $G(\cdot) > 0$, its frequency is scaled up.

Next, the phase of the adaptive oscillator is dependent on the adjusted frequency as $\dot{\phi} = \omega$ where the phase ϕ varies between 0 and 2π and grows uniformly within one cycle. The phase of the adaptive oscillator is referred to in order to compute a stride percentage as $stride\ \% = \frac{\phi}{2\pi} \times 100$. Then, the trajectory of the robot can be adjusted by normalizing a reference trajectory (e.g., the individualized gait trajectory in the subsequent development) with the stride percentage based on a lookup table. The normalized trajectory can now match the actual gait of the patient, i.e., starting from the same gait event under the same frequency. Therefore, the synchronization between the robot and the patient is realized. The aforementioned workflow is illustrated in Fig. 5.3.

5.1.4 Individualized Gait Generation

A cloud-based management platform is established for BEAR-H (see Fig. 5.4), and the platform structure is based on the Internet of Things (IoT). It can receive and store the real-time training data of patients and then allow doctors to master the situation and progress of rehabilitation. With some intelligent diagnosis algorithms, doctors can quickly provide feedback to patients via remote access, laying the foundation for home rehabilitation. Additionally, the established cloud-based platform provides multiple interfaces.

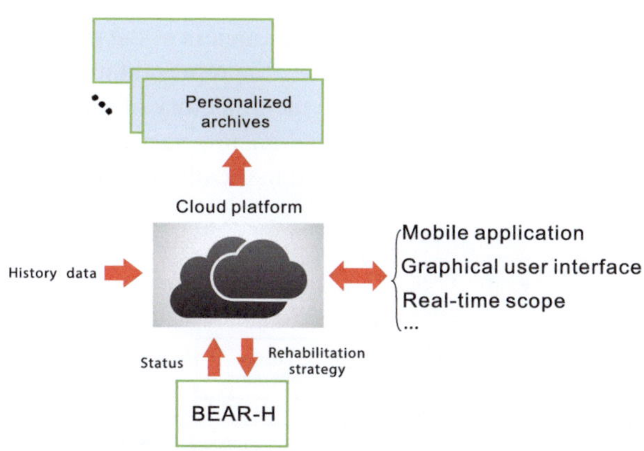

Fig. 5.4 The cloud-computing platform provides the foundation for an intelligent diagnosis, where the patient can carry out rehabilitation with BEAR-H at home, while the doctor can remotely evaluate the quality. Different interfaces are integrated at the platform, e.g., the mobile application (where family members can check if there is any safety issue), the graphical user interface (where the engineer from the company can monitor if there is any failure), and the real-time data plotting (where the doctor can conclude if the rehabilitation is valid)

(1) The doctor can use the mobile application to check the training record and training report of patients or control the robot to adjust the strategy of rehabilitation before the next course of treatment.
(2) As the platform visualizes all of the real-time data, the doctor can closely monitor the robot's motions (e.g., the trajectory of the robot's joints and level of assistance) and then provide immediate feedback.
(3) The platform allows the user to revisit the history data (which are stored as personalized archives) and hence have an overview of the whole procedure of rehabilitation; the platform can also be used as the interface for remote maintenance and fault diagnosis.

Additionally, the platform stores healthy subjects' data, that is, the body parameters (e.g., height, age, wrist width, etc.) and the gait trajectories under different walking speeds. By referring to the large dataset, a neural network (NN) is constructed in BEAR-H for the generation of individualized gaits under the structure shown in Fig. 5.5. The overall structure consists of four modules: *Input*, *Feature Extraction*, *Mapping*, and *Output*, and the workflow can be summarized as follows.

(1) The NN is constructed to realize a probabilistic mapping from the body parameters of patients to the features of the gait trajectory. The features are designed as a series of coefficients of the Fourier transform of the gait trajectory, which can then be reconstructed by computing the Fourier inversion with reference to the predicted features (i.e., the output of NN).
(2) The training dataset is collected from healthy subjects. Specifically, the gait trajectory of the subject is recorded using anthropometry methods and set as the ground truth. The multi-scale body parameters (e.g., height, weight, leg length, etc.) are measured and then fed into the NN. The NN outputs the predicted gait features, which are compared with the ground truth to update the weights and hence train the NN.
(3) The well-trained NN receives the body parameters of the patient and then generates an individualized gait trajectory (i.e., the representation with gait features) for a specific subject, corresponding to the trajectory of the healthy subject with similar body parameters. As the body parameters vary for different subjects, the individualized gait also varies and is automatically adjusted by the trained NN.

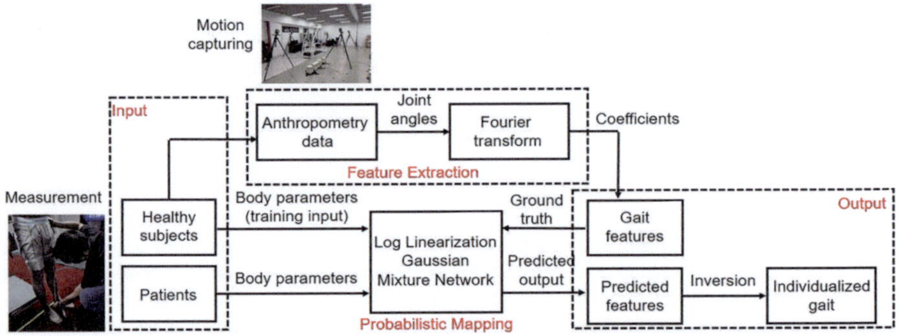

Fig. 5.5 The overall structure of gait generation consists of four modules: *Input*, *Feature Extraction*, *Mapping*, and *Output*

Note that the backbone network is based on a Log Linearization Gaussian Mixture Network (LLGMN), and the training and implementation phases are illustrated in Fig. 5.6. Therefore, the individualized gait is actually generated by estimating the posterior probability of the gait in a data-driven way, given the specific body parameters of a patient. Such a formulation explores the underlying relationship between the body parameters and gait; hence, it can better suit the patient to guarantee his/her natural motion after rehabilitation.

Note that the subject-specific gait is related to EMG signals and the strength of a human limb. However, a stroke patient has significantly different EMG signals and strength data, compared with the situation when he/she was in a healthy condition. Therefore, that information is not used as the input to retrieve the original healthy gait pattern.

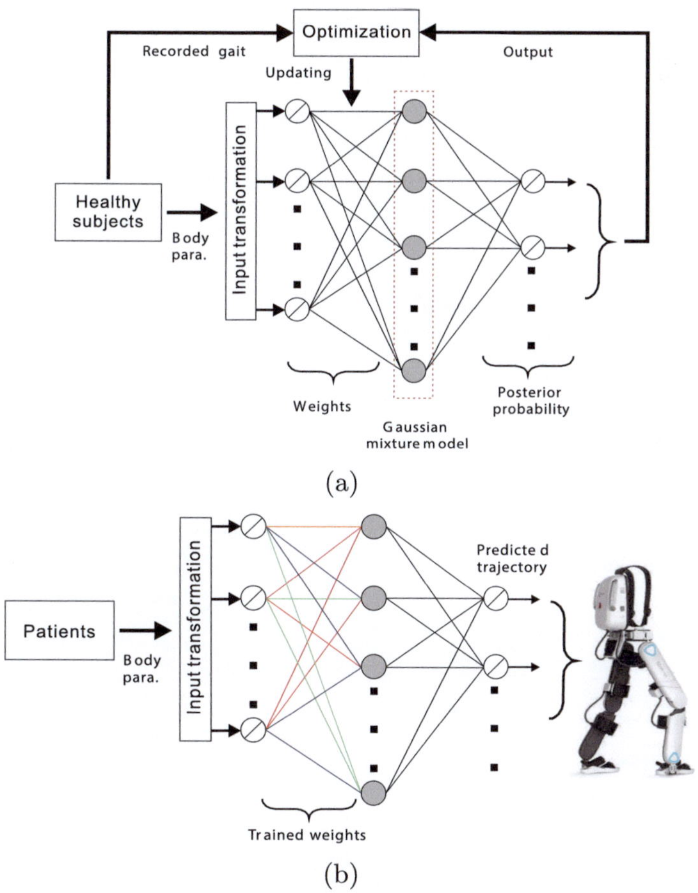

Fig. 5.6 BEAR-H can generate an individualized gait according to the body parameters of patients (e.g., height, weight, leg length): **a** *training phase*: both the gait trajectory and body parameters of healthy subjects are recorded to train the LLGMN; **b** *implementation*: the trained network receives the body parameters of a patient and then predicts the gait (i.e., corresponding to the gait of a healthy subject with similar body parameters), which is then used to drive the motion of BEAR-H

5.1.5 Results

Clinical trials are carried out to study the performance of BEAR-H, where the patient wears the robot and walks in a straight line in a hospital environment (Fig. 5.7). Before clinical trials, the patients are informed about the purposes of the study and asked to sign informed consent. Then, some preliminary training is performed to help patients get used to wearing the robot, which ensures that the patient has the ability to participate in the subsequent formal trials; the preliminary training lasts for three to five days. Next, the formal trial is performed for five days each week, and the patient walks by wearing the robot twice per day, with each walking trial lasting for around 30 mins.

A total of 30 patients were involved in the clinical trials, including 23 males and 7 females. The patients are selected according to the following criteria.

(1) The subject can walk with some assistance;
(2) The subject's age is between 18 and 75 years old, including both male and female subjects;
(3) The subject has not been previously diagnosed with hemiplegia, and the *duration* is between two weeks and six months;
(4) The upper limb can function properly and is able to hold the stand (which is used as the walking assistant device);

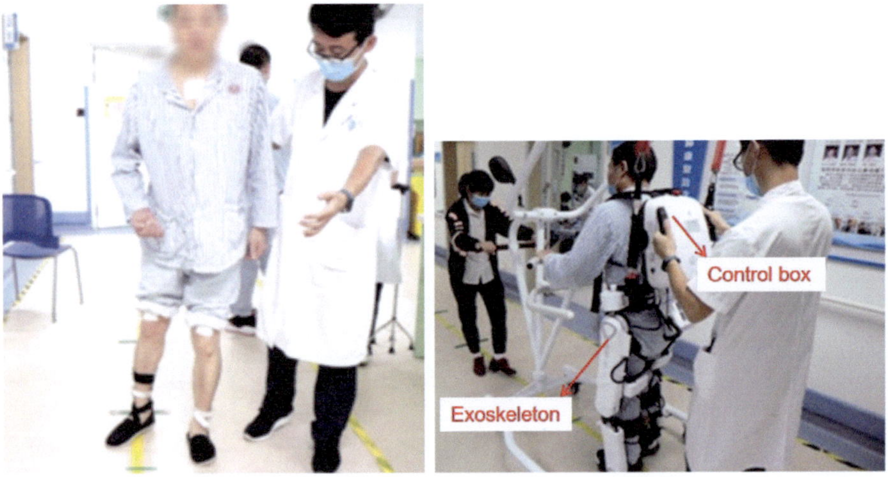

(a) (b)

Fig. 5.7 Clinical trials have been performed to validate the performance of the robot: **a** Hemiplegia patients were involved in the clinical trials; **b** The patient is wearing the robot and then walking on the ground, where the metabolic data before and after walking is compared to study the performance of BEAR-H

Table 5.5 Statistical summary of patients

PARAMETER		COUNT	MEAN	STD	MIN	MAX
Age (years)		30	50.67	12.81	22	72
Height (cm)		30	167.27	6.87	150	176
Weight (kg)		30	64.87	8.80	45	80
Duration (days)		30	89.40	46.23	16	162
Causes	CH	10				
	CI	20				

(5) The walking is not stable, and the speed is significantly slower than that of healthy subjects;
(6) He or she can understand commands from the doctor and clearly acknowledges all of the training details.

At the same time, the patients with the following conditions will be excluded.

(1) His or her joints can only move within a significantly limited range;
(2) The fracture has not fully recovered, or the osteoporosis is significant;
(3) There are other serious symptoms involved.

The statistical summary of their body parameters and medical condition is given in Table 5.5. In Table 5.5, the *duration* represents the interval since the subject was diagnosed (for example, the duration of 15 means the subject was diagnosed 15 days ago), and there are two types of hemiplegia causes: CH (cerebral hemorrhage) and CI (cerebral infarction). From Table 5.5, it can be seen that the age is between 22 (young) and 72 (old), the height is between 150 cm (short) and 176 cm (tall), and the weight is between 45 kg (thin) and 80 kg (fat). Hence, the selected patients are representative in the sense that patients with different body parameters have been considered. The selection also covers different medical conditions, ranging from new cases (e.g., 15 days ago) to long-standing ones (e.g., 162 days ago). In addition, the number of CI is double that of CH, which is reasonable and matches the occurrence probability in clinical trials.

The formal trial is carried out as follows. (1) BEAR-H is tested to ensure that the system functions properly; (2) the conditions of the subject are evaluated by measuring blood pressure and heart rate, and the trial proceeds if all readings are normal; (3) The clinical staff helps the patient to wear the robot, fasten the safety belt of the stand, and then input the body parameters of the patient; (4) the patient starts to walk once the robot is activated; (5) during the walking phase, the doctor and clinical staff continue to monitor the procedure and take action to stop the trial if any safety issues arise; (6) after the walking phase is completed, the patient takes off the robot, and the staff measures the blood pressure and heart rate of the patient again.

Comparative studies were performed in the clinical trials, where all the 30 patients were equally divided into two groups: Group 1 includes patients from No. 1 to No.

15, who were subject to the traditional manual rehabilitation, that is, an experienced therapist holds and assists the patient during the walking test; Group 2 includes patients from No. 16 to No. 30, who were subject to robotic rehabilitation, by wearing BEAR-H with minimal manual assistance (the doctor intervenes only when there might be a safety issue). Both groups have a similar distribution of body parameters and medical conditions, as given in Table 5.5.

The performance of BEAR-H was studied in the Six-Minute Walk Test (6MWT) [9], where the patient wears the robot and then walks continuously for six minutes. During the walking phase, the doctor accompanies the patient with minimal communication (i.e., the doctor talks to the patient to encourage him or her), and the patient can rest between the two tests. The trials were performed in a closed corridor so other pedestrians could not interrupt. The walking path is straight and solid, and the total length is more than 30 m. There is a returning sign at the end of the path, and there is also a series of reminders every five meters.

In the first experiment, the heart rate of the patient before and after 6MWT is recorded and compared. For example, the variation of 10% of heart rate implies the patient's heart rate after walking increases by 10%, with respect to that before walking. Note that smaller increases in heart rate imply a better effect, where assistance from the doctor/robot benefits the patient's motion, such that the patient is relaxed while walking. Significant increases in heart rate are a sign of ineffective training, indicating that the patient is not comfortable with assistance from the doctor/robot and needs to exert additional effort while walking. The experimental results are shown in Fig. 5.8, where the blue bar denotes the heart rate before walking, and the red bar denotes the heart rate after walking. The variation of heart rate for Group 1 (i.e., manual rehabilitation) is summarized in Table 5.6, which is between 5.71% and 14.63%, and the average change of heart rate equals 8.89%. The variation of heart rate for Group 2 (i.e., robotic rehabilitation) is summarized in Table 5.7, which is between 1.22% and 8.05%, and the average change of heart rate equals 4.10%, less than Group 1 and hence proving the effectiveness of the robot.

In the second experiment, the distance the patient can walk within six minutes is compared before and after training, and the further distance implies a faster walking speed and hence a better effect of rehabilitation, and vice versa. Experimental results for both groups are shown in Fig. 5.9, represented as the improvement of distance (i.e., the difference between the distance the patient can walk after training and that before it). Note that the training does not function for patient No. 18, who does not make progress. This is common in clinical trials, mainly due to the specific condition of patients (e.g., other diseases or extreme discomfort). From Fig. 5.9, it is obtained that the average improvement with manual rehabilitation is 61.07 m, and that with robotic rehabilitation is 65.51 m (after the removal of data for No. 18), proving the effectiveness of robotic rehabilitation.

In the third experiment, the effect of training is assessed by referring to the Fugl-Meyer Assessment for Lower Extremity (FMA-LE) [10], which has become a standard tool for the evaluation of post-stroke recovery. To carry out FMA-LE, the doctor asks the patient to perform a series of tasks and then assesses his/her difficulty in performing such tasks. The performance before and after training is compared, and a

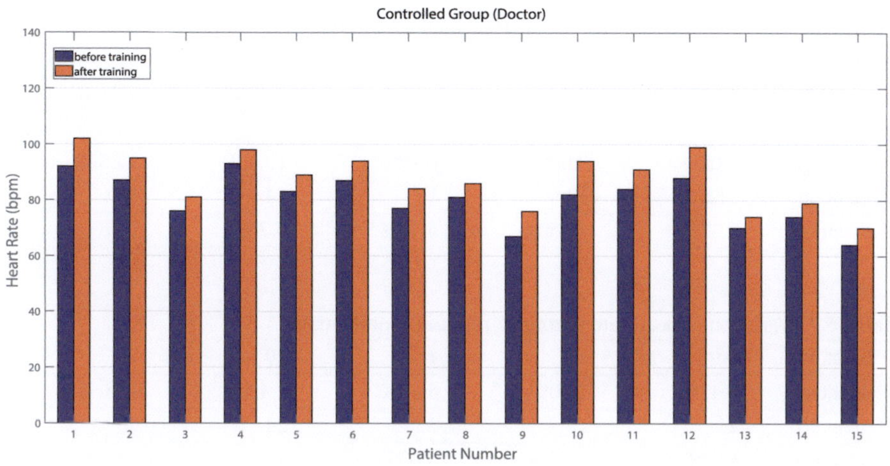

(a)

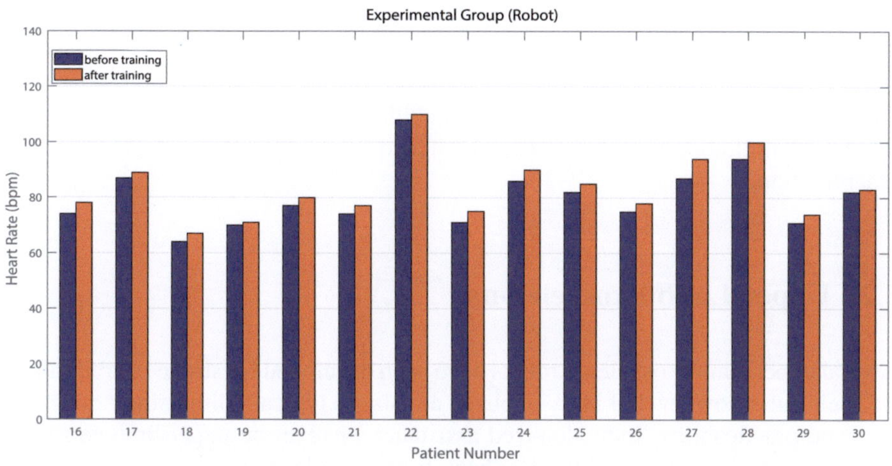

(b)

Fig. 5.8 Variation of heart rates: **a** Manual rehabilitation; **b** Robotic rehabilitation

higher assessment mark implies a better effect of rehabilitation, and vice versa. Experimental results of FMA-LE are shown in Fig. 5.10, represented as the improvement of FMA-LE marks after the training. Note that FMA-LE marks each task with 3 points, that is,

(1) Score 0: The patient cannot perform the task;
(2) Score 1: The patient can perform the task partially;
(3) Score 2: The patient can perform the task fully.

and the full mark of FMA-LE is 34. Similarly, the rehabilitation does not function for a few patients (i.e., No. 3, No. 14, and No. 28). After the removal of invalid data,

Table 5.6 Variation of heart rate (Group 1: manual rehabilitation)

Patient no.	1	2	3	4	5
Percentage	10.87%	9.20%	6.58%	5.10%	7.23%
Patient no.	6	7	8	9	10
Percentage	8.05%	9.09%	6.17%	13.43%	14.63%
Patient no.	11	12	13	14	15
Percentage	8.33%	12.50%	5.71%	6.76%	9.38%

Table 5.7 Variation of heart rate (Group 2: robotic rehabilitation)

Patient no.	16	17	18	19	20
Percentage	5.41%	2.30%	4.69%	1.43%	3.90%
Patient no.	21	22	23	24	25
Percentage	4.05%	1.85%	5.63%	4.65%	3.66%
Patient no.	26	27	28	29	30
Percentage	4.00%	8.05%	6.38%	4.23%	1.22%

the average improvement with manual rehabilitation is 4.38 $marks$, and that with robotic rehabilitation is 5.21 $marks$, which also proves the effectiveness of robotic rehabilitation.

5.2 Upper-Limb Exoskeletons

Several upper-limb exoskeleton robots have been developed for stroke rehabilitation, but their rather low level of individualized assistance typically limits their effectiveness and practicability. Individualized assistance involves an upper-limb exoskeleton robot continuously assessing feedback from a stroke patient and then precisely adjusting interaction forces to suit specific conditions and online changes.

This section describes the development of a new upper-limb exoskeleton robot with a novel online generative capability that allows it to provide individualized assistance to support the rehabilitation training of stroke patients. Specifically, the upper-limb exoskeleton robot exploits generative models to customize the fine and fit trajectory for the patient, as medical conditions, responses, and comfort feedback during training generally differ between patients. This generative capability is integrated into the two working modes of the upper-limb exoskeleton robot: an active mirroring mode for patients who retain motor abilities on one side of the body and a passive following mode for patients who lack motor ability on both sides of the body.

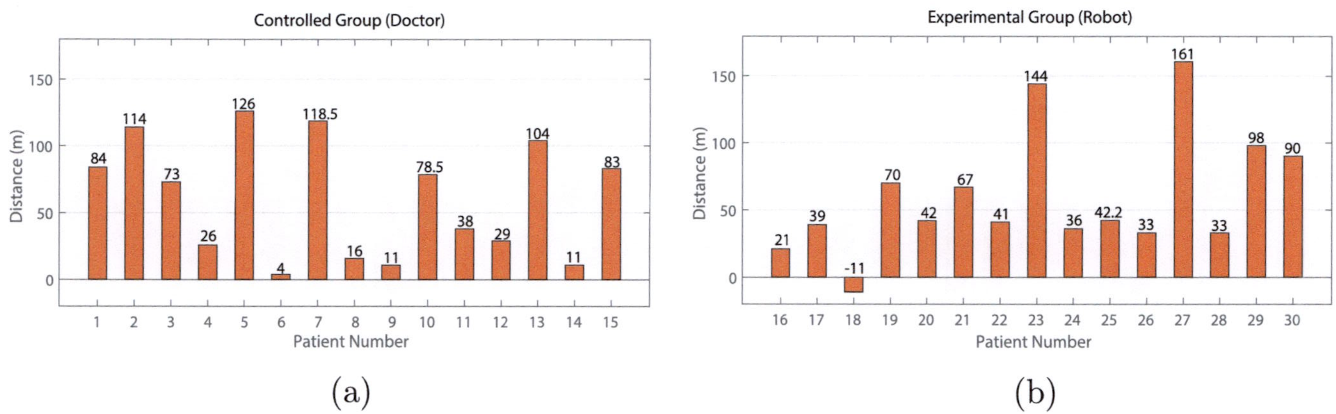

Fig. 5.9 Improvement of walking distance: **a** Manual rehabilitation; **b** Robotic rehabilitation

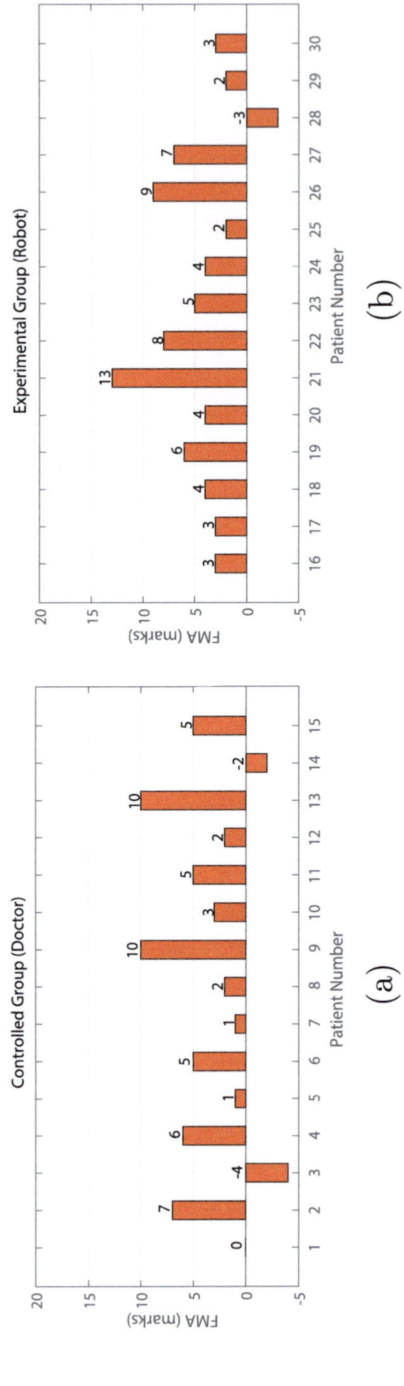

Fig. 5.10 Improvement of FMA-LE marks: **a** Manual rehabilitation; **b** Robotic rehabilitation

Table 5.8 Comparison of upper-limb exoskeletons

Exoskeleton	Joints	DOFs	Sensors	Actuation
[11]	Shoulder, Elbow	4A+1P	Encoders	Direct
[12]	Shoulder, Elbow, Forearm	9A	Encoders, Force	Compliant
[13]	Shoulder, Elbow, Forearm	5A	Encoders, Force	Cable-driven
Ours	Shoulder, Elbow, Forearm	5A+1P	Encoders, Force	Cable + Compliant

A-active, P-passive

5.2.1 Overall Structure

Compared with other upper-limb exoskeleton robots, the upper-limb exoskeleton robot has three other attractive features. First, it has six degrees of freedom (DoFs), namely five active DoFs and one passive DoF, to assist the shoulder and the elbow joints and cover the full range of upper-limb movement. Second, most of its active joints are driven by series elastic actuators (SEAs) and a cable mechanism, which absorb energy and have low inertia. These compliantly driven high DoFs provide substantial flexibility and ensure hardware safety, but require an effective controller. Thus, based on the singular perturbation approach, a model-based impedance controller is proposed to fully exploit the advantages of the hardware. Third, the safety of the upper-limb exoskeleton robot is guaranteed by its hardware and software. Regarding hardware, its SEAs are tolerant to impacts and have high backdrivability. Regarding software, online trajectory refinement is performed to regulate the assistance provided by the upper-limb exoskeleton robot, and an anomaly detection network is constructed to detect and relax physical conflicts between the upper-limb exoskeleton robot and the patient. The comparison between the developed robot and others is illustrated in Table 5.8.

This study examined a prototype cable-driven upper-limb exoskeleton robot that was co-developed by Tsinghua University and Shenzhen MileBot Robotics Co., Ltd. A computer-aided design model of the upper-limb exoskeleton robot is illustrated in Fig. 5.11. The upper-limb exoskeleton robot is equipped with one passive and five active joints that allow refinement of the upper-arm and forearm lengths within ranges of 20 and 40 cm, respectively. The passive DoF, labeled as Joint 0, accommodates the eccentric movements of the shoulder joint and thereby eliminates the limitations of the upper-limb motion range and increases the space for patient motion training. The active joints are numbered 1–5 and serve specific functions, as described below.

(1) Joint 1: shoulder abduction and adduction
(2) Joint 2: shoulder flexion and extension
(3) Joint 3: upper-arm internal and external rotation
(4) Joint 4: elbow flexion and extension
(5) Joint 5: forearm internal and external rotation

Joint 1 and 2 consist of direct-drive joint modules (RJSIIT-17-RevB2 and RJSIIT-17-RevB5, respectively), each of which is outfitted with force sensors and supported by the frame. Joint 3–5 utilize harmonic deceleration servo motors (AK80-64) that are configured as cable-driven SEAs and equipped with torque sensors (TK17-191151) and encoders (3590S-2-104L for the internal and external rotation joints, and QY2204-SSI for the elbow joint). Additionally, these joints incorporate two potentiometers for measuring spring compression during motion. The detailed technical specifications of the upper-limb exoskeleton robot are provided in Table 5.9. Motor drivers are implemented to control Joint 3–5, and together with two joint modules, facilitate communication with the main board through a controller area network. The main board receives commands from a personal computer (PC) via a universal asynchronous receiver/transmitter.

The upper-limb exoskeleton robot employs a hierarchical control system. At the low level, the main board executes motion control and environmental perception, enabling precise force control and real-time acquisition of sensory data. At a high level, a Linux system on the PC operates a motion planning module and performs computationally intensive tasks, such as dynamic calculation and network inference. All programs are integrated into Robot Operating System (ROS) [14] and run concurrently.

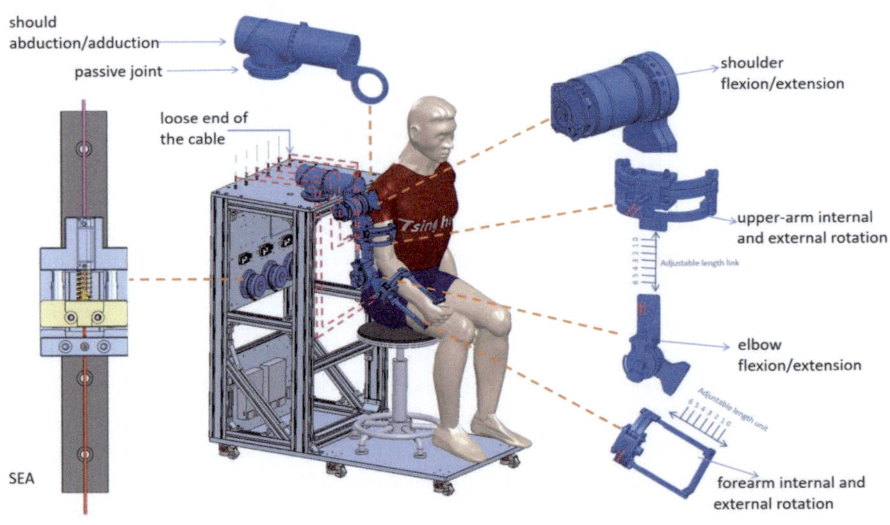

Fig. 5.11 Overview of the developed upper-limb exoskeleton robot, which is cable-driven and consists of five active joints (Joint 1–5) and one passive joint (Joint 0)

Table 5.9 Specifications of the upper-limb exoskeleton

Property	Value
Continuous torque	49 N·m @ Joint 1 and 2
	48 N·m @ Joint 3, 4, and 5
Backdrivability	Less than 0.7 N·m @ 10°/s
Weight	7.78 kg excluding the frame
Control frequency	1000 Hz

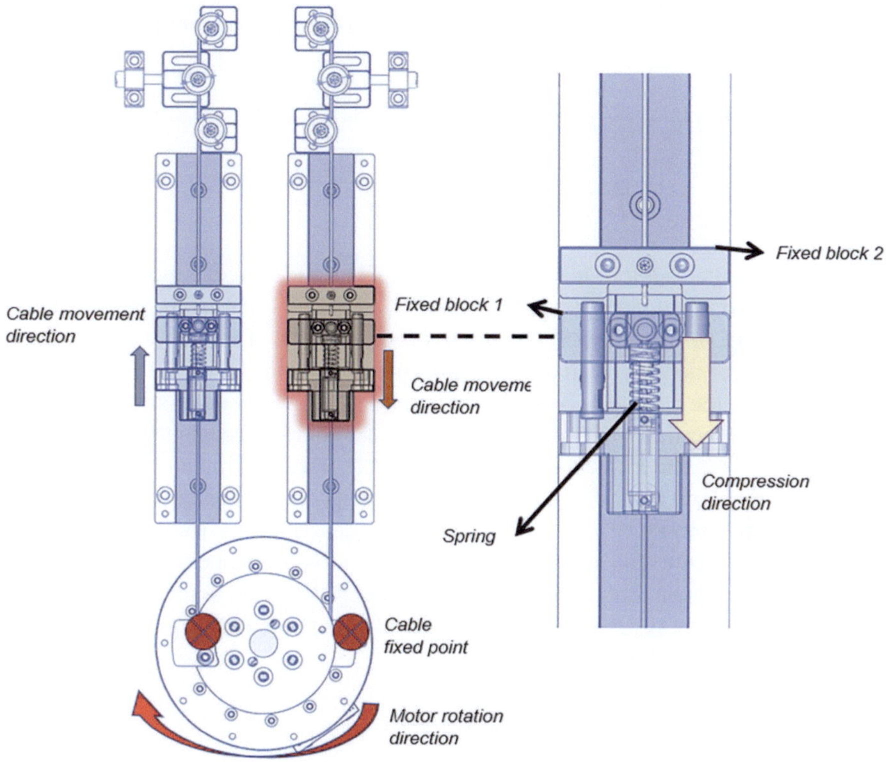

Fig. 5.12 The operating principle of the self-developed SEA

The joints that undergo significant spatial movement, namely Joints 3, 4, and 5, utilize our self-developed SEAs equipped with cable-driven mechanisms. A working principle diagram of a SEA is provided in Fig. 5.12. Each joint motor is outfitted with two sets of pulleys, with each set corresponding to a different rotational direction. Each end of the motor output shaft is affixed with a steel cable that is connected to a potentiometer to constitute the SEA. The potentiometer assembly includes two fixed blocks, namely fixed block 1 and fixed block 2, and a spring component. The cable from the motor end (depicted in red) is connected via a spring to fixed block 1, while another cable (depicted in blue) connects the moving joint components to fixed block 2. As the motor applies torque, fixed block 1 moves along a sleeve attached to fixed block 2, thereby compressing the spring and enabling the measurement of tension within the cable. Upon spring compression, the blocks form a unified structure that moves in synchrony with and in the same direction as the joint, thereby facilitating overall movement.

For safety, the upper-limb exoskeleton robot is equipped with an emergency stop button that physically disconnects power during urgent situations. Additional safety measures are incorporated at the software level, including measures that limit the range of motion, prevent self-collision, restrict joint velocity, and automatically cease operation in cases of excessive interaction force or torque.

The dynamic model of the developed upper-limb exoskeleton robot can be defined as follows [15]:

$$\boldsymbol{M}(\boldsymbol{q})\ddot{\boldsymbol{q}} + \boldsymbol{C}(\dot{\boldsymbol{q}}, \boldsymbol{q})\dot{\boldsymbol{q}} + \boldsymbol{g}(\boldsymbol{q}) = \boldsymbol{S}_1\boldsymbol{u} + \boldsymbol{S}_2^{\mathrm{T}}\boldsymbol{K}(\boldsymbol{\theta} - \boldsymbol{S}_2\boldsymbol{q}) + \boldsymbol{\tau}_e + \boldsymbol{S}_2^{\mathrm{T}}\boldsymbol{\tau}_f, \tag{5.5}$$

$$\boldsymbol{B}\ddot{\boldsymbol{\theta}} + \boldsymbol{K}(\boldsymbol{\theta} - \boldsymbol{S}_2\boldsymbol{q}) = \boldsymbol{S}_2\boldsymbol{u}, \tag{5.6}$$

where $\boldsymbol{q} \in \Re^5$ is the vector of joint positions; $\boldsymbol{\theta} \in \Re^3$ is the vector of motor-rotor-shaft positions; and $\boldsymbol{M}(\boldsymbol{q}) \in \Re^{5\times5}$, $\boldsymbol{C}(\dot{\boldsymbol{q}}, \boldsymbol{q})\dot{\boldsymbol{q}} \in \Re^{5\times5}$, and $\boldsymbol{g}(\boldsymbol{q}) \in \Re^5$ are the inertia matrix, centripetal and Coriolis torques, and gravitational torques of the robot, respectively. The selection matrices $\boldsymbol{S}_1 = \mathrm{diag}(1, 1, 0, 0, 0) \in \Re^{5\times5}$ and $\boldsymbol{S}_2 = [\boldsymbol{0}, \boldsymbol{I}_3] \in \Re^{3\times5}$ are used to separate the cable-driven joints. In addition, $\boldsymbol{K} \in \Re^{3\times3}$ is the stiffness matrix, $\boldsymbol{B} \in \Re^{3\times3}$ is the inertia matrix of the motor, $\boldsymbol{\tau}_e \in \Re^5$ is the physical interaction torque vector, $\boldsymbol{\tau}_f \in \Re^3$ is the disturbance due to the cable transmission and the friction of the joint system, and $\boldsymbol{u} \in \Re^5$ is the control torque applied to the actuator.

The dynamic model of the overall system is described by (5.5) and (5.6) and exhibits different time scales. Specifically, the SEA subsystem operates on a fast time-scale, while the upper-limb exoskeleton robot subsystem functions on a slow time-scale. To control this system effectively, the control input is formulated according to singular perturbation theory [16], as follows:

$$\boldsymbol{u} = \boldsymbol{u}_f + \boldsymbol{u}_s, \tag{5.7}$$

where $\boldsymbol{u}_f$ is the fast time-scale control term for stabilizing the model defined by (5.6), and $\boldsymbol{u}_s$ is the slow time-scale control term for stabilizing the model defined by (5.5). A representative form of $\boldsymbol{u}_f$ is expressed as follows:

$$\boldsymbol{u}_f = -\boldsymbol{S}_2^{\mathrm{T}}\boldsymbol{K}_v(\dot{\boldsymbol{\theta}} - \boldsymbol{S}_2\dot{\boldsymbol{q}}), \tag{5.8}$$

where $\boldsymbol{K}_v \in \Re^{3\times3}$ is a diagonal and positive-definite matrix.

Substituting (5.7) and (5.8) into (5.6) yields

$$\boldsymbol{B}\ddot{\boldsymbol{\theta}} + \boldsymbol{K}(\boldsymbol{\theta} - \boldsymbol{S}_2\boldsymbol{q}) + \boldsymbol{K}_v(\dot{\boldsymbol{\theta}} - \boldsymbol{S}_2\dot{\boldsymbol{q}}) = \boldsymbol{S}_2\boldsymbol{u}_s, \tag{5.9}$$

which can be rewritten as follows by defining $\boldsymbol{\tau}_o = \boldsymbol{K}(\boldsymbol{\theta} - \boldsymbol{S}_2\boldsymbol{q})$:

$$\boldsymbol{B}\ddot{\boldsymbol{\tau}}_o + \boldsymbol{K}_v\dot{\boldsymbol{\tau}}_o + \boldsymbol{K}\boldsymbol{\tau}_o = \boldsymbol{S}_2\boldsymbol{u}_s - \boldsymbol{B}\boldsymbol{S}_2\ddot{\boldsymbol{q}}. \tag{5.10}$$

By introducing $\boldsymbol{K} = \boldsymbol{K}_1/\varepsilon^2$ and $\boldsymbol{K}_v = \boldsymbol{K}_2/\varepsilon$, with ε being a small positive parameter, (5.10) can be written as follows:

$$\varepsilon^2\boldsymbol{B}\ddot{\boldsymbol{\tau}}_o + \varepsilon\boldsymbol{K}_2\dot{\boldsymbol{\tau}}_o + \boldsymbol{K}_1\boldsymbol{\tau}_o = \boldsymbol{K}_1(\boldsymbol{S}_2\boldsymbol{u}_s - \boldsymbol{B}\boldsymbol{S}_2\ddot{\boldsymbol{q}}). \tag{5.11}$$

When $\varepsilon = 0$, the solution of (5.11) is $\bar{\boldsymbol{\tau}}_o = \boldsymbol{S}_2\boldsymbol{u}_s - \boldsymbol{B}\boldsymbol{S}_2\ddot{\boldsymbol{q}}$.

If the fast time-scale is set as $\gamma = \frac{t}{\varepsilon}$, $\bar{\boldsymbol{\tau}}_o$ is achieved at $\gamma \to \infty$. $\bar{\boldsymbol{\tau}}_o$ remains constant at $\varepsilon = 0$. Next, we introduce a new variable $\boldsymbol{\eta} = \boldsymbol{\tau}_o - \bar{\boldsymbol{\tau}}_o$ to rewrite (5.11) on the fast time-scale, as follows:

$$\boldsymbol{B}(\frac{\mathrm{d}^2\boldsymbol{\eta}}{\mathrm{d}\gamma^2}) + \boldsymbol{K}_2(\frac{\mathrm{d}\boldsymbol{\eta}}{\mathrm{d}\gamma}) + \boldsymbol{K}_1\boldsymbol{\eta} = \mathbf{0}, \tag{5.12}$$

which defines the *boundary-layer system.*

Substituting the solution of (5.12) into (5.5) affords a *quasi-steady-state system* that captures the slow dynamics of the overall system. These dynamics can be expressed as follows:

$$(\boldsymbol{M}(\boldsymbol{q}) + \bar{\boldsymbol{B}})\ddot{\boldsymbol{q}} + \boldsymbol{C}(\dot{\boldsymbol{q}}, \boldsymbol{q})\dot{\boldsymbol{q}} + \boldsymbol{g}(\boldsymbol{q}) = \boldsymbol{u}_s + \boldsymbol{\tau}_e + \boldsymbol{S}_2^{\mathrm{T}}\boldsymbol{\tau}_f, \tag{5.13}$$

where $\bar{\boldsymbol{B}} = \boldsymbol{S}_2^{\mathrm{T}}\boldsymbol{B}\boldsymbol{S}_2$ is the projected motor inertia matrix.

According to singular perturbation theory, the stability of the overall system is guaranteed if the *boundary-layer system* and the *quasi-steady-state system* are both exponentially stable.

The overall dynamic model described by (5.5) and (5.6) is a high-order system, where (5.5) represents the rigid-joint side, and (5.6) represents the compliant actuator side. It is non-trivial to stabilize and control such a system.

An upper-limb exoskeleton robot needs to guide the patient to perform repetitive motions via close interaction to help the patient regain motor function. Thus, the upper-limb exoskeleton robot is controlled to track a time-varying trajectory in accordance with the following impedance model:

$$\boldsymbol{M}_d(\ddot{\boldsymbol{q}} - \ddot{\boldsymbol{q}}_d) + \boldsymbol{C}_d(\dot{\boldsymbol{q}} - \dot{\boldsymbol{q}}_d) + \boldsymbol{K}_d(\boldsymbol{q} - \boldsymbol{q}_d) = \boldsymbol{\tau}_e, \tag{5.14}$$

where $\boldsymbol{q}_d \in \Re^n$ is the vector of the desired trajectory, and $\boldsymbol{M}_d, \boldsymbol{C}_d, \boldsymbol{K}_d \in \Re^{n\times n}$ are the desired inertia, the desired damping, and the desired stiffness matrices, respectively, which are diagonal and positive-definite. Tracking the desired trajectory in accordance with the impedance model allows the patient to deviate from the trajectory and hence provides a certain level of compliance to improve safety.

The proposed dual-mode training framework is illustrated in Fig. 5.13. Both the anomaly detector and intention predictor are designed using trained generative models based on a dataset of motions of healthy individuals. These models are designed to capture the motion intentions of the unaffected side of the patient and evaluate the comfort and naturalness of the current patient–robot interaction. Individualization is facilitated by employing two methods, i.e., employing the anomaly score to guide the refinement of a desired trajectory that complies with dynamic constraints, and employing ProMPs to integrate various assistive trajectories into a personalized assistance distribution. Additionally, the safety of the motion process is enhanced by adjusting the impedance parameters according to the anomaly score, thereby ensuring a secure and effective rehabilitation environment.

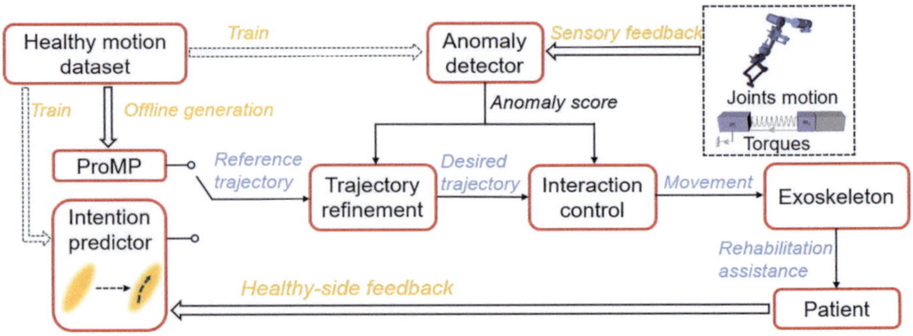

Fig. 5.13 Workflow of dual-mode training. In active mirroring mode, the predicted motion intention serves as the reference trajectory. In the passive following mode, the reference trajectory is generated offline by fitting ProMPs to the data obtained from the demonstrations of healthy individuals. In both modes, the primary goal is to deliver the most personalized assistance possible. Additionally, the motion data from healthy individuals is used to train the intention predictor and anomaly detector, and the anomaly score is used to guide trajectory refinement and the adjustment of impedance parameters

5.2.2 *Online Trajectory Refinement*

This section introduces the new trajectory generation method for the upper-limb exoskeleton robot. This method is based on the generative models and can be refined online to suit the patient while maintaining safety and individualized features, and hence maintaining effectiveness. First, in both training modes (i.e., active mirroring and passive following modes), we introduce a general integrated vector $\boldsymbol{x}_d = [\boldsymbol{q}_d^\top, \dot{\boldsymbol{q}}_d^\top]^\top$, which encapsulates a sequence of discretized trajectory points within the joint space, thereby parameterizing the trajectory. The following quadratic programming problem is formulated to describe the planning process:

$$
\begin{aligned}
\min_{\boldsymbol{x}_d,\boldsymbol{u}_d,s} \; & \sum_{i=t}^{t+N_p} \left[\|\boldsymbol{q}_d^{(i)} - \boldsymbol{q}_r^{(i)}\|_{\boldsymbol{Q}}^2 + \|\boldsymbol{u}_d^{(i)}\|_{\boldsymbol{R}}^2 + (s^{(i)})^2 \right], \\
s.t. \; \; & \boldsymbol{x}_d^{(t+1)} = \begin{bmatrix} \boldsymbol{I} & \boldsymbol{I}\Delta t \\ \boldsymbol{0} & \boldsymbol{I} \end{bmatrix} \boldsymbol{x}_d^{(t)} + \begin{bmatrix} \boldsymbol{0} \\ \boldsymbol{I}\Delta t \end{bmatrix} \boldsymbol{u}_d^{(t)} \\
& s^{(t+1)} = s^{(t)} + \begin{bmatrix} \boldsymbol{0} \\ -(\frac{\partial f}{\partial \boldsymbol{\tau}_e})^\top \boldsymbol{K}_a \Delta t \end{bmatrix} \boldsymbol{x}_d^{(t)} \\
& \qquad - (\frac{\partial f}{\partial \boldsymbol{\tau}_e})^\top \boldsymbol{C}_a \boldsymbol{u}_d^{(t)} \Delta t \\
& \boldsymbol{x}_d \in \mathcal{X}, \boldsymbol{u}_d \in \mathcal{U}
\end{aligned}
\tag{5.15}
$$

where N_p is the predictive horizon, $\boldsymbol{Q}$ and $\boldsymbol{R}$ are symmetric positive-definite weighting matrices, $\boldsymbol{u}_d$ is the acceleration of the desired trajectory, $\boldsymbol{C}_a$ and $\boldsymbol{K}_a$ are the varying impedance parameters to be defined later, and $f(\cdot)$ is a functional representation of the anomaly detection network. In addition, the superscript indicates the time step, and $\mathcal{X}$ and $\mathcal{U}$ are the sets of trajectory space constraints and acceleration constraints, respectively. Thus, (5.15) enables various constraints and safety measures to be included in both training modes.

5.2.3 Active Training Mode

In cases of mild motor impairment, such as that observed in early-stage stroke, patients retain functional motion capabilities on their unaffected side. Thus, the upper-limb exoskeleton robot must accurately estimate the motion intentions of the patient's unaffected side to provide mirrored support for the patient's affected side. This mirroring process enables an affected limb to emulate the movement of an unaffected limb, thereby enhancing the recovery of motor function.

The intention estimation system inputs a series of N historical observations, defined as $\boldsymbol{o}_{(i)} = \{\boldsymbol{q}_{h(i)}^{(t)} \in \Re^5 | t = -N_o, -N_o + 1, \cdots, 0\}, \forall i \in \{1, 2, \cdots, N\}$, where N_o is the number of past time steps, and $\boldsymbol{q}_{h(i)}^{(t)}$ is the position of the unaffected side of the body at time step t. The system also generates a prediction of the motion intention, expressed as $\boldsymbol{p}_{(i)} = \{\boldsymbol{q}_{h(i)}^{(t)} \in \Re^5 | t = 1, 2, \cdots, N_p\}$. To estimate the potential distribution of plausible trajectories, we introduce a probabilistic predictor based on denoising diffusion probabilistic models (DDPMs) [17], as DDPMs can effectively describe the uncertain nature of human motion intention. For clarity, we omit the subscript (i) in the following subsections and refer to the past and future trajectories as $\boldsymbol{o}$ and $\boldsymbol{p}$, respectively.

Intention Predictor: The role of the intention predictor is to produce probabilistic forecasts of future trajectories. As such, the intention predictor is informed by an initial, ambiguous predicted trajectory $\boldsymbol{p}_T$, which represents a noise-perturbed estimate of the future trajectory $\boldsymbol{p}_0$ after T diffusion steps, where T is the pre-defined maximum number of diffusion steps. The predictor is designed to infer the reverse diffusion process, articulated as the sequence $(\boldsymbol{p}_T, \boldsymbol{p}_{T-1}, \cdots, \boldsymbol{p}_0)$. This sequence is mathematically structured as a Markov chain that is characterized by Gaussian transition probabilities and refines the prediction by reducing the initial uncertainty. Furthermore, the intention predictor incorporates historical observations encoded into a context vector $\boldsymbol{c}$, which is synthesized via a neural network parametrized by ϕ. This context vector informs the trajectory generation process. The principle of the intention predictor is depicted in Fig. 5.14. Within this framework, the upper-limb exoskeleton robot is endowed with the ability to precisely estimate the future trajectory based on a simple sampling from the initial noisy prediction. Importantly, a future trajectory represents the patient's motion intention and is not based on direct measurements of the patient's unaffected limb. Thus, this approach generates a self-mirrored trajectory that closely aligns with the patient's original movement patterns, thereby representing a personalized trajectory.

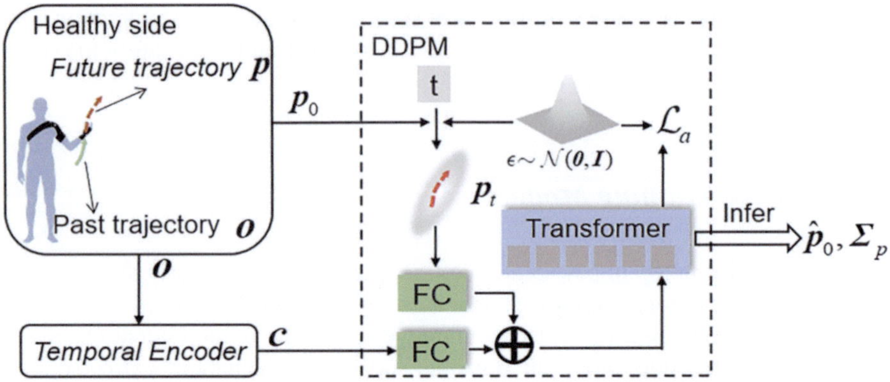

Fig. 5.14 Overall structure of the intention predictor: In the training phase, the past trajectory $\boldsymbol{o}$ is encoded into a context vector $\boldsymbol{c}$, which is used in conjunction with the future trajectory $\boldsymbol{p}$ to train the diffusion model. In the inference phase, only the context vector is employed to guide the denoising process, which is used to predict the future trajectory

The posteriors associated with diffusion and reverse processes are given as follows:

$$q(\boldsymbol{p}_t|\boldsymbol{p}_{t-1}) = \mathcal{N}(\boldsymbol{p}_t; \sqrt{1-\beta_t}\boldsymbol{p}_{t-1}, \beta_t \boldsymbol{I}), \tag{5.16}$$

$$q_\psi(\boldsymbol{p}_{t-1}|\boldsymbol{p}_t, \boldsymbol{c}) = \mathcal{N}(\boldsymbol{p}_{t-1}; \boldsymbol{\mu}_\psi(\boldsymbol{p}_t, \boldsymbol{c}), \tilde{\beta}_t \boldsymbol{I}), \tag{5.17}$$

where $\beta_1, \beta_2, \cdots, \beta_T$ is the variance schedule used to adjust the injected noise. The adjusted variance, $\tilde{\beta}_t$, is defined as $\tilde{\beta}_t = \frac{1-\bar{\alpha}_{t-1}}{1-\bar{\alpha}_t}\beta_t$, where this formulation is derived by adopting the definitions $\alpha_t = 1 - \beta_t$ and $\bar{\alpha}_t = \prod_{i=1}^{t} \alpha_i$.

The objective of the predictor is to maximize the log-likelihood, denoted as $\mathbb{E}[\log q_\psi(\boldsymbol{p}_0)]$. Given that direct computation of this metric is infeasible, a variational lower bound is utilized as a surrogate for the training loss:

$$\begin{aligned}\mathcal{L}_{vlb}(\phi, \psi) = &\sum_{t=2}^{T} D_{KL}\left(q(\boldsymbol{p}_{t-1}|\boldsymbol{p}_t) \| q_\psi(\boldsymbol{p}_{t-1}|\boldsymbol{p}_t, \boldsymbol{c})\right) \\ &- \log q_\psi(\boldsymbol{p}_0|\boldsymbol{p}_1, \boldsymbol{c}),\end{aligned} \tag{5.18}$$

where $D_{KL}(\cdot)$ signifies the Kullback–Leibler divergence function.

In the proposed methodology, the conditional probability $q(\boldsymbol{p}_{t-1}|\boldsymbol{p}_t)$ is reformulated as $q(\boldsymbol{p}_{t-1}|\boldsymbol{p}_t, \boldsymbol{p}_0)$. This reformulation allows the following closed-form expression to be devised:

$$q(\boldsymbol{p}_{t-1}|\boldsymbol{p}_t, \boldsymbol{p}_0) = \mathcal{N}(\boldsymbol{p}_{t-1}; \tilde{\boldsymbol{\mu}}_t(\boldsymbol{p}_t, \boldsymbol{p}_0), \tilde{\beta}_t \boldsymbol{I}). \tag{5.19}$$

By iteratively employing parameterization techniques in the diffusion process (5.16), the mean of the posterior distribution in (5.19) and the reverse process (5.17) can be articulated as follows:

Algorithm 5.1 Preemptive Tuning for the Reference Trajectory

Given: $\hat{\boldsymbol{p}}_0, \hat{\boldsymbol{\Sigma}}_p, \boldsymbol{q}, \varepsilon$
$\boldsymbol{q}_r^{(0)} = \text{Saturate}(\hat{\boldsymbol{q}}_h^{(0)}, \boldsymbol{q}_{\min}, \boldsymbol{q}_{\max})$
for $t = 1$ **to** N_p **do**
  for $i = 1$ **to** 5 **do**
    $\delta_i^{(t)} = \min(\|\hat{q}_{hi}^{(t-1)} - q_{\min,i}\|, \|q_{\max,i} - \hat{q}_{hi}^{(t-1)}\|)$
    if $\frac{\hat{\sigma}_{hi}^{(t)}}{\hat{\sigma}_{hi}^{(t)} + (\delta_i^{(t)})^2} \leq \varepsilon$ **then**
      $q_{ri}^{(t)} = \hat{q}_{hi}^{(t)}$
    else
      $q_{ri}^{(t)} = q_{ri}^{(t-1)}$
    end if
  end for
end for
Output: $\boldsymbol{q}_r^{(0)}$ **to** $\boldsymbol{q}_r^{(N_p)}$

$$\tilde{\boldsymbol{\mu}}_t(\boldsymbol{p}_t, \boldsymbol{p}_0) = \frac{1}{\sqrt{\alpha_t}}(\boldsymbol{p}_t - \frac{\beta_t}{\sqrt{1-\bar{\alpha}_t}}\boldsymbol{\epsilon}_t), \tag{5.20}$$

$$\boldsymbol{\mu}_\psi(\boldsymbol{p}_t, \boldsymbol{c}) = \frac{1}{\sqrt{\alpha_t}}(\boldsymbol{p}_t - \frac{\beta_t}{\sqrt{1-\bar{\alpha}_t}}\boldsymbol{\epsilon}_\psi(\boldsymbol{p}_t, \boldsymbol{c})), \tag{5.21}$$

where $\boldsymbol{\epsilon} \sim \mathcal{N}(\boldsymbol{0}, \boldsymbol{I})$, and the loss function applied in active mirroring mode is further simplified as follows

$$\mathcal{L}_a(\phi, \psi) = \mathbb{E}_{t,\boldsymbol{p}_0,\boldsymbol{\epsilon}}[\|\boldsymbol{\epsilon} - \boldsymbol{\epsilon}_{\phi,\psi}(\boldsymbol{p}_t, \boldsymbol{o})\|^2]. \tag{5.22}$$

Once the reverse process is trained, multiple trajectories are sampled by utilizing the predictor to infer from the initial noise. For simplicity and conservatism, it is assumed that the future of each joint is predicted in an independent manner. Consequently, the entire inference and statistical process is encapsulated in the following equation:

$$\hat{\boldsymbol{p}}_0, \hat{\boldsymbol{\Sigma}}_p = f_a(\boldsymbol{o}), \tag{5.23}$$

where $\hat{\boldsymbol{p}}_0 = \{\hat{\boldsymbol{q}}_h^{(t)} \in \Re^5 | t = 1, 2, \cdots, N_p\}$ represents the estimated motion intention, and $\hat{\boldsymbol{\Sigma}}_p = \{diag(\hat{\boldsymbol{\sigma}}_h^{(t)}) \in \Re^{5\times5} | t = 1, 2, \cdots, N_p\}$ represents the standard deviation of the prediction.

Preemptive Tuning: Given the distribution of a predicted future trajectory, it is imperative to preemptively tune the reference trajectory to ensure compliance with established constraints. Specifically, for each joint i, denoted by subscript, the feasible trajectory must hold the following probabilistic inequality:

$$\mathcal{P}(q_{hi}^{(t)} \notin \mathcal{Q}_i) < \varepsilon, \tag{5.24}$$

where $\mathcal{Q}_i$ is the pre-defined permissible motion range $[q_{\min,i}, q_{\max,i}]$. It is stipulated that $\hat{q}_{hi}^{(0)} = q_i$, where q_i is the current position of joint i of the upper-limb exoskeleton robot. In accordance with Cantelli's inequality, the predicted distribution is characterized by:

$$\mathcal{P}(q_{hi}^{(t)} \geq \hat{q}_{hi}^{(t)} + \delta_i^{(t)}) \leq \frac{\hat{\sigma}_{hi}^{(t)}}{\hat{\sigma}_{hi}^{(t)} + (\delta_i^{(t)})^2}, \tag{5.25}$$

$$\delta_i^{(t)} = \min(\|\hat{q}_{hi}^{(t-1)} - q_{\min,i}\|, \|q_{\max,i} - \hat{q}_{hi}^{(t-1)}\|). \tag{5.26}$$

Now, we introduce the following tuning law for prediction:

$$q_{ri}^{(t)} = \begin{cases} \hat{q}_{hi}^{(t)}, & \text{if } \frac{\hat{\sigma}_{hi}^{(t)}}{\hat{\sigma}_{hi}^{(t)} + (\delta_i^{(t)})^2} \leq \varepsilon, \\ q_{ri}^{(t-1)}, & \text{otherwise.} \end{cases} \tag{5.27}$$

Here, $q_{ri}^{(t)}$ is the reference trajectory of joint i at time-step t. It updates to the predicted motion intention $\hat{q}_{hi}^{(t)}$ when the movement trend is within the established boundaries. However, if a potential boundary violation is detected, it maintains trajectory from the previous time-step to prevent the limit being crossed. The aforementioned process is also detailed in Algorithm 5.1. The adjusted reference trajectory $\boldsymbol{q}_r$ offers an accurate estimate of the intended motions of the unaffected side of the body, guaranteeing that safety considerations are adequately addressed. By resolving (5.15), it is feasible to generate a trajectory for the upper-limb exoskeleton robot that is not only smooth and characterized by minimal acceleration but also ensures the safety of the desired future trajectory.

5.2.4 Passive Training Mode

In cases of significant motor impairment, the upper-limb exoskeleton robot is engaged in passive following training. This mode of training is essential for assisting the patient to follow a pre-defined trajectory. Moreover, to promote rehabilitation, it is crucial to individualize the assistance in the regulation of assist-as-needed. The workflow of passive following training is depicted in Fig. 5.13. In this mode, motion data from the upper limbs of healthy subjects are collected to facilitate the training of two distinct modules.

Anomaly Detector: To individualize assistance, a real-time criterion is needed for evaluating the assistive trajectory. We use an anomaly detector to quantify the comfort of wear and the effectiveness of rehabilitation. This anomaly detector is based on a diffusion model architecture and identifies irregular patterns in upper-limb movements, as illustrated in Fig. 5.15. Subsequently, the anomaly detector computes a score that quantifies the deviation between the current human-robot interaction and a natural interaction condition, thereby guiding the individualization of assistance. In this approach, sensory feedback at a given timestep i, captured through a sliding

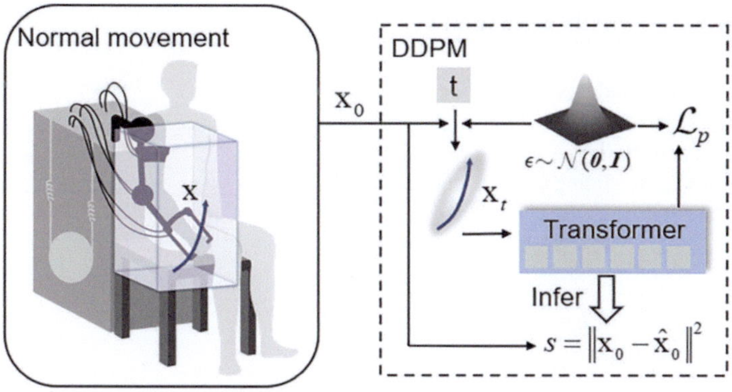

Fig. 5.15 Illustration of the anomaly detector: The transparent rectangular prism delineates the feasible workspace, encompassing the range within which upper-limb motion is categorized as normal and safe. These movements incorporate position, velocity, and interaction torque information, which are collectively depicted as a curved blue arrow and represented by $\mathbf{x}$

window mechanism, is fed into the anomaly detector, denoted as $\mathbf{x}^{(i)} \in \Re^{L_s N_c}$, where L_s is the width of the sliding window, and N_c is the number of data channels.

As mentioned, the anomaly detector is based on a diffusion model, wherein the reverse diffusion process of duration T^p is delineated as the sequence $(\mathbf{x}_{T^p}, \mathbf{x}_{T^p-1}, \cdots, \mathbf{x}_0)$. The associated diffusion and reverse diffusion processes are defined as follows:

$$q(\mathbf{x}_t|\mathbf{x}_{t-1}) = \mathcal{N}(\mathbf{x}_t; \sqrt{1-\beta_t^p}\mathbf{x}_{t-1}, \beta_t^p \boldsymbol{I}), \tag{5.28}$$

$$q_\Psi(\mathbf{x}_{t-1}|\mathbf{x}_t) = \mathcal{N}(\mathbf{x}_{t-1}; \boldsymbol{\mu}_\Psi(\mathbf{x}_t), \tilde{\beta}_t^p \boldsymbol{I}), \tag{5.29}$$

where $\beta_1^p, \beta_2^p, \cdots, \beta_{T^p}^p$ are variance schedules utilized to modulate the level of noise injected during the process. The adjusted variance, $\tilde{\beta}_t^p$, is calculated using the formula $\tilde{\beta}_t^p = \frac{1-\bar{\alpha}_{t-1}^a}{1-\bar{\alpha}_t^p}\beta_t^p$, and by exploiting the definitions $\alpha_t^p = 1 - \beta_t^p$ and $\bar{\alpha}_t^p = \prod_{i=1}^t \alpha_i^p$.

In alignment with the concept outlined in the *intention predictor*, the loss function used to train the diffusion model in passive following mode is specified as follows

$$\mathcal{L}_p(\Psi) = \mathbb{E}_{t,\mathbf{x}_0,\epsilon}[\|\epsilon - \epsilon_\Psi(\mathbf{x}_t)\|^2]. \tag{5.30}$$

Once the reverse process has been effectively learned from the dataset comprising upper-limb movements, the anomaly detector proficiently filters the noise from contaminated sensory data to yield a clarified output. This capability facilitates the generation of a refined sensory input through the anomaly detector. This input is employed to compute the anomaly score, as delineated in Algorithm 5.2, where $\epsilon_p, z_p \sim \mathcal{N}(\mathbf{0}, \boldsymbol{I})$ and $\nu \in [1, T^p]$ is a constant parameter. For the sake of brevity, the methodology for calculating the anomaly score is encapsulated by the following function:

$$s = f(\boldsymbol{q}, \dot{\boldsymbol{q}}, \boldsymbol{\theta}, \dot{\boldsymbol{\theta}}, \boldsymbol{\tau}_e). \tag{5.31}$$

Algorithm 5.2 Anomaly Detection

Require: $\boldsymbol{q}, \dot{\boldsymbol{q}}, \boldsymbol{\theta}, \dot{\boldsymbol{\theta}}, \boldsymbol{\tau}_e$

Initialize sliding window queue
for each new data point **do**
  $\mathbf{x}_{now}^{(i)} \leftarrow (\boldsymbol{q}, \dot{\boldsymbol{q}}, \boldsymbol{\theta}, \dot{\boldsymbol{\theta}}, \boldsymbol{\tau}_e)$
  Enqueue $\mathbf{x}_{now}^{(i)}$ to sliding window queue
  if sliding window queue is full **then**
    $\mathbf{x}_0 \leftarrow$ GetWindowData
    $\epsilon_p \leftarrow$ Sample
    $\hat{\mathbf{x}}_\nu \leftarrow \mathbf{x}_0\sqrt{\bar{\alpha}_\nu^p} + \epsilon_p\sqrt{1-\bar{\alpha}_\nu^p}$
    for $t = \nu, \cdots, 1$ **do**
      $z_p \leftarrow$ Sample()
      $\hat{\mathbf{x}}_{t-1} \leftarrow \frac{1}{\sqrt{\alpha_t^p}}(\hat{\mathbf{x}}_t - \frac{1-\alpha_t^p}{\sqrt{1-\bar{\alpha}_t^p}}\epsilon_\Psi(\hat{\mathbf{x}}_t)) + \tilde{\beta}_t^p z_p$
    end for
    $s \leftarrow ||\mathbf{x}_0 - \hat{\mathbf{x}}_0||^2$
    Output: s
    Dequeue from sliding window queue
  end if
end for

The anomaly detector integrates diffusion models and thus is adept at capturing the inherent spatiotemporal patterns and stochastic motion tendencies of upper-limb movements through the analysis of sensor data. Consequently, anomaly scores can be computed in real-time for human-robot interactions. These scores serve as indicators of the comfort levels and the naturalness of the assistance provided by the upper-limb exoskeleton robot.

Reference Generation: To customize assistance during passive following training, historical upper-limb trajectories, coupled with online trajectory refinement as outlined in (5.15), are integrated to develop a probabilistic model. In this context, ProMPs are employed to encode a set of trajectories into a probabilistic framework [18], which is capable of generating similar references through sampling. The application of ProMPs for trajectory sampling is particularly suited for analyzing the repetitive movements encountered in passive following training. This suitability is because of the probabilistic model's effective accommodation of sensor noise, human uncertainty, and individual biases.

To implement the ProMPs, we express the trajectory by means of the weight vector $\boldsymbol{\omega} \in \Re^{Dn\times 1}$, where D is the number of basis functions, and n is the number of active joints, such that

$$\boldsymbol{y}_t = \begin{bmatrix} \boldsymbol{q}_{1,t}^{\mathsf{T}} & \cdots & \boldsymbol{q}_{n,t}^{\mathsf{T}} \end{bmatrix}^{\mathsf{T}} = \begin{bmatrix} \boldsymbol{\Phi}_t \\ \dot{\boldsymbol{\Phi}}_t \end{bmatrix} \omega + \epsilon_y, \tag{5.32}$$

$$\boldsymbol{q}_{i,t} = \begin{bmatrix} q_{i,t} & \dot{q}_{i,t} \end{bmatrix}^{\mathsf{T}}, \tag{5.33}$$

$$p(\boldsymbol{\tau}_y|\boldsymbol{w}) = \prod_t \mathcal{N}(\boldsymbol{y}_t|\boldsymbol{\Phi}_t\omega, \boldsymbol{\Sigma}_y), \tag{5.34}$$

where $\boldsymbol{q}_{i,t} \in \Re^2$ represents the composite vector of the ith joint at time step t, $\boldsymbol{\epsilon}_y \sim \mathcal{N}(\mathbf{0}, \boldsymbol{\Sigma}_y)$ represents zero-mean i.i.d. Gaussian noise, $\boldsymbol{\tau}_y$ is the trajectory over the demonstration, and $\boldsymbol{\Phi}_t \in \Re^{n\times Dn}$, chosen as a Gaussian form, is the time-variant basis matrix.

Given the assumption that $\boldsymbol{\omega}$ follows a normal distribution, $\boldsymbol{\omega} \sim \mathcal{N}(\boldsymbol{\omega}|\boldsymbol{\mu}_\omega^{(k)}, \boldsymbol{\Sigma}_\omega^{(k)})$, a new trajectory at time step t can be modeled as follows:

$$p(\boldsymbol{y}_t; \boldsymbol{\mu}_\omega^{(k)}, \boldsymbol{\Sigma}_\omega^{(k)}) = \int \mathcal{N}(\boldsymbol{y}_t|\boldsymbol{\Phi}_t\boldsymbol{\omega}, \boldsymbol{\Sigma}_y)\mathcal{N}(\boldsymbol{\omega}|\boldsymbol{\mu}_\omega^{(k)}, \boldsymbol{\Sigma}_\omega^{(k)})d\boldsymbol{\omega}. \tag{5.35}$$

Therefore, the reference trajectory is given as follows:

$$\boldsymbol{q}_r(t) = [q_{i,t}, \cdots, q_{n,t}] \tag{5.36}$$

To facilitate the generation of personalized assistance, the following cost function is introduced to evaluate the performance of the provided reference trajectory:

$$\mathcal{S}(\boldsymbol{q}_r) = \int_{T_r} \{\|\boldsymbol{q}_d(\boldsymbol{q}_r) - \boldsymbol{q}\|_Q^2 + s^2\}dt, \tag{5.37}$$

where T_r signifies the duration of the assistance, and $\boldsymbol{q}_d$ denotes the desired trajectory obtained from online refinement via (5.15).

Given that each sampled trajectory can be evaluated by its cost, it becomes feasible to attribute an information-theoretic weight to them, reflecting the performance across all k trajectories [19]:

$$w^{(i)} = \frac{1}{\eta_k} \exp(-\frac{1}{\lambda_p}\mathcal{S}(\boldsymbol{q}_r^{(i)})), \tag{5.38}$$

where $\boldsymbol{q}_r^{(i)}$ is the reference trajectory from the ith sampling, η_k is a normalization constant, and λ_p is a small positive constant.

Considering the performance of the trajectory, the parameters $\boldsymbol{\mu}_\omega^{(k)}$ and $\boldsymbol{\Sigma}_\omega^{(k)}$ are deduced from the k collected trajectories, as follows:

$$\boldsymbol{\omega}^{(i)} = (\boldsymbol{\Phi}^\top\boldsymbol{\Phi})^{-1}\boldsymbol{\Phi}^\top\boldsymbol{Y}^{(i)}, \tag{5.39}$$

$$\boldsymbol{\mu}_\omega^{(k)} = \sum_i w^{(i)}\boldsymbol{\omega}^{(i)}, \tag{5.40}$$

$$\boldsymbol{\Sigma}_\omega^{(k)} = \sum_i w^{(i)}(\boldsymbol{\omega}^{(i)} - \boldsymbol{\mu}_\omega^{(k)})(\boldsymbol{\omega}^{(i)} - \boldsymbol{\mu}_\omega^{(k)})^\top, \tag{5.41}$$

where $\boldsymbol{\Phi} \in \Re^{N_r n\times Dn}$ is a matrix comprised of block diagonal matrices $\boldsymbol{\Phi}_t$, stacked vertically in accordance with sampling number N_r, and $\boldsymbol{Y}^{(i)} \in \Re^{N_r n}$ corresponds to the ith gathered trajectory. This process facilitates the construction of the probabilistic model.

The adjusted trajectory is integrated into the upper-limb exoskeleton robot to assist the patient. Subsequently, the patient's actual movement during the training sessions is captured and utilized to iteratively refine the probabilistic model (5.35). Specifically, $\boldsymbol{\mu}_\omega^{(k)} \leftarrow \boldsymbol{\mu}_\omega^{(k+1)}$ and $\boldsymbol{\Sigma}_\omega^{(k)} \leftarrow \boldsymbol{\Sigma}_\omega^{(k+1)}$, which is then used to facilitate the planning of subsequent trajectories, as illustrated in Algorithm 5.3. The variable N_s is the number of times that free exploration is performed based on the pre-collected demonstrations. The refinement of assistance is governed by the anomaly score, reflecting prior healthy movement behavior. This score serves as an indicator of movement comfort and naturalness during rehabilitation, ensuring that the trajectory adjustments intrinsically enhance the quality of assistance.

Algorithm 5.3 Reference Trajectory Generation

Require: $\boldsymbol{Y}^{(k+1)}$
1: **if** $k \leq N_s$ **then**
2: Initialize $\boldsymbol{\mu}_\omega^{(k+1)}$, $\boldsymbol{\Sigma}_\omega^{(k+1)}$ with pre-collected data.
3: **else**
4: Calculate cost $\mathcal{S}(\boldsymbol{q}_r^{(k+1)})$ using Eq. (5.37)
5: **for** all collected $k+1$ trajectories **do**
6: Calculate cost $w^{(i)}$, $\boldsymbol{\omega}^{(i)}$ using Eqs. (5.38) and (5.39)
7: **end for**
8: Calculate distribution $\mathcal{N}(\boldsymbol{\omega}|\boldsymbol{\mu}_\omega^{(k+1)}, \boldsymbol{\Sigma}_\omega^{(k+1)})$ using Eqs. (5.40) and (5.41)
9: **end if**
10: $\boldsymbol{\omega} \leftarrow$ Sample($\mathcal{N}(\boldsymbol{\omega}|\boldsymbol{\mu}_\omega^{(k+1)}, \boldsymbol{\Sigma}_\omega^{(k+1)})$)
11: $\boldsymbol{q}_r(t) \leftarrow \boldsymbol{\Phi}_t\boldsymbol{\omega}$

For a specific training task involving human-robot interaction, we assume that the patient intends to synchronize movement with the upper-limb exoskeleton robot. However, physical factors such as the randomness of human motion intentions may impact the tracking performance of the upper-limb exoskeleton robot. Given the challenge of explicitly accounting for such disturbances within the dynamics, we model this interference as additional noise in trajectory planning. The dynamics are described as follows:

$$\boldsymbol{x}_p^{(t+1)} = g_p(\boldsymbol{x}_p^{(t)}, \boldsymbol{q}_r), \tag{5.42}$$

$$\boldsymbol{q}_r \sim \mathcal{N}(\boldsymbol{q}_{rp}, \boldsymbol{\Sigma}_p), \tag{5.43}$$

where $\boldsymbol{x}_p = [\boldsymbol{x}_d^\top, s]^\top$ is an augmented state vector, $g_p(\cdot)$ is a nonlinear time-variant function that integrates online refinement, the impedance controller, and deterministic components of human motion intention with human-robot interaction. The mean of the assistance distribution is denoted by $\boldsymbol{q}_{rp}$, and Σ_p encapsulates the overall stochastic disturbance, which includes the randomness of human motion intentions and the sampling variability of generative models.

According to [19], to minimize the following cost

$$\hat{\mathcal{S}}(\boldsymbol{q}_r) = \int_{T_r} \{\|\boldsymbol{q}_d(\boldsymbol{q}_r) - \boldsymbol{q}\|_Q^2 + s^2 + \frac{\lambda_p}{2}\boldsymbol{q}_{rp}^\top \Sigma_p^{-1} \boldsymbol{q}_{rp}\} dt, \tag{5.44}$$

where the optimal control input is structured in a cost-decoupled manner as follows:

$$\hat{w}(\boldsymbol{q}_r) = \frac{1}{\hat{\eta}_k} \exp(-\frac{1}{\lambda_p}(\hat{\mathcal{S}}(\boldsymbol{q}_r) + \gamma_p \sum \tilde{\boldsymbol{q}}_{rp}^{\mathsf{T}} \Sigma_p^{-1} \boldsymbol{q}_r)), \tag{5.45}$$

$$\hat{\boldsymbol{q}}_{rp}^* = \mathbb{E}[\hat{w}(\boldsymbol{q}_r)\boldsymbol{q}_r]. \tag{5.46}$$

Here, $\gamma_p = \lambda_p(1 - \alpha_p)$ is the decoupled temperature parameter, defined with $\alpha_p \in [0, 1]$, and $\tilde{\boldsymbol{q}}_{rp}$ is the mean of the initial trajectory estimates.

In this approach, the optimal assistance is derived from the expected value of the current trajectory distribution. Each sampling iteration is based on the previously improved trajectory distribution, which may lead to inconsistencies in the sampling space. To address these potential inconsistencies and ensure adequate exploration, we opt for a large sample size N_s in Algorithm 5.3. Additionally, we set $\alpha_p = 1$ to mitigate the influence of inconsistencies in $\tilde{\boldsymbol{q}}_{rp}$ across different samples and thereby enhance the robustness and reliability of the optimal control solution.

Given that the last term in (5.44) is unexpected in the rehabilitation process, the ideal set is $\lambda_p = 0$, leading to the following definition of optimal assistance:

$$\hat{\boldsymbol{q}}_{rp}^* = \arg\min_{\boldsymbol{q}_r} \hat{\mathcal{S}}(\boldsymbol{q}_r). \tag{5.47}$$

In this scenario, the optimal assistance is identified as the sampled trajectory with the lowest cost, and all other sampled outcomes are disregarded. However, this formulation disregards the distribution of the optimal assistance, rendering the iterative improvement mechanism nonviable. Thus, to maintain the feasibility of iterative improvements and also minimize the impact of the undesirable term, the parameter λ_p, which governs the tightness of the solution, is set to a small value. This adjustment ensures that $\hat{\mathcal{S}}(\boldsymbol{q}_r) \to \mathcal{S}(\boldsymbol{q}_r)$, and the optimal weight formula in (5.45) simplifies to the configuration used in (5.38). This setting balances the need to minimize the undesired term with the need to maintain a practical and effective iterative improvement process.

The optimal assistance distribution is identified via a coarse-to-fine approach. Initially, the sample distribution is established based on pre-defined demonstrations, which primarily guide the exploratory phase using the initial samples. Subsequent iterations improve the assistance distribution by progressively narrowing the sampling space and utilizing previously explored optimal trajectory values. The probability model focuses on the best-performing trajectories to provide optimal assistance, which effectively addresses the various uncertainties that the patient may encounter during the task. Incorporation of the patient's motion data ensures that subsequent trajectories become increasingly individualized and thus align increasingly more with the patient's medical needs. Hence, this method enables the upper-limb exoskeleton robot to exploit online interactions for the dual purpose of enhancing interaction safety and improving the efficacy of passive following training.

5.2.5 Interaction Control

The joint below the shoulder joints is cable-driven, which decreases the inertia associated with movement and thus increases comfort. Given the presence of these joints, there is substantial friction, which can hinder the movement of joints and therefore should be considerably compensated for.

To achieve this, the upper-limb exoskeleton robot first performs movements in the absence of the patient, where $\boldsymbol{\tau}_e = \mathbf{0}$. In addition, the disturbance torque can be parameterized as follows [20]:

$$\begin{aligned}\boldsymbol{\tau}_f &= (\boldsymbol{a}_f + \boldsymbol{b}_f \odot e^{-\boldsymbol{c}_f \odot \dot{\boldsymbol{q}}} + \boldsymbol{d}_f \odot \dot{\boldsymbol{q}}) \odot \boldsymbol{sgn}(\dot{\boldsymbol{q}}) \\ &\approx (\bar{\boldsymbol{a}}_f + \bar{\boldsymbol{b}}_f \odot \dot{\boldsymbol{q}} + \bar{\boldsymbol{c}}_f \odot \dot{\boldsymbol{q}} \odot \dot{\boldsymbol{q}}) \odot \boldsymbol{sgn}(\dot{\boldsymbol{q}}) = \boldsymbol{Y}(\dot{\boldsymbol{q}})\boldsymbol{\zeta},\end{aligned} \tag{5.48}$$

where $\boldsymbol{a}_f, \boldsymbol{b}_f, \boldsymbol{c}_f, \boldsymbol{d}_f$ are the unknown parameters, $\bar{\boldsymbol{a}}_f, \bar{\boldsymbol{b}}_f, \bar{\boldsymbol{c}}_f$ are derived from the Taylor expansion as simplifications for the model, $\odot$ denotes the Kronecker product, $\boldsymbol{sgn}(\cdot)$ represents a sign function, $\boldsymbol{Y}(\cdot)$ represents a regressor matrix, and $\boldsymbol{\zeta}$ represents the vector of model parameters. The approximation presented in (5.48) is reasonable because the velocities of the joints of the upper-limb exoskeleton robot remain rather low during a given rehabilitation process.

As the upper-limb exoskeleton robot is equipped with force sensors on all of its active joints, friction can be directly measured and recorded in the absence of a patient. Together with the recorded joints' velocities, the parameters of the friction model are learned via polynomial fitting. The estimated friction is represented as follows

$$\hat{\boldsymbol{\tau}}_f = \boldsymbol{Y}(\dot{\boldsymbol{q}})\hat{\boldsymbol{\zeta}}. \tag{5.49}$$

Once the friction is estimated, a variable impedance model is proposed. This model must be capable of identifying the human–robot interaction condition and addressing the conflict during rehabilitation. These capabilities enable the model to regulate the action of the upper-limb exoskeleton robot.

To regulate the impedance model, a weighting function [21] is introduced to consider the anomaly score. This function is mathematically defined as follows:

$$w(s) = \lambda_1 \tanh(-\frac{s}{\chi_1} + \chi_2) + \lambda_2, \tag{5.50}$$

where λ_1 and λ_2 are positive constants that determine the range and median of the weighting function, respectively, χ_1 is a constant that normalizes the anomaly score into a specified small range, and χ_2 is the offset of the weighting function from the origin of the coordinates along the positive horizontal axis. Based on this function, the desired impedance model is redefined as follows:

$$\boldsymbol{C}_d(\dot{\boldsymbol{q}} - \dot{\boldsymbol{q}}_d) + \boldsymbol{K}_d(\boldsymbol{q} - \boldsymbol{q}_d) = \frac{1}{w(s)}\boldsymbol{\tau}_e. \tag{5.51}$$

Multiplying both sides of (5.51) by $w(s)$ yields

$$\boldsymbol{C}_a(t)(\dot{\boldsymbol{q}} - \dot{\boldsymbol{q}}_d) + \boldsymbol{K}_a(t)(\boldsymbol{q} - \boldsymbol{q}_d) = \boldsymbol{\tau}_e, \tag{5.52}$$

where $\boldsymbol{C}_a(t) \triangleq w(s)\boldsymbol{C}_d$ and $\boldsymbol{K}_a(t) \triangleq w(s)\boldsymbol{K}_d$, are the time-varying apparent impedance parameters. These parameters are utilized to deduce the relationship between the desired trajectory and the interaction torque, as outlined in (5.15). The function of the mechanism is such that an increase in the anomaly score leads to a decrease in $w(s)$, thereby reducing the magnitude of the impedance parameters. This reduction ensures that there is an increase in the passiveness with which the upper-limb exoskeleton robot responds to any detected conflicts. Conversely, when the anomaly score is low, the impedance parameters revert to their original values, thereby maintaining the level of assistance provided by the upper-limb exoskeleton robot.

Subsequently, an impedance vector is introduced as follows

$$\begin{aligned} \boldsymbol{z} &= \dot{\boldsymbol{q}} - \dot{\boldsymbol{q}}_r \\ &= \dot{\boldsymbol{q}} - \dot{\boldsymbol{q}}_d + \boldsymbol{C}_d^{-1}\boldsymbol{K}_d(\boldsymbol{q} - \boldsymbol{q}_d) - \frac{1}{w(s)}\boldsymbol{C}_d^{-1}\boldsymbol{\tau}_e, \end{aligned} \tag{5.53}$$

where

$$\dot{\boldsymbol{q}}_r = \dot{\boldsymbol{q}}_d - \boldsymbol{C}_d^{-1}\boldsymbol{K}_d(\boldsymbol{q} - \boldsymbol{q}_d) + \frac{1}{w(s)}\boldsymbol{C}_d^{-1}\boldsymbol{\tau}_e \tag{5.54}$$

is a reference vector. According to (5.53), the convergence of $\boldsymbol{z} \to \boldsymbol{0}$ implies the realization of the desired impedance model (5.51).

The overall control input is designed as defined in (5.7), with the fast time-scale control term defined as in (5.8). Next, the slow time-scale control term is established by using the estimated friction $\hat{\boldsymbol{\tau}}_f$ to stabilize the dynamics expressed in (5.13) and achieve the desired impedance model, as follows:

$$\begin{aligned} \boldsymbol{u}_s = &- \boldsymbol{K}_z\boldsymbol{z} - \boldsymbol{S}_2^{\mathsf{T}}\hat{\boldsymbol{\tau}}_f - \boldsymbol{\tau}_e - k_g \cdot \boldsymbol{sgn}(\boldsymbol{z}) \\ &+ (\boldsymbol{M}(\boldsymbol{q}) + \bar{\boldsymbol{B}})\ddot{\boldsymbol{q}}_r + \boldsymbol{C}(\dot{\boldsymbol{q}}, \boldsymbol{q})\dot{\boldsymbol{q}}_r + \boldsymbol{g}(\boldsymbol{q}), \end{aligned} \tag{5.55}$$

where $\boldsymbol{sgn}(\cdot)$ is the sign function and is defined as follows:

$$\operatorname{sgn}(z) = \begin{cases} 1, & z > 0 \\ 0, & z = 0 \\ -1, & z < 0 \end{cases} \tag{5.56}$$

where k_g is a positive constant, and $\boldsymbol{K}_z \in \Re^{n\times n}$ is a diagonal and positive-definite matrix.

To prove the stability of the whole system, we substitute Eqs. (5.53) and (5.55) into (5.13), resulting in the following system dynamics:

$$(\boldsymbol{M}(\boldsymbol{q}) + \bar{\boldsymbol{B}})\dot{\boldsymbol{z}} + \boldsymbol{C}(\dot{\boldsymbol{q}}, \boldsymbol{q})\boldsymbol{z} = -\boldsymbol{K}_z\boldsymbol{z} - \boldsymbol{S}_2^{\mathsf{T}}\tilde{\boldsymbol{\tau}}_f - k_g\boldsymbol{sgn}(\boldsymbol{z}), \tag{5.57}$$

where $\tilde{\boldsymbol{\tau}}_f = \hat{\boldsymbol{\tau}}_f - \boldsymbol{\tau}_f$ represents the estimation error of the friction.

Subsequently, we propose the following candidate Lyapunov function:

$$V = \frac{1}{2}z^T(\boldsymbol{M}(\boldsymbol{q}) + \bar{\boldsymbol{B}})z. \tag{5.58}$$

By differentiating (5.58) with respect to time and substituting the dynamics with those from (5.57), we derive the following expression:

$$\dot{V} = -z^T \boldsymbol{K}_z z - z^T \boldsymbol{S}_2^{\mathrm{T}} \tilde{\boldsymbol{\tau}}_f - k_g z^T \boldsymbol{sgn}(z). \tag{5.59}$$

Assuming that $\|\boldsymbol{S}_2^{\mathrm{T}} \tilde{\boldsymbol{\tau}}_f\| \leq \kappa$, the upper bound for $\dot{V}$ is derived as follows:

$$\dot{V} \leq -z^T \boldsymbol{K}_z z - (k_g - \kappa)\|z\|, \tag{5.60}$$

If k_g is adequately large such that $k_g > \kappa$, the inequality simplifies to

$$\dot{V} \leq -z^T \boldsymbol{K}_z z < 0, \tag{5.61}$$

Given that $V > 0$ and $\dot{V} < 0$, the *quasi-steady-state system* is exponentially stable. Considering that the *boundary-layer system* can be made intrinsically stable by appropriate tuning of $\boldsymbol{K}_1$ and $\boldsymbol{K}_2$, the stability of the closed-loop system is assured according to [22], ensuring convergence to the desired impedance vector.

5.2.6 Results

The proposed dual-mode trajectory refinement method was implemented within the upper-limb exoskeleton robot to assess the effectiveness of both rehabilitation modes. Figure 5.16 depicts the experimental configuration in which the main board controlled the impedance of the upper-limb exoskeleton robot. This board was interfaced via a serial port connection with a PC, which was outfitted with an Intel i5-13490F CPU and an RTX 4060Ti graphics card. The PC executed the anomaly detection module, calculated the reference trajectory $\boldsymbol{q}_r$, and dynamically updated the desired trajectory $\boldsymbol{q}_d$ online. Subjects performed rehabilitation training in a passive following mode or an active mirroring mode. In the latter mode, the subject wore a brace equipped with markers on the unaffected side of the body to facilitate engagement in rehabilitation exercises. Optical motion capture equipment (NOKOV) was used to accurately capture motion intentions.

Implementation of the variable impedance controller required knowledge of the dynamic model. Thus, the dynamic parameters were computed analytically in real time using the open-source Orocos Kinematics and Dynamics Library. In this computation, the upper-limb exoskeleton robot is deconstructed into a sequence of links and joints to formulate a model defining its physical configuration, encompassing characteristics such as the length, mass, and inertia of each link. Subsequently, the forward dynamics are derived using Newton–Euler equations based on this model.

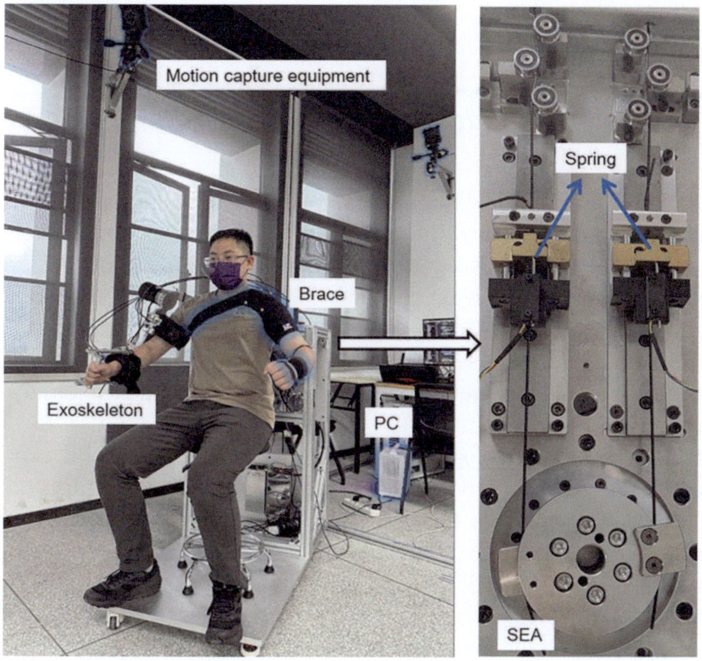

Fig. 5.16 The experimental setup in active mirroring mode used motion capture equipment to obtain feedback from the unaffected side of the body. The SEA was constructed by connecting the motor output and the joint end with a spring and cable

The intention predictor and the anomaly detection module required motion data, which was collected from the upper-limb exoskeleton robot during operation by healthy subjects. To facilitate free and natural movement, the upper-limb exoskeleton robot was set to operate in a transparent mode during the data collection phase [23]. Specifically, the controller was designed as follows [24]:

$$\begin{aligned} \boldsymbol{u}_0 =& (\boldsymbol{M}(\boldsymbol{q}) + \bar{\boldsymbol{B}})\ddot{\boldsymbol{q}}_0 + \boldsymbol{C}(\dot{\boldsymbol{q}}, \boldsymbol{q})\dot{\boldsymbol{q}} \\ &+ \boldsymbol{g}(\boldsymbol{q}) - \boldsymbol{S}_2^{\mathsf{T}}\hat{\boldsymbol{\tau}}_f - \boldsymbol{\tau}_e + \boldsymbol{u}_f, \end{aligned} \tag{5.62}$$

$$\ddot{\boldsymbol{q}}_0 = \frac{1}{\gamma_0}(\boldsymbol{M}(\boldsymbol{q}) + \bar{\boldsymbol{B}})^{-1}\boldsymbol{\tau}_e, \tag{5.63}$$

where γ_0 represents a parameter controlling the magnitude of virtual mass, and $\ddot{\boldsymbol{q}}_0$ denotes the desired acceleration. In this transparent mode, healthy subjects or patients were able to maneuver the upper-limb exoskeleton robot effortlessly, without experiencing significant discomfort.

The next section presents the results of five experiments, which are described below.

(1) *Intention Predictor*: This experiment aimed to validate the efficacy and accuracy of the proposed intention predictor. A diverse set of intention prediction methods were trained using the collected data, and then a comparative analysis was performed to demonstrate the superiority of the proposed intention predictor.

Table 5.10 Motion range of joints

Joint	1	2	3	4	5
Min(°)	−90	−45	−30	−5	−30
Max(°)	30	115	30	120	30

(2) *Anomaly Detection*: This experiment aimed to evaluate the performance of the anomaly detection network. Its detection accuracy was demonstrated in various simulated anomaly scenarios, including movements outside the normal range, stroke-induced convulsions, and human-robot interaction conflicts. Furthermore, comparative studies were conducted to illustrate that the proposed anomaly detector, which is based on a diffusion model, exhibits detection accuracy that is significantly better than that exhibited by other anomaly detection methods.

(3) *Interaction Control*: This experiment aimed to assess the dynamic capabilities of the system and its ability to reject disturbances. Thus, a trajectory-tracking task was conducted. In addition, the effectiveness of the proposed variable impedance controller was validated in a scenario involving anomalies.

(4) *Active Mirroring Training*: This experiment aimed to assess the effectiveness of the online trajectory refinement. Thus, ablation studies were conducted. In addition, how the anomaly score influences the trajectories generated by this refinement process was examined. Furthermore, an active mirroring mode was implemented, and the motion capture system was used to verify that the proposed method effectively constrains assistive trajectories and maintains safety throughout the rehabilitation process.

(5) *Passive Following Training*: This experiment aimed to validate the improvements in rehabilitation facilitated by passive following training. An ablation study was conducted to demonstrate that the online trajectory refinement significantly enhances movement naturalness and task performance throughout the training process. Moreover, a clinical trial was performed with stroke patients to obtain evidence that this rehabilitation framework significantly aids in the recovery of motor functions.

Two able-bodied participants with no prior experience with upper-limb exoskeleton robots were recruited for motion data collection. Table 5.10 presents the motion range of the upper-limb exoskeleton robot during data collection. This motion range covers the main spatial areas of daily activities in the upper limbs.

Table 5.11 presents an overview of the participants' statistical attributes. The participants signed a written informed consent form prior to the experimental sessions. Next, the participants underwent a 6-min pre-training phase to familiarize themselves with moving while wearing the upper-limb exoskeleton robot. During the data collection phase, no specific movement guidelines were imposed. Therefore, the participants were permitted to move the upper-limb exoskeleton robot at their preferred speed, i.e., rapidly or slowly, or to maintain it in a stationary position. The collected data were compiled into a dataset that was utilized for training the intention predictor and anomaly detection modules.

Table 5.11 Statistical information of the subjects

Subject	Gender	Age (y)	Weight (kg)	Height (cm)
1	Male	30	87	181
2	Male	24	70	172

Table 5.12 Prediction results for different methods

Method	FDE (°)	ADE (°)	MAE (°)	RMSE (°)
Forward-int.	0.171	0.102	0.055	0.066
CNN-LSTM	0.165 ↑	0.099 ↑	0.056 ↓	0.067 ↓
Ours	0.096 ↑	0.061 ↑	0.033 ↑	0.040 ↑

Intention Predictor: To achieve real-time estimation of human motion intention, the collected motion data were utilized to train the intention predictor. To enhance the generalizability of the trained model, we partitioned the collected motion data into training, testing, and validation sets in an 8:2:1 ratio during the training process. We employed Trajectron++ [25] as the backbone for extracting trajectory features and selected a transformer architecture to manage the diffusion process. The parameters for the intention predictor were set as follows: $N_o = 5$, $N_p = 7$, and $T = 100$. Denoising diffusion implicit models (DDIMs) [26] were employed in the inference process. Intention prediction relies on historical observation data, and the predicted trajectory length was rather short. Therefore, we selected forward integration and a convolutional neural network–long short-term memory (CNN-LSTM)-based neural network that previously demonstrated excellent performance in estimating robot joint behavior [27] as the baselines for the experiments.

Furthermore, we adopted final displacement error (FDE), average displacement error (ADE), mean absolute error (MAE), and root-mean-square error (RMSE) as the evaluation metrics to measure the accuracy of the predictions. FDE and ADE were computed on a two-dimensional plane defined by Joint 2 and Joint 4, which possess the largest range of motion and exhibit the most frequent movements. MAE and RMSE were averaged across all active joints. The predictions for the validation set are presented in Table 5.12. These results suggest that in the considered real-time intention prediction task, the trajectory prediction performance of the CNN-LSTM-network-based method was about the same as the forward integration method. However, our intention predictor based on a diffusion model significantly outperformed the forward integration method. Specifically, our intention predictor showed a 43.9% lower FDE, a 40.2% lower ADE, a 40.0% lower MAE, and a 39.4% lower RMSE than the forward integration method.

A comparative analysis was conducted to validate the applicability of our intention predictor across different movement conditions in four tasks. Two tasks were based on motion data that were collected while the participants were wearing the upper-limb exoskeleton robot, including data on free movements similar to the training data and arm-swinging sinusoidal movements. The other two tasks were based on simulated smooth trajectories, namely circular and lemniscate trajectories. Not all tested trajectories were included in the pre-collected dataset employed in the training

Table 5.13 Prediction results for different tasks

Task	Method	FDE (°)	ADE (°)	MAE (°)	RMSE (°)
Free move	Forward integration	0.209	0.125	0.076	0.090
	CNN-LSTM	0.163	0.098	0.060	0.073
	Ours	0.111 (↑)	0.071 (↑)	0.042 (↑)	0.052 (↑)
Sinusoidal	Forward integration	0.731	0.379	0.243	0.297
	CNN-LSTM	0.739	0.387	0.248	0.302
	Ours	0.616 (↑)	0.331 (↑)	0.212 (↑)	0.258 (↑)
Circular	Forward integration	0.078	0.037	0.024	0.029
	CNN-LSTM	0.101	0.054	0.031	0.038
	Ours	0.066 (↑)	0.029 (↑)	0.018 (↑)	0.024 (↑)
Lemniscate	Forward integration	0.111	0.049	0.032	0.041
	CNN-LSTM	0.095	0.045	0.028	0.036
	Ours	0.086 (↑)	0.035 (↑)	0.022 (↑)	0.029 (↑)

phase. Table 5.13 reports the performance of the intention predictor in the four tasks, averaged across five active joints. It can be seen that our intention predictor exhibits the best performance in all metrics.

Figure 5.17 illustrates trajectories from steps 1, 4, and 7 of trajectory prediction, together with the actual reference trajectory. For clarity, we present the experimental results in a two-dimensional plane in which Joint 2 is the x-axis and Joint 4 is the y-axis. Moreover, the right side of each subplot shows an expanded view of the trajectory prediction details within the black box. It can be seen that the trajectory prediction at step 1 aligns well with the actual trajectory, while the trajectory predictions at steps 4 and 7 deviate slightly from the real trajectory. These results demonstrate that as the number of prediction steps increases, the prediction accuracy slightly decreases and the corresponding prediction variance increases. This trend is explicitly considered in the preemptive tuning algorithm. Overall, the above-mentioned results confirm that our intention predictor reliably forecasts upper-limb joint movements and certain regular motions by capturing the dynamic trends of trajectories based on historical observations. The free-move task was directly related to human movements in practical applications, as it simulated real-life scenarios. In contrast, the sinusoidal, circular, and lemniscate tasks were designed to test the generalization performance of our intention predictor in contexts other than rehabilitation training.

Anomaly Detection: The collected motion data, including interaction information, were integrated to train our anomaly detector, which operated in real time to evaluate safety and the naturalness of motion. Therefore, these data served as a repository of safe and natural interaction feedback, enabling the anomaly detector to discern their latent relationship and subsequently identify abnormal interactions during human–robot interaction. The parameters were $L_s = 100$ and $N_c = 21$, denoting the history

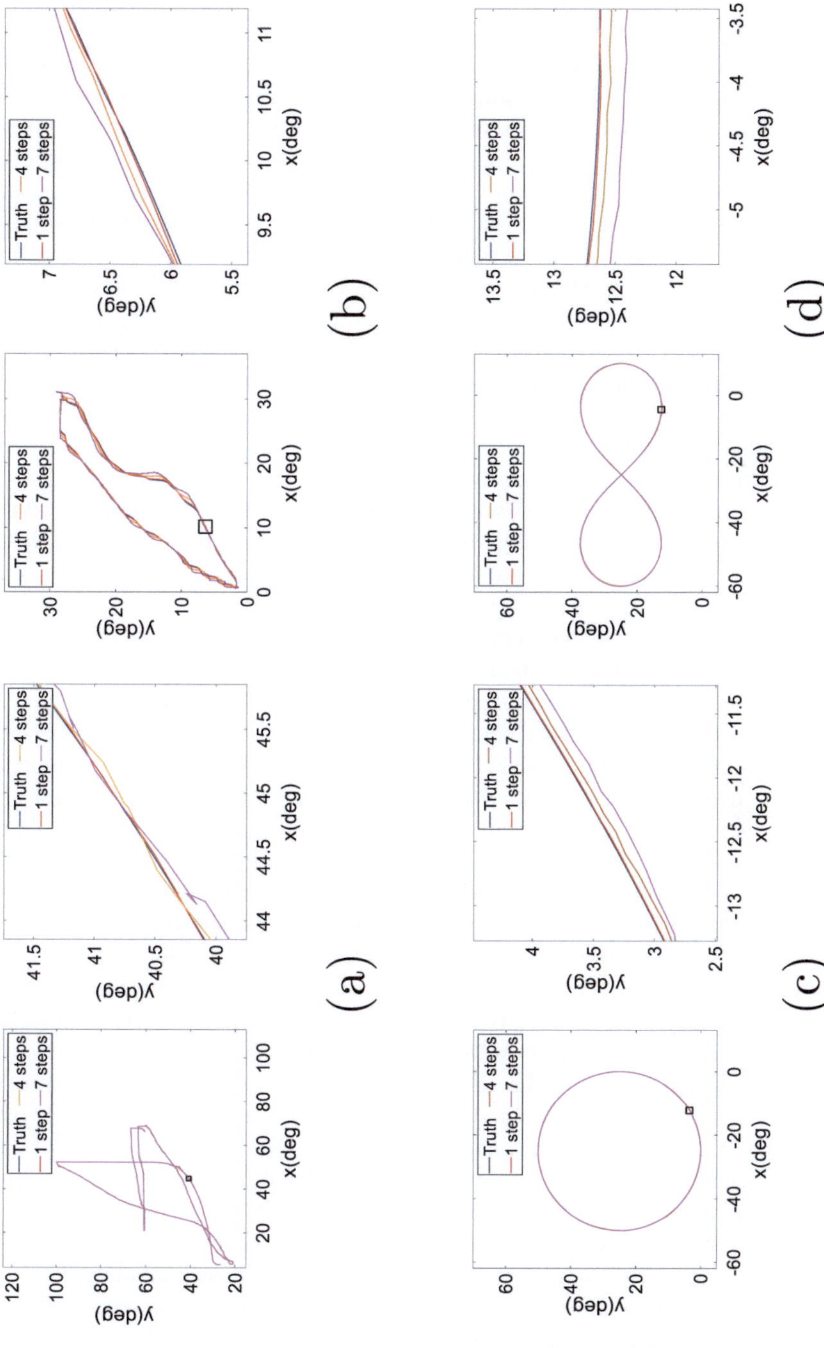

Fig. 5.17 Predicted trajectories for various steps across four distinct tasks. **a** Free move; **b** sinusoidal; **c** circular; and **d** lemniscate tasks

of joint and motor motion data, in addition to interaction torque. The diffusion model parameters during both training and inference phases were $T^p = 100$ and $\nu = 60$. DDIMs were also employed during inference.

During the experiment, a participant was required to move while wearing the upper-limb exoskeleton robot to simulate various anomalies. Three abnormal scenarios were considered: an excessive movement scenario (involving joint movements beyond the normal range), a balance deviation scenario (involving deviations from the relative balance position), and a simulated stroke tremors scenario. In the excessive movement scenario, the participant initially maintained the upper-limb exoskeleton robot within the normal movement range and then lowered the arm to simulate the anomaly. As the arm was gradually lowered, there was a decrease in the shoulder joint angle that progressively crossed the motion boundary, and thus, the anomaly level increased gradually. Subsequently, the arm was raised back to the normal motion range. The results are presented in Fig. 5.18 and reveal that as the joint positions gradually approached and then crossed the boundary, the anomaly score increased. These results demonstrate that our anomaly detector effectively detected deviations from the normal motion range (i.e., the motion range of the collected data).

In the simulated stroke tremors scenario, the participant initially maintained the upper-limb exoskeleton robot in the rest phase. Subsequently, the participant shook the entire arm of the upper-limb exoskeleton robot to simulate tremors. The corresponding anomaly score and joint positions are depicted in Fig. 5.19. It is evident that during the simulated stroke tremors, the anomaly score generally increased, and while the tremors persisted, it remained rather high.

In the balance deviation scenario, the participant initially maintained the upper-limb exoskeleton robot in a static equilibrium position that represented the balance state in which the upper-limb exoskeleton robot offered assistance. The participant was then instructed to manipulate the upper-limb exoskeleton robot upward and downward to simulate misalignment during assistance. The results are presented in Fig. 5.20 and reveal that there was a marked increase in anomaly scores whenever deviations occurred, regardless of the direction (i.e., regardless of whether the patient's movement trajectory was above or below the predetermined trajectory). This increase in scores indicates that the above-mentioned anomalies were detected.

The performance of our anomaly detector was experimentally assessed using a VAE-based anomaly detector [21] as a baseline. Specifically, detection performance in the stroke tremor scenario was evaluated using a receiver operating characteristic (ROC) curve, which illustrates the ability of a classification model to differentiate between classes. The areas under the ROC curve for the considered models are presented in Table 5.14, and the calculated anomaly scores are depicted in Fig. 5.21. In the figure, it can be seen that the scores generated by the proposed anomaly detector markedly increased as the anomaly occurred. Furthermore, the scores generated by the VAE-based anomaly detector failed to return to a normal level once the anomaly ceased. These results indicate that compared with the VAE-based anomaly detector, our anomaly detector is more adaptive in classifying anomalies in different joint configurations, owing to the superior generative performance of the diffusion model.

Interaction Control: During rehabilitation training, the upper-limb exoskeleton robot is required to regulate human-robot interaction in a desired impedance model

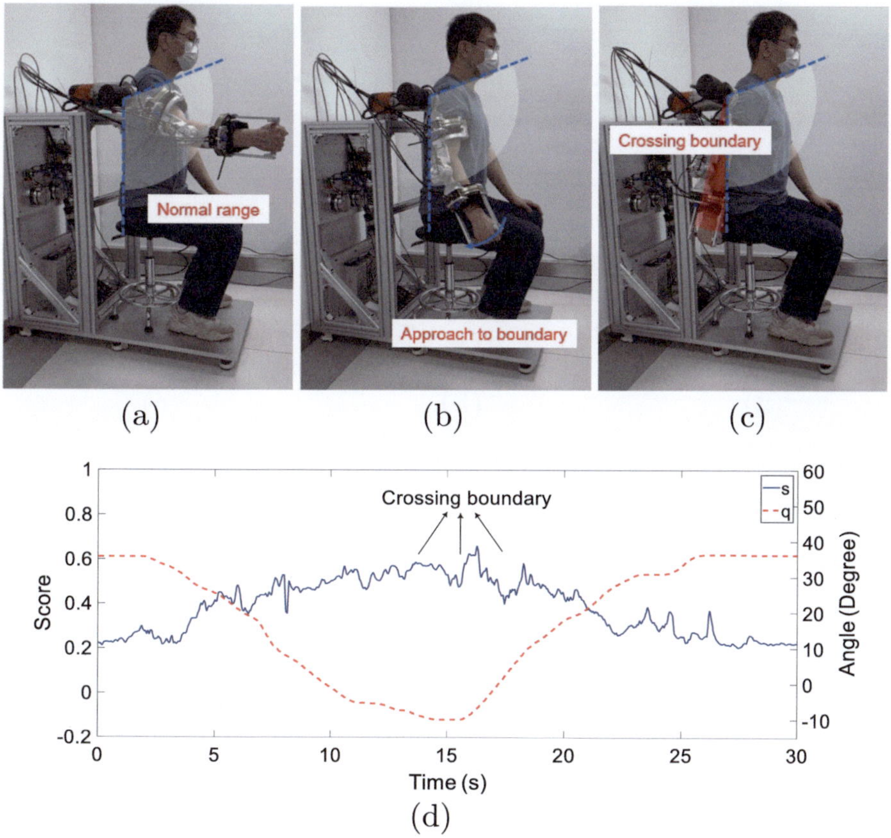

Fig. 5.18 **a–c** Snapshots of anomaly due to excessive movements; **d** anomaly score and joint position during excessive movements. The white sectors represent areas where activity is normal, while the red sectors represent areas where activity is anomalous

with consideration of friction compensation. To this end, the upper-limb exoskeleton robot was operated to move slowly in the absence of a participant, and friction was quantified. That is, the difference between the readings from the potentiometers at the motor output and the measurements from the torque sensor at the joint end was calculated. In addition, joint velocities were recorded. Subsequently, polynomial fitting was applied based on the simplified friction model (5.48) to afford the estimated values for friction parameters presented in Table 5.15.

Next, we implemented our impedance controller in the upper-limb exoskeleton robot, devoid of a patient, to track a pre-defined sinusoidal trajectory involving all active joints. This experiment aimed to validate the accuracy of the friction compensation and evaluate the dynamic performance of the controller. The impedance control parameters were set as summarized in Table 5.16. The experimental results demonstrate that the impedance controller effectively compensated for friction, enabling the upper-limb exoskeleton robot to accurately follow the desired trajectory, as illustrated in Fig. 5.22. Specifically, the RMSEs for the joints during this trajectory tracking task were 0.613° for Joint 1, 0.728° for Joint 2, 0.997° for Joint 3, 2.143° for Joint 4, and 0.948° for Joint 5.

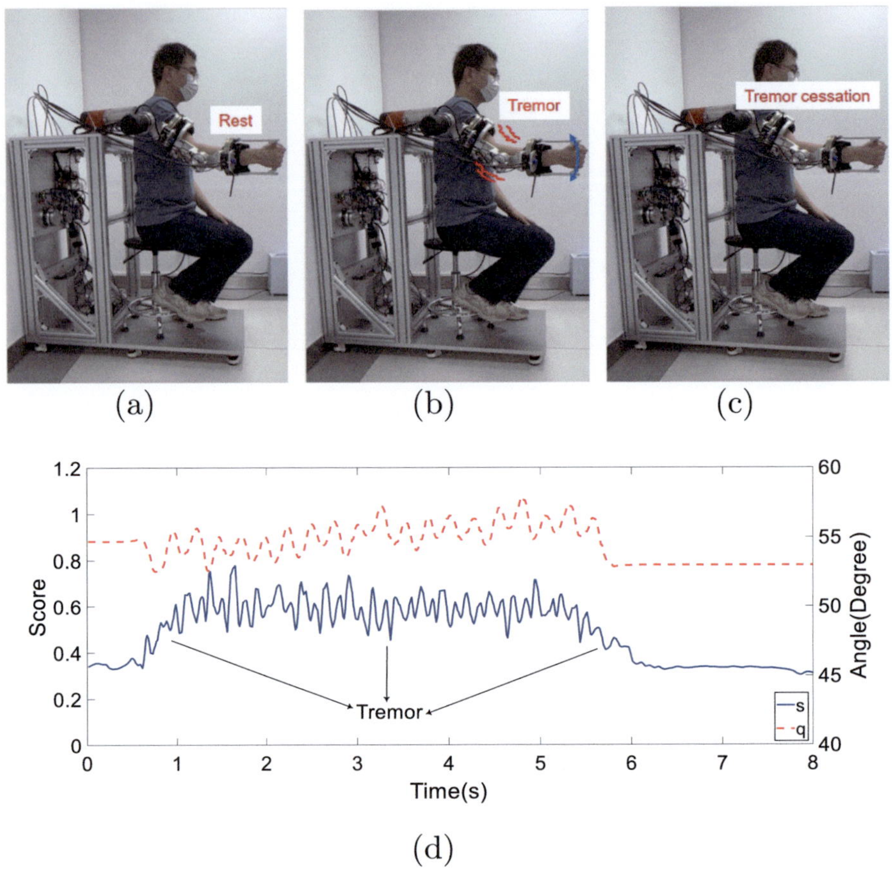

Fig. 5.19 **a–c** Snapshots of conflict due to simulated stroke tremors; **d** anomaly score and joint position during simulated stroke tremors

Table 5.14 Comparison of AUCs of the two anomaly detection methods

Method	VAE	Ours
AUC	0.865	0.999

Additionally, we assessed the performance of our impedance controller under conditions involving anomalies. This was achieved by having the upper-limb exoskeleton robot guide the patient in trajectory tracking while the patient held the upper-limb exoskeleton robot in positions that simulated anomalies. The results of this experiment are depicted in Fig. 5.23. It can be seen that throughout the experiment, the impedance vector remained close to zero, indicating that the desired variable impedance model was effectively maintained despite human involvement and the occurrence of anomalies.

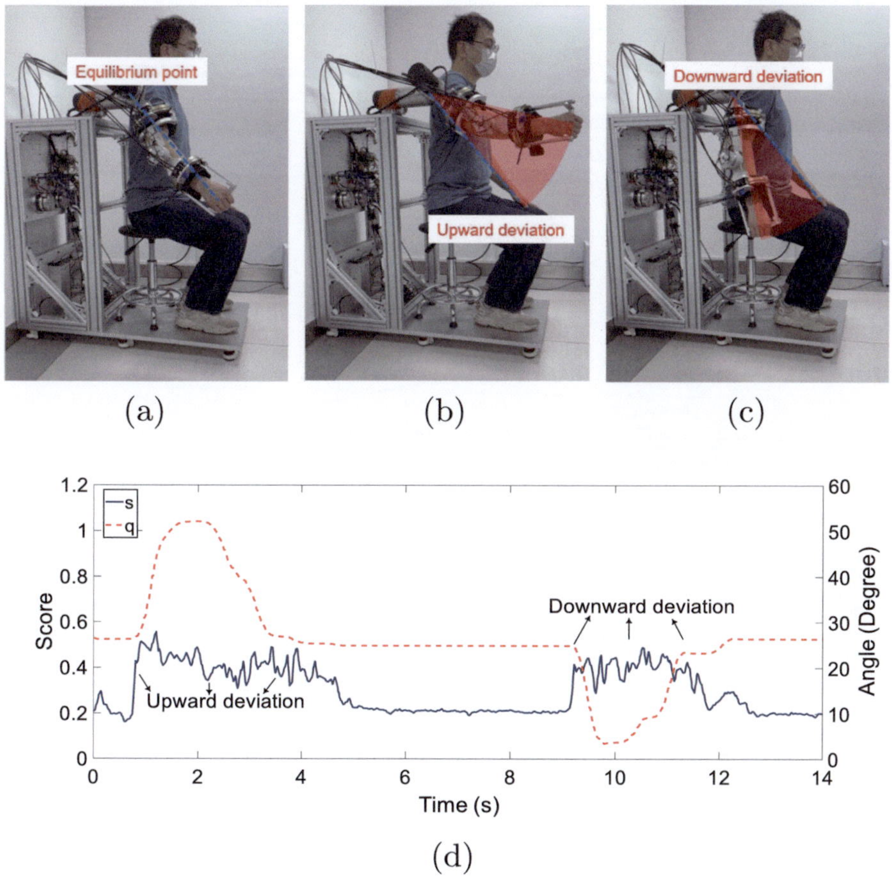

Fig. 5.20 **a–c** Snapshots of conflict due to balance deviation; **d** anomaly score and joint position during balance deviation. The red sectors represent areas where anomalies occurred

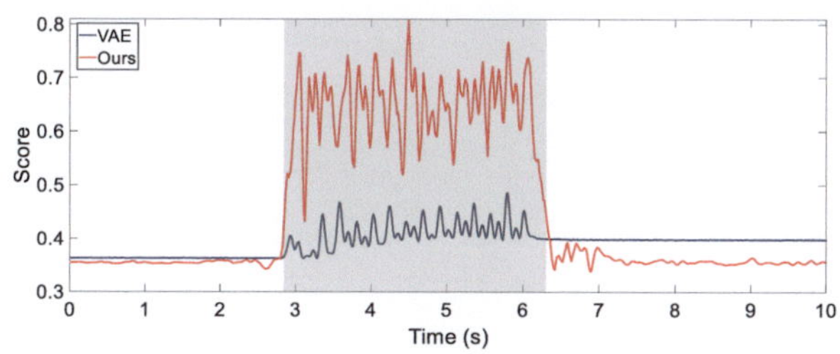

Fig. 5.21 Anomaly score curves for the two different anomaly detection methods, with shaded areas indicating the areas where anomalies occurred

Active Mirroring Training: In active mirroring training, the movements of the upper-limb exoskeleton robot were aligned with the motion intentions of the unaf-

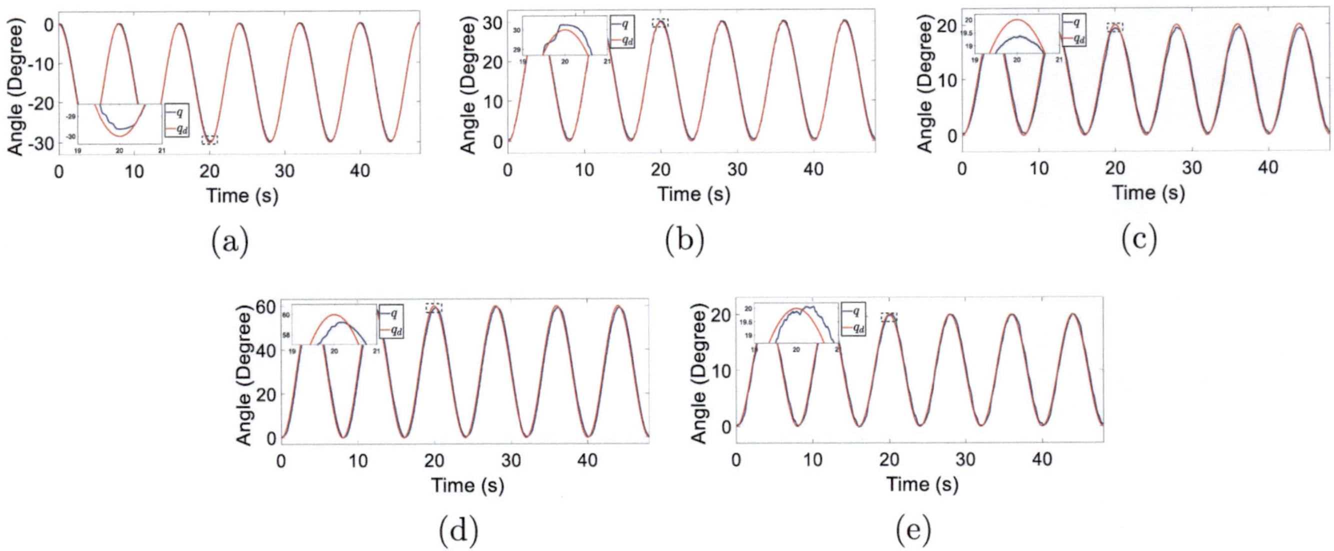

Fig. 5.22 a–e Trajectory tracking performance across Joint 1 through 5 of the exoskeleton

Table 5.15 Estimated friction model

	$\bar{a}_f$	$\bar{b}_f$	$\bar{c}_f$
Joint 3	0.822	−2.132	0.557
Joint 4	1.718	11.441	−3.028
Joint 5	0.636	−1.212	1.005

Table 5.16 Parameters of VIC

C_d	K_d	λ_1	χ_1	χ_2	λ_2	K_v	K_z
$10\boldsymbol{I}_5$	$50\boldsymbol{I}_5$	0.5	0.04	8.75	1.5	$1.1\boldsymbol{I}_3$	diag(1.5, 0.6, 0.7, 4, 1.8)

fected side of the body of the patient (as illustrated in Fig. 5.16). In addition, online trajectory refinement was used to smoothen movement commands and enhance safety through dynamic constraints. The details of the experiments are given below.

(1) To evaluate the capability of the proposed method to handle unexpected external impact, we conducted an experiment on Joint 2 of the upper-limb exoskeleton robot's shoulder.
(2) We simulated sudden external disturbances by abruptly altering the desired position during the experiments.
(3) To represent the lower constraint of the online trajectory refinement, we imposed a lower bound on the position command at 20°, thereby simulating a tendency to exceed the acceptable movement range.
(4) During comparative trials, we eliminated motion command inconsistencies caused by feedback from the unaffected side of the body by deactivating the positional feedback on this side. Therefore, we instead relied on the proprioceptive sensors of the upper-limb exoskeleton robot (specifically, its encoders) and conducted the experiment without a patient.

The position commands and joint positions with and without the implementation of online trajectory refinement, respectively, are illustrated in Fig. 5.24. This figure reveals that without online trajectory refinement, the joint position surpassed the set movement boundary. However, when online trajectory refinement was activated, the joint movement responded effectively to the dynamic constraints and remained close to the boundary. Specifically, without trajectory refinement, the movement exceeded the established boundary by approximately 9°. In contrast, with trajectory refinement, the excess was significantly reduced to just 0.5°. The trajectory refinement is designed to adjust the trajectory to adhere to constraints and to mitigate the extent to which joint position violates movement boundaries. Therefore, even with trajectory refinement, the joint position may slightly cross the boundary, as observed with the 0.5° transgression in this experiment.

Next, we experimented with Joint 1 to verify that the proposed method identified anomaly regions and guided the planned trajectory towards safer areas with lower anomaly scores than the current area. Ideally, this joint's normal operational range

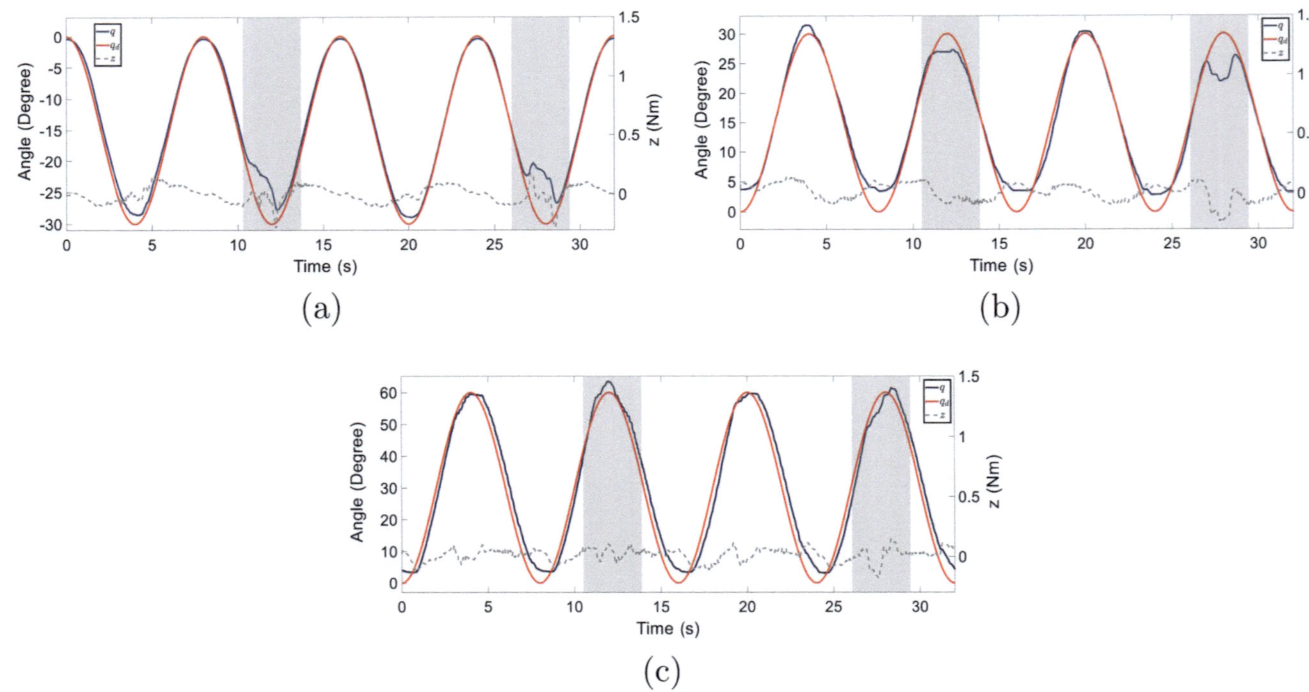

Fig. 5.23 **a–c** Performance of variable impedance controller for Joint 1, 2, and 4 of the upper-limb exoskeleton robot in the presence of a patient. The shaded areas indicate the periods when anomalies occurred

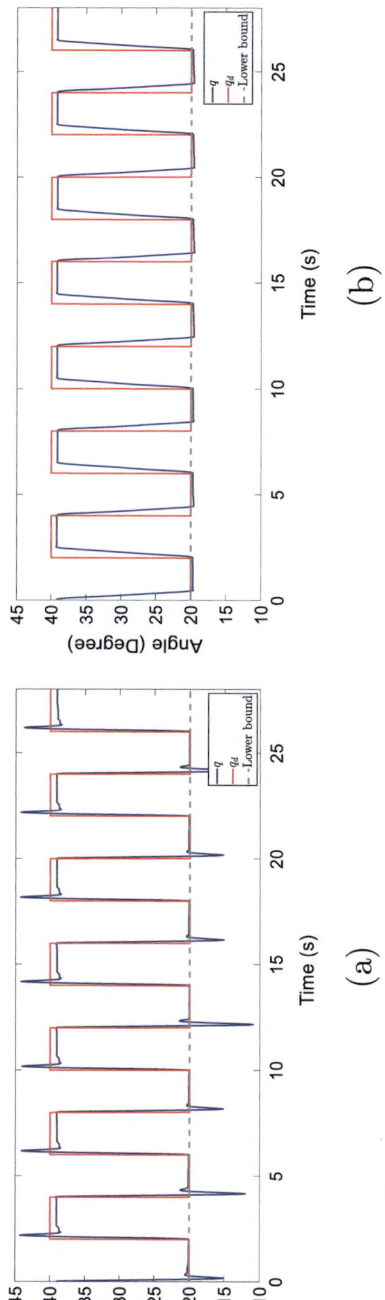

Fig. 5.24 Test of ability to handle unexpected external impacts on Joint 2 **a** without online trajectory refinement; and **b** with online trajectory refinement

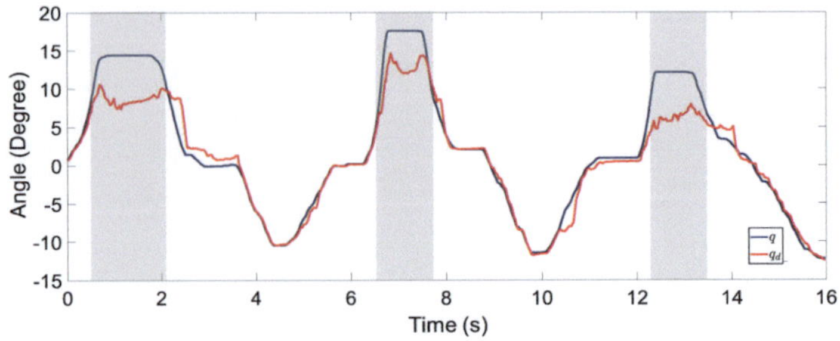

Fig. 5.25 Performance of the system in guiding the upper-limb exoskeleton robot to areas with lower anomaly scores than other areas through online trajectory refinement, with shaded areas indicating anomalies caused by abnormal adduction angles

should not significantly exceed zero. In particular, the shoulder adduction angle should remain small during typical activities. Thus, as a motion capture system may yield inaccurate adduction angle estimates due to marker obstruction, we instead utilized the encoder feedback from the upper-limb exoskeleton robot to obtain adduction angle estimates. The results are displayed in Fig. 5.25. In the figure, it can be seen that as the shoulder adduction angle increased, the motion commands generated by the online trajectory refinement decreased the adduction angle of movements rather than ensuring that they strictly adhered to the target trajectory. Moreover, when the joint angle returned to the normal range, i.e., when the shoulder moved into the abduction space, the trajectory refinement resumed its focus on aligning the trajectory with the target trajectory within the dynamic constraints. These results confirm that the proposed method is capable of tracking changes in human motion intention within the normal activity range and refining the trajectory accordingly.

For active mirroring training, we evaluated our method by using the upper-limb exoskeleton in a healthy subject by mapping the motion of the left arm (i.e., the healthy side) to that of the right arm (i.e., the mock stroke side). Acquiring precise angles of human upper limbs through an optical motion capture system requires a custom-made suit, which was beyond the scope of this study. Instead, we employed a self-manufactured brace embedded with key markers and gloves fitted with markers to estimate the angles at three limb joints, specifically those corresponding to Joint 1, 2, and 4. However, due to brace deformation and marker obstruction, the accuracy of the upper-limb joint angles was limited to within a certain range. Thus, to ensure safety and demonstrate the capability of our method to impose constraints on upper-limb movements, we set position constraints for Joint 1, 2, and 4 as $[-40^\circ, 10^\circ]$, $[-10^\circ, 80^\circ]$, and $[0^\circ, 60^\circ]$, respectively.

As shown in Fig. 5.26, the desired trajectory was consistently aligned with the trajectory of the human limb $\boldsymbol{q}_h$ and was refined according to dynamic constraints. Moreover, when the unaffected side of the body moved rapidly to a position outside the established movement boundary, the proposed trajectory refinement was swiftly adjusted. That is, a predetermined maximum speed was implemented to prevent the set joint position boundaries from being exceeded. When the movements of the joints on the unaffected side of the body remained within these dynamic constraints, the upper-limb exoskeleton robot tracked the trajectory of the healthy limb. Simultaneously and throughout the training session, our variable impedance controller effectively sustained the human-robot interaction within the desired impedance model.

Passive Following Training: We conducted a series of experiments to validate the efficiency of the proposed individualization framework on the passive following training task. These experiments were centered on a typical task in upper-limb rehabilitation: raising the upper limb to a fixed point. This task requires coordination between the shoulder and elbow, specifically involving Joint 1, 2, and 4. During the cost calculation, the anomaly score was scaled to have the same magnitude as the tracking error, and the hyperparameter λ_p was set to 0.003. Moreover, as the exploration part of individualized assistance largely consists of initial trajectory sampling based on a demonstration, we set $N_s = 40$. The training process was designed to stop when the RMSE of the mean trajectory of the distribution between two consecutive iterations decreased to less than 0.1° or when the maximum number of iterations was reached. Furthermore, the sampling space was greatly reduced by the demonstration data. The number of sample sets was considered to be sufficient for exploration of the current task based on practical experience.

First, we conducted ablation studies with a healthy patient present to validate the efficacy of the online refinement module in passive following training scenarios. In addition, five restarts were executed to bypass local optima. The results of these studies are depicted in Fig. 5.27. Within the figure, the black dashed line illustrates the trajectory that achieved the lowest cost during the exploration phase, while the red solid line represents the mean of the assistive trajectory distribution. Without online trajectory refinement, the experiment spanned 67 iterations, whereas with online trajectory refinement, the experiment spanned 55 iterations. The experimental results reveal that without online refinement in passive following training individualization, the converged assistive trajectory distribution exhibited a large variance, and the trajectory with the lowest cost significantly deviated from the mean of the distribution. This indicates that the trajectory distribution with similar costs was rather wide. In contrast, with online refinement in passive following training individualization, the converged assistive trajectory distribution had a smaller variance, suggesting that there was a more concentrated set of trajectories with similar costs. This concentration is largely attributable to the use of sensor feedback and anomaly scores in the online refinement, thereby facilitating real-time adjustments to the assistive trajectory. This process effectively reduced the uncertainty of motion intentions and minimized conflicts in human-robot interaction. Thus, the precision with which the cost function evaluated and distinguished the performance of different trajectories was increased, leading to the overall effectiveness of the training being enhanced.

The effectiveness of the individualized assistance was assessed using four metrics: the tracking RMSE, the anomaly score, the cost (5.37), and the EMG signal level of the biceps brachii. The results of this evaluation are displayed in Fig. 5.28. To allow comparison, all metrics were averaged over one cycle and normalized relative

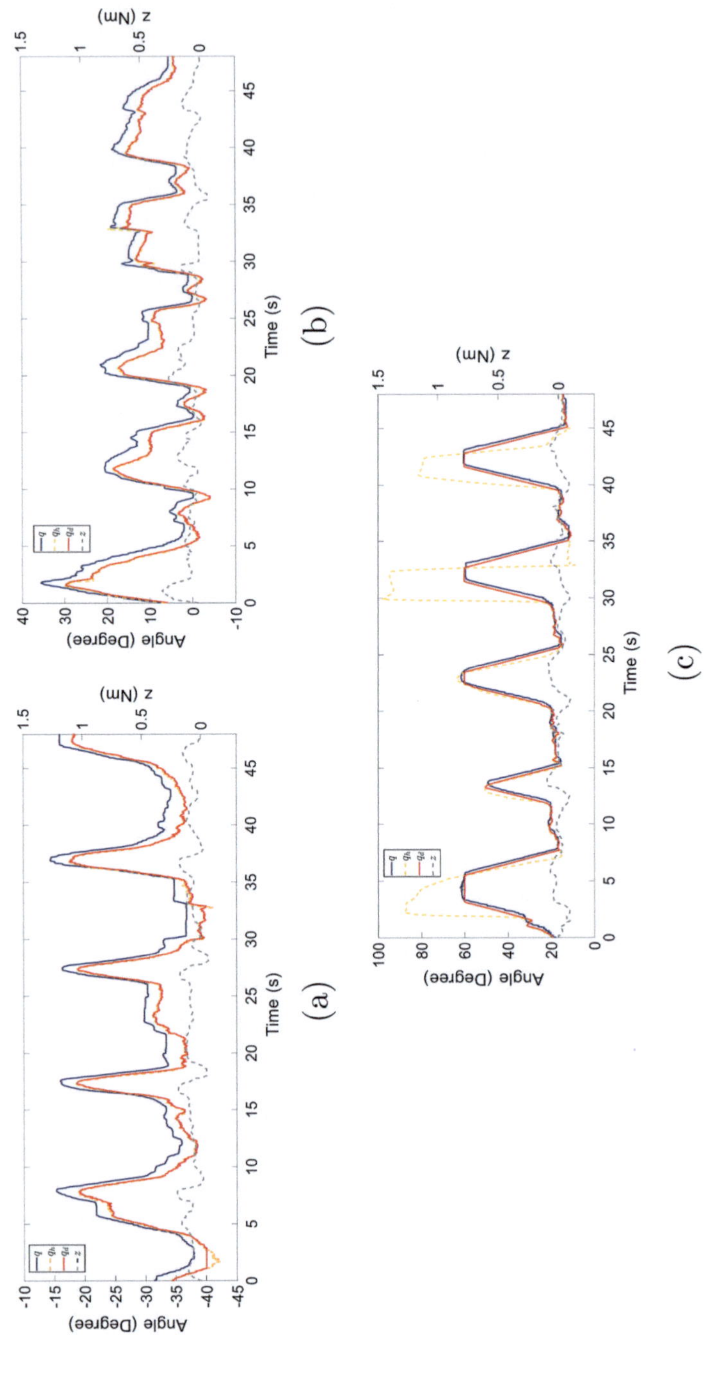

Fig. 5.26 a–c Performance of active mirroring training on Joints 1, 2, and 4 of the upper-limb exoskeleton robot

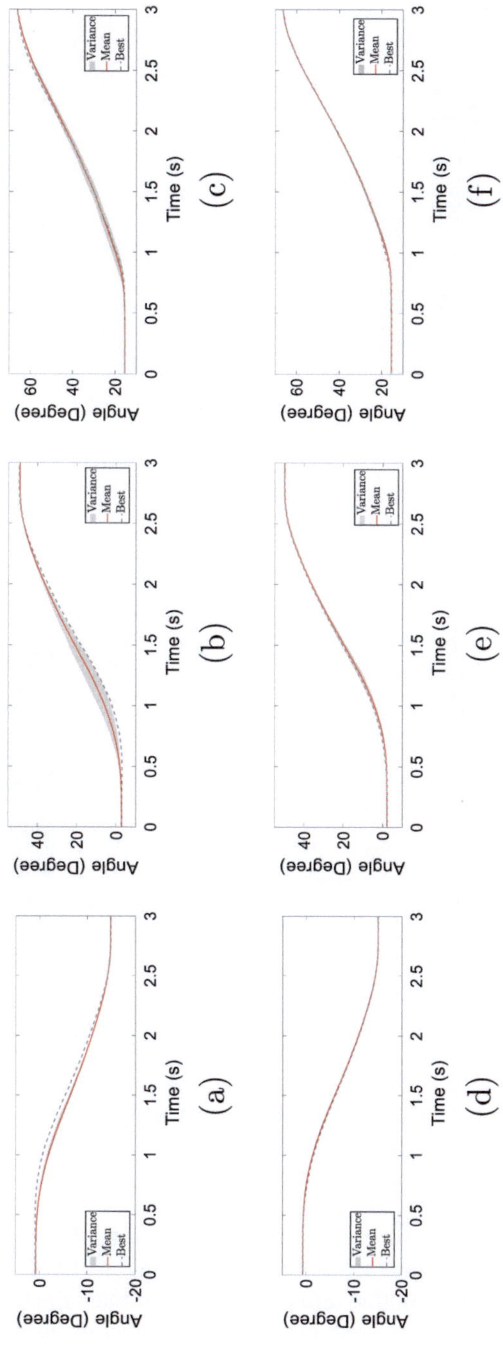

Fig. 5.27 Optimized assistance distribution **a–c** without online refinement and **d–f** with online refinement

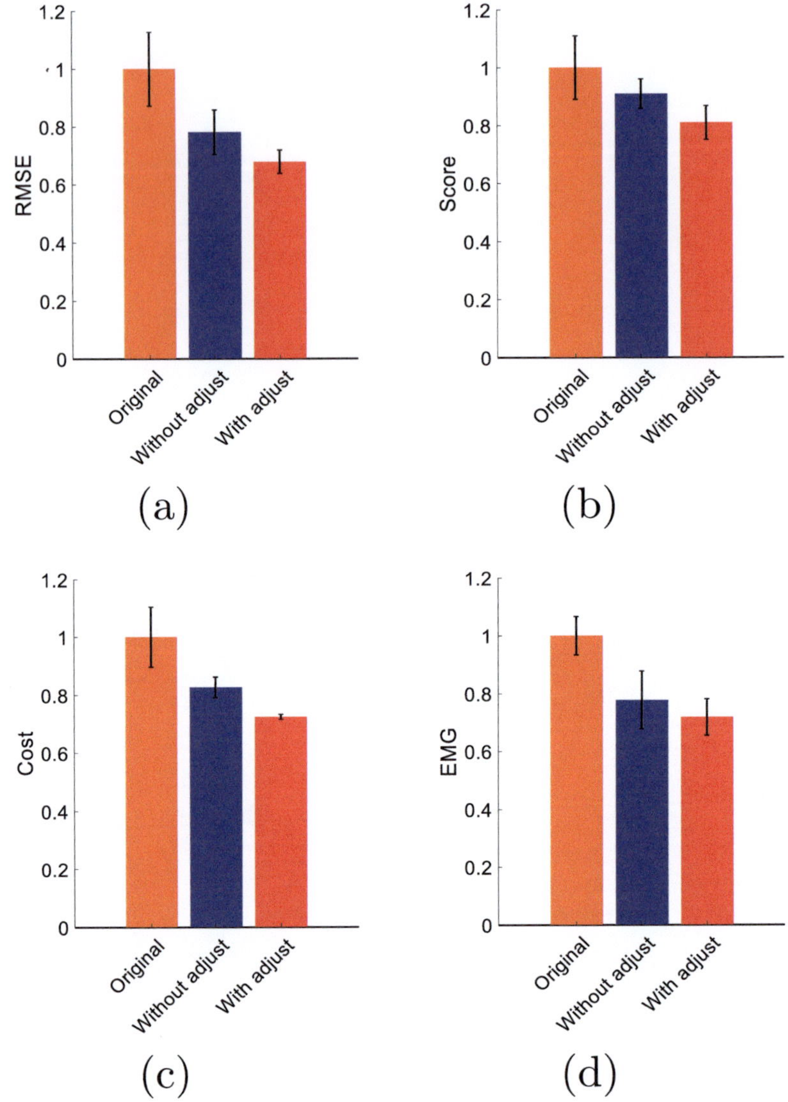

Fig. 5.28 a–d Evaluation results of passive following training showing the metrics: RMSE, anomaly score, cost, and EMG signal level of the biceps brachii. All metrics were averaged over one cycle and normalized relative to the original assistance

to the original assistance. Compared with the other assistance (i.e., the mean trajectory of the demonstrations and the optimized assistance without online refinement), the assistance incorporating online refinement demonstrated the best performance in improving tracking accuracy and reducing the anomaly score during motion. Furthermore, implementing online refinement resulted in the lowest cost following passive

Table 5.17 Attributes of participants

Grp.	Subj.	Gen.	Age (y)	Wt. (kg)	Ht. (cm)	Arm L. (cm)	Dx.
CG	1	Male	46	70	176	61	CH
	2	Male	60	65	167	57	Stroke
	3	Male	62	75	170	58	Stroke
EG	4	Male	44	67	169	59	Stroke
	5	Male	51	77	177	64	Stroke
	6	Male	66	66	167	57	Stroke

training and a reduction in the EMG signal level. These outcomes substantiate the efficacy of the proposed individualization framework for passive following training.

We also recruited seven patients for clinical trials to demonstrate the effectiveness of the proposed method. As one participant was transferred to another hospital, only six participants completed the trial. All the patients had a healthy side that did not move naturally, due to the effects of their stroke. Hence, passive following training was employed. The participants were allocated to either a control group or an experimental group, and their details are summarized in Table 5.17. Both groups engaged in regular daily rehabilitation exercises, and the experimental group participated in an additional 14 days of passive following training.

The rehabilitation task was set to be the same as the previously described task of raising the arm, involving the coordinated movement of Joint 1, 2, and 4, as illustrated in Fig. 5.29. Interactive information was recorded during the limb-lifting phase, which was considered as the training task, while the limb-lowering phase, which began at $t = 3.5$ s, was used to return to the initial position. The motor abilities of both groups were evaluated and scored by professional healthcare personnel using specific evaluation metrics. These metrics were the muscle tone level and the Fugl-Meyer assessment (FMA) score [10]. A low level of muscle tone and a high FMA score are indicative of good upper-limb motor ability. Both groups were subjected to a motor function assessment before treatment. During the treatment phase, the experimental group participated in daily passive-following training sessions, each of which lasted approximately 15 min. Both groups underwent reassessment 2 weeks after the start of the treatment. The experimental results normalized to the initial evaluation are displayed in Fig. 5.29. In this clinical trial, muscle tone levels were assessed using a rating scale with grades 0, 1, 1+, 2, 3, and 4, where grade 0 indicates normal muscle tone. For the purposes of numerical analysis, grade 1+ is quantified as 1.5. The evaluation results are documented in Table 5.18. All assessments were exclusively focused on the upper limb. In the upper limb segment of the FMA, the maximum score, indicating normal function, is 66 points. Unlike the control group, the experimental group exhibited significant improvements in all metrics compared with their initial assessments before the treatment. Thus, compared with the results of the control group, the results of the experimental group indicate that the passive

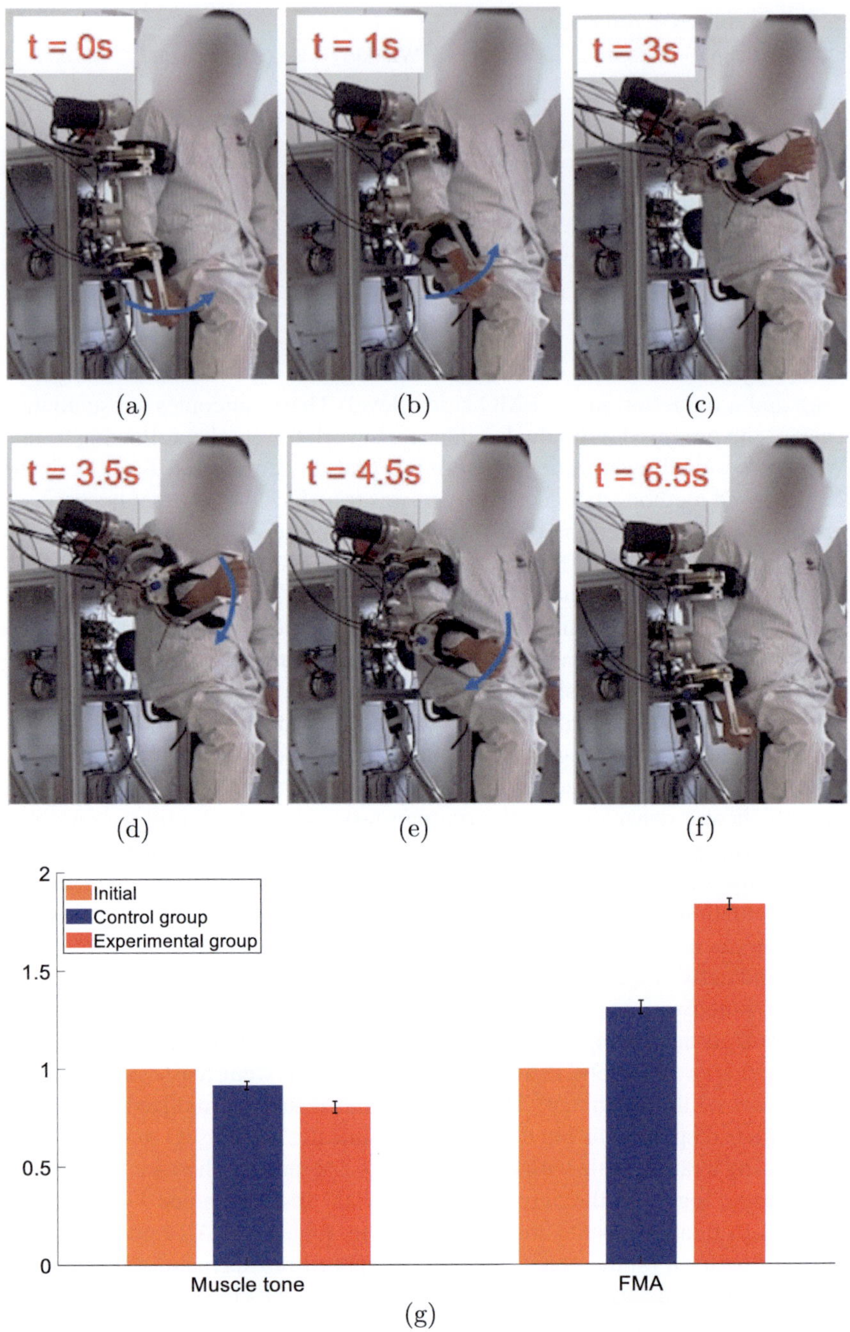

Fig. 5.29 **a–f** Snapshots of upper-limb movement during training, with red arrows indicating the direction of movement; **g** comparative results of the clinical trial normalized to the initial evaluation

Table 5.18 Detailed Clinical Evaluation

Grp.	Subj.	Init. Tone	Post Tone	Init. FMA	Post FMA
CG	1	1.5	1.5	8	12
	2	1.5	1.5	31	35
	3	2	1.5	23	30
EG	4	1.5	1	8	16
	5	2	1.5	19	35
	6	2	2	3	5

following training with the upper-limb exoskeleton robot accelerated the recovery of motor functions. Therefore, this training could enhance the effectiveness of treatment for conditions such as stroke and cerebral hemorrhage.

5.3 Summary and Notes

The integration of exoskeletons in rehabilitation training offers significant advantages, including reducing the workload of healthcare professionals and enabling easily customizable training programs tailored to individual patients. Unlike traditional rehabilitation methods that rely heavily on therapist intervention, robotic exoskeletons ensure consistent, repetitive motion patterns essential for neuromuscular recovery. Additionally, they allow for precise monitoring and adaptive control, ensuring patients receive optimal support throughout their rehabilitation process. The focus of rehabilitation differs between the upper and lower limbs. Lower-limb rehabilitation emphasizes restoring walking ability and balance, while upper-limb rehabilitation targets fine motor skills and functional reach movements essential for daily activities.

Existing lower-limb rehabilitation exoskeletons primarily rely on direct-driven actuation [5], often providing assistance to only a single joint, such as the hip [28] or knee [29], limiting their adaptability across different gait impairments. Furthermore, their generalization capability across different patients remains poor as they struggle to accommodate varying anthropometric characteristics and neurological conditions. In contrast, our developed system features multi-joint actuation for comprehensive lower-limb rehabilitation, integrating intelligent functionalities such as gait recognition and synchronization, cloud-computing-based diagnosis, and individualized gait generation. This advanced system was evaluated on 30 stroke patients, demonstrating its effectiveness in promoting motor recovery and improving walking ability.

Similarly, existing upper-limb rehabilitation exoskeletons often adopt either SEA [30] or cable-driven mechanisms [31] but struggle to balance dynamic performance and compliance. Many systems provide assistance only to a single joint, such as the shoulder [32] or elbow [33], limiting their ability to support complex, multi-joint movements required for functional rehabilitation. To address these challenges, our proposed exoskeleton features six degrees of freedom and employs a cable-driven serial elastic actuation system, which enhances movement compliance while reducing inertia effects during motion. Additionally, it offers both active and passive rehabilitation modes, making it adaptable for stroke patients with varying levels of motor impairment, ultimately delivering personalized rehabilitation assistance to optimize recovery outcomes.

References

1. L. Liu, D. Wang, K.L. Wong, Y. Wang, Stroke and stroke care in china: huge burden, significant workload, and a national priority. Stroke **42**(12), 3651–3654 (2011)
2. Milebot, "Milebot official website," n.d. Accessed: YYYY-MM-DD
3. A. Tsukahara, Y. Hasegawa, K. Eguchi, Y. Sankai, Restoration of gait for spinal cord injury patients using hal with intention estimator for preferable swing speed. IEEE Trans. Neural Syst. Rehabil. Eng. **23**(2), 308–318 (2014)
4. L. Brenner, Exploring the psychosocial impact of ekso bionics technology. Arch. Phys. Med. Rehabil. **97**(10), e113 (2016)
5. A. Esquenazi, M. Talaty, A. Packel, M. Saulino, The rewalk powered exoskeleton to restore ambulatory function to individuals with thoracic-level motor-complete spinal cord injury. Am. J. Phys. Med. Rehabil. **91**(11), 911–921 (2012)
6. J. Zhang, T. Wang, Q. Zhao, S. Liu, Impacts of a lower limb exoskeleton robot on the muscle strength of tibialis anterior muscle in stroke patients, in *E3S Web of Conferences*, vol. 185 (EDP Sciences, 2020), p. 03036
7. G.A. Pratt, M.M. Williamson, Series elastic actuators, in *Proceedings 1995 IEEE/RSJ International Conference on Intelligent Robots and Systems. Human Robot Interaction and Cooperative Robots*, vol. 1 (IEEE, 1995), pp. 399–406
8. G. Chen, P. Qi, Z. Guo, H. Yu, Gait-event-based synchronization method for gait rehabilitation robots via a bioinspired adaptive oscillator. IEEE Trans. Biomed. Eng. **64**(6), 1345–1356 (2016)
9. J. Ye, H. Wu, L. Wu, J. Long, Y. Zhang, G. Chen, C. Wang, X. Luo, Q. Hou, Y. Xu, An adaptive method for gait event detection of gait rehabilitation robots. Front. Neurorobot. **14**, 38 (2020)
10. A.R. Fugl-Meyer, L. Jääskö, I. Leyman, S. Olsson, S. Steglind, A method for evaluation of physical performance. Scand. J. Rehabil. Med. **7**(1), 13–31 (1975)
11. Q. Wu, X. Wang, F. Du, Development and analysis of a gravity-balanced exoskeleton for active rehabilitation training of upper limb. Proc. Inst. Mech. Eng. C J. Mech. Eng. Sci. **230**(20), 3777–3790 (2016)
12. Y. Zimmermann, J. Song, C. Deguelle, J. Läderach, L. Zhou, M. Hutter, R. Riener, P. Wolf, Human–robot attachment system for exoskeletons: Design and performance analysis. IEEE Trans. Robot. (2023)
13. Y. Mao, S.K. Agrawal, Design of a cable-driven arm exoskeleton (carex) for neural rehabilitation. IEEE Trans. Rob. **28**(4), 922–931 (2012)
14. M. Quigley, K. Conley, B. Gerkey, J. Faust, T. Foote, J. Leibs, R. Wheeler, A.Y. Ng et al., Ros: an open-source robot operating system, in *ICRA Workshop on Open Source Software*, vol. 3 (Japan, Kobe, 2009), p .5
15. A. Albu-Schäffer, C. Ott, G. Hirzinger, A unified passivity-based control framework for position, torque and impedance control of flexible joint robots. Int. J. Robot. Res. **26**(1), 23–39 (2007)
16. M.W. Spong, Modeling and control of elastic joint robots (1987)
17. J. Ho, A. Jain, P. Abbeel, Denoising diffusion probabilistic models. Adv. Neural. Inf. Process. Syst. **33**, 6840–6851 (2020)

18. A. Paraschos, C. Daniel, J.R. Peters, G. Neumann, Probabilistic movement primitives, *Advances in Neural Information Processing Systems*, vol. 26 (2013)
19. G. Williams, N. Wagener, B. Goldfain, P. Drews, J.M. Rehg, B. Boots, E.A. Theodorou, Information theoretic mpc for model-based reinforcement learning, in *2017 IEEE International Conference on Robotics and Automation (ICRA)* (IEEE, 2017), pp. 1714–1721
20. C.C. De Wit, H. Olsson, K.J. Astrom, P. Lischinsky, A new model for control of systems with friction. IEEE Trans. Autom. Control **40**(3), 419–425 (1995)
21. X. Zhang, Y. Shu, Y. Chen, G. Chen, J. Ye, X. Li, X. Li, Multi-modal learning and relaxation of physical conflict for an exoskeleton robot with proprioceptive perception, in *2023 IEEE International Conference on Robotics and Automation (ICRA)* (IEEE, 2023), pp. 10490–10496
22. A.N. Tikhonov, Systems of differential equations containing small parameters in the derivatives. Matematicheskii Sbornik **73**(3), 575–586 (1952)
23. Y. Chen, S. Miao, G. Chen, J. Ye, C. Fu, B. Liang, S. Song, X. Li, Learning to assist different wearers in multitasks: efficient and individualized human-in-the-loop adaptation framework for lower-limb exoskeleton. IEEE Trans. Robot. (2024)
24. Y. Zimmermann, E. B. Küçüktabak, F. Farshidian, R. Riener, M. Hutter, Towards dynamic transparency: Robust interaction force tracking using multi-sensory control on an arm exoskeleton, in *2020 IEEE/RSJ International Conference on Intelligent Robots and Systems (IROS)* (IEEE, 2020), pp. 7417–7424
25. T. Salzmann, B. Ivanovic, P. Chakravarty, M. Pavone, Trajectron++: Dynamically-feasible trajectory forecasting with heterogeneous data, in *Computer Vision–ECCV 2020: 16th European Conference, Glasgow, UK, August 23–28, 2020, Proceedings, Part XVIII 16* (Springer, 2020), pp. 683–700
26. J. Song, C. Meng, S. Ermon, Denoising diffusion implicit models (2020). arXiv:2010.02502
27. T.-Y. Kim, S.-B. Cho, Predicting residential energy consumption using cnn-lstm neural networks. Energy **182**, 72–81 (2019)
28. K. Seo, J. Lee, Y. Lee, T. Ha, Y. Shim, Fully autonomous hip exoskeleton saves metabolic cost of walking, in *2016 IEEE International Conference on Robotics and Automation (ICRA)* (IEEE, 2016), pp. 4628–4635
29. N.V. Divekar, G.C. Thomas, A.R. Yerva, H.B. Frame, R.D. Gregg, A versatile knee exoskeleton mitigates quadriceps fatigue in lifting, lowering, and carrying tasks. Sci. Robot. **9**(94), eadr8282 (2024)
30. A. Ebrahimi, D. Gröninger, R. Singer, U. Schneider, Control parameter optimization of the actively powered upper body exoskeleton using subjective feedbacks, in *2017 3rd International Conference on Control, Automation and Robotics (ICCAR)* (IEEE, 2017), pp. 432–437
31. J.C. Perry, J. Rosen, S. Burns, Upper-limb powered exoskeleton design. IEEE/ASME Trans. Mechatron. **12**(4), 408–417 (2007)
32. A. Nasr, S. Bell, J. McPhee, Optimal design of active-passive shoulder exoskeletons: A computational modeling of human-robot interaction. Multibody Syst. Dyn. **57**(1), 73–106 (2023)
33. T. Chen, R. Casas, P.S. Lum, An elbow exoskeleton for upper limb rehabilitation with series elastic actuator and cable-driven differential. IEEE Trans. Rob. **35**(6), 1464–1474 (2019)

Chapter 6
Emerging Technologies and Future Directions

In previous chapters, we have primarily focused on the generation of assistive trajectories for exoskeletons and the implementation of interaction control between the exoskeleton and the wearer. These efforts addressed fundamental aspects of motor control and user safety, ensuring that exoskeletons can provide consistent and effective assistance during rehabilitation or augmentation tasks. However, the discussion so far has largely revolved around low-level control strategies and trajectory optimization, which, while essential, do not fully explore the potential of integrating advanced technologies for higher-level planning and decision-making.

This chapter aims to expand the scope by examining how emerging technologies, particularly Large Language Models (LLMs), can enhance the task-level planning and adaptability of exoskeleton robots. Unlike previous sections that focused on trajectory generation and interaction control, this chapter delves into how LLMs can improve the semantic understanding and decision-making capabilities of exoskeletons, enabling them to respond to complex, varied scenarios in a generalized manner. The motivation for this chapter arises from the need to address challenges in generalization and adaptability. Many real-world applications require exoskeletons to seamlessly interact with dynamic environments and diverse user intentions. For instance, recognizing various terrains in gait assistance, distinguishing between objects of different properties during manipulation, and adapting to collaborative tasks in crowded or unpredictable environments all demand a higher level of contextual awareness.

To deliver comprehensive assistance to humans, the development of large language models (LLMs) has introduced advanced reasoning capabilities and a deep understanding of semantic information to robotic systems. These models enable logical analysis and step-by-step responses to described tasks [1] or directly generate motion commands from the action space [2]. However, in the context of exoskeleton robots, effectively integrating human intention with the planned actions derived from LLMs remains an open challenge. Current approaches primarily utilize LLMs as natural language processing modules, facilitating voice-based input for motion control [3]. For instance, verbal commands can adjust exoskeleton joint angles [4]. While functional, these methods often overlook the rich semantic information embedded

X. Li and H. Yu, *Compliantly-Driven Robots with Application to Exoskeletons*,
https://doi.org/10.1007/978-981-96-9544-7_6

in task descriptions provided by language. Such semantic differences are critical for tailoring assistance to be both comprehensive and generalized across varied scenarios. Leveraging LLMs to fully utilize this semantic information holds the potential to significantly enhance the breadth and adaptability of exoskeleton assistance.

By taking the advantages of emerging technologies such as LLMs, this chapter introduces a semantic-aware pipeline for upper-limb exoskeletons that is capable of assisting the wearer in a safe manner, allowing the wearer's motion intentions to intervene and adjust the current assistance strategy during the assistance process.

6.1 LLM-Driven Exoskeleton

The integration of LLMs into exoskeleton systems presents promising opportunities for enhancing the generalization capabilities of robotic assistance. However, achieving seamless and adaptive assistance for upper-limb exoskeletons remains a complex challenge due to the following factors:

(1) Unlike manipulators, where LLMs are implemented in isolation [1], exoskeletons operate in collaboration with human users. This requires the exoskeleton to identify the wearer's motion intentions before activating assistance. Furthermore, the system must incorporate a replanning mechanism to address conflicts between the human's intention and the provided assistance.
(2) Since the exoskeleton augments the wearer's movements, task planning must incorporate human factors. Semantic information from the task environment needs to be accurately interpreted and used to reason solutions that align with human needs and perspectives.
(3) During movement execution, the exoskeleton must thoroughly account for potential risks associated with limb movements. Real-time regulation of human-robot interaction is essential to ensure safety throughout the assistance process.

To tackle these challenges, we will introduce a semantic-aware pipeline for upper-limb assistance, as illustrated in Fig. 6.2. This approach aims to leverage LLMs to enhance the adaptability and intelligence of exoskeleton systems by enabling them to understand and respond to task descriptions effectively. The proposed pipeline is structured around two main phases: task planning and task execution. During the planning phase, an Intent-integrative Grasping mechanism allows the exoskeleton to operate in a transparent mode, facilitating reliable grasping that aligns with the wearer's intentions. This stage also incorporates the anomaly detector introduced in Chap. 4, where the anomaly score serves as an unsupervised metric to evaluate the quality of human-robot interaction. When the anomaly score exceeds a predefined threshold, it triggers a replanning mechanism to adjust the task strategy accordingly (Fig. 6.1).

The execution phase builds upon the trajectory refinement techniques discussed in Chap. 5, where the assistive trajectory is generated in accordance with the semantic constraints derived from the task description. For example, if the task involves

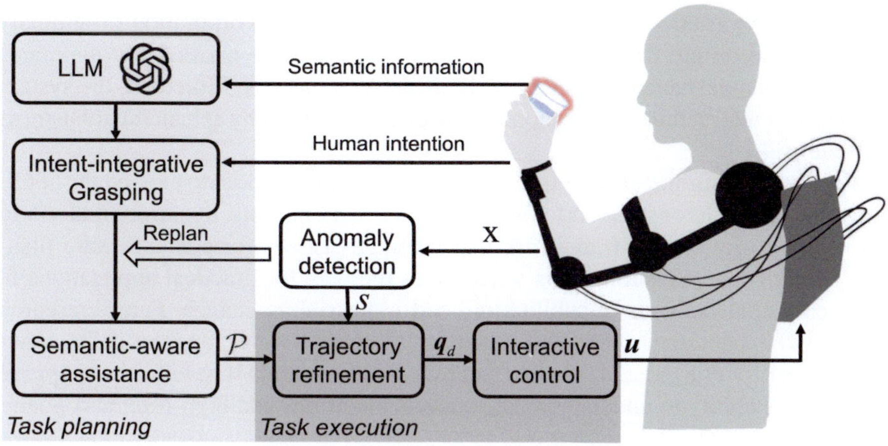

Fig. 6.1 Block diagram of the proposed method: Semantic information is extracted by the LLM to enable task planning. Human intention intervenes during the grasping phase and subsequently receives assistance from the exoskeleton. The anomaly score s, computed from sensory feedback $\mathbf{x}$, guides online trajectory refinement and serves as a trigger for replanning, enhancing robustness to external anomalies. The reliable interactive control is ensured by an impedance controller designed to account for the cable-driven SEA mechanism

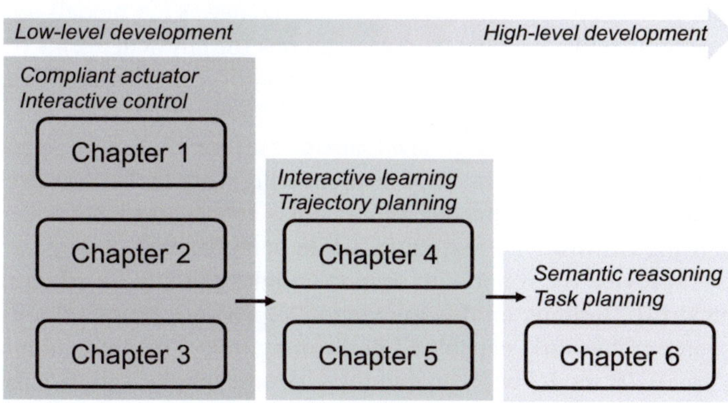

Fig. 6.2 The conceptual flow of all the chapters

carrying a heavy object, a higher impedance parameter is applied to ensure stability, whereas handling fragile items requires a lower speed constraint to prevent accidental damage. The assistive trajectory is then implemented through an impedance controller, enabling precise and adaptive upper-limb assistance. The integration of LLMs ensures that the exoskeleton can automatically interpret task descriptions and generate appropriate assistive strategies based on the semantic information provided.

The proposed approach provides a comprehensive framework that addresses the generalization challenges associated with exoskeleton assistance. By utilizing LLMs for automated task planning and execution, the system achieves a high level of adaptability across varied scenarios and user intentions. The incorporation of anomaly detection and online trajectory refinement ensures that the exoskeleton maintains safety and responsiveness throughout the assistance process. Moreover, the system enables user intervention when anomalies are detected, thereby enhancing robustness and safety.

The research significance of this approach lies in its potential to expand the application of exoskeletons beyond controlled environments into dynamic, real-world settings. By providing a framework that leverages LLMs for context-aware planning and adaptive execution, this work paves the way for practical applications in home-care environments, rehabilitation, and industrial assistance. Future research will aim to expand the set of motion patterns to cover a wider range of daily activities and incorporate external sensors, such as cameras and IMUs, to further improve the system's ability to understand the wearer's intentions and high-level task goals.

6.2 Vision-Language-Action Model Towards Exoskeleton

With the recent advancements in LLMs, a new paradigm known as the Vision-Language-Action (VLA) model has emerged at the intersection of language reasoning, visual understanding, and embodied intelligence. VLA models are designed to generate generalizable and context-aware actions for physical agents by integrating natural language instructions and visual perception. In contrast to traditional pipeline-based robotic systems, VLA models unify perception, cognition, and control, allowing robots to interpret high-level semantic information and respond to open-ended tasks in dynamic environments. The pioneering work in this domain, Google's RT-2, introduced a VLA architecture that enables robots to identify and manipulate objects using LLM-driven reasoning, while being trained on large-scale internet datasets to provide the robot with semantic understanding based on real-world human context [2]. Building on this foundation, OpenVLA proposed a lightweight and open-source framework, capable of performing object manipulation tasks with a relatively small 7B model while maintaining strong generalization across unseen items [5]. More recently, $\pi 0$ [6] advanced the VLA framework by incorporating flow match techniques to construct an auxiliary action generation network that leverages both high-level semantic goals and robot proprioception feedback, significantly improving action robustness and context-awareness. The VLA paradigm represents a promising direction for robotics by enabling agents to operate in unstructured and semantically rich environments, thereby expanding their adaptability and autonomy across a broader range of real-world applications.

Integrating VLA models into exoskeleton systems presents both conceptual opportunities and domain-specific constraints. Unlike traditional applications of VLA in robotics, where manipulation is typically executed by end effectors such as grippers, exoskeletons function within the joint space of the human body, necessitating fine-grained alignment with user limb motion. This introduces higher control complexity and demands stronger embodiment to interpret human intention. While recent progress in embodied AI has shown the capability of systems to perform complex multimodal tasks by fusing visual inputs and language commands, its application in exoskeletons is still emerging. However, current embodied AI technologies already demonstrate their value in healthcare-related domains such as physical assistance and rehabilitation, with exoskeletons used to support locomotion in individuals with spinal cord injuries [7]. By extending the reasoning and perception capabilities of VLA models to the domain of physical human-robot collaboration, such systems could support more responsive and personalized assistance strategies. Rather than relying on predefined trajectories, VLA-driven exoskeletons could dynamically generate assistive actions, adapting to novel user tasks such as picking up unknown items in a kitchen, assisting with dressing, or supporting arm movements during grooming or eating. While current VLA models demonstrate strong capabilities in language and vision-based reasoning, they typically do not account for physical interaction forces. However, when applied to exoskeletons, incorporating such contact forces is essential, as they are a critical component of the multimodal information required for safe and responsive human-robot interaction. By integrating this additional layer of physical awareness, VLA-driven exoskeletons can better interpret user intent and adapt assistance strategies. This advancement could shift exoskeletons from structured rehabilitation tools to versatile assistive agents capable of supporting a wide range of daily activities and home-care scenarios.

6.3 Summary and Notes

In this chapter, we propose a semantic-aware framework for exoskeleton robots that provides comprehensive upper-limb assistance through an intent-integrative and safety-assured framework. The assistance process incorporates anomaly detection and online trajectory refinement to ensure safe and adaptive human-robot interaction. Furthermore, this chapter explores the potential of integrating Vision-Language-Action models into exoskeleton systems, highlighting their ability to unify semantic understanding and embodied intelligence. By leveraging language and vision inputs, VLA-driven exoskeletons can move beyond predefined behaviors to generate context-aware assistance. This approach offers a promising path towards highly adaptive, generalized support in diverse daily-life scenarios, ultimately transforming exoskeletons into intelligent, versatile companions.

The book *Compliantly-Driven Robots with Application to Exoskeletons* is devoted to exploring how compliant actuation technologies can be systematically integrated into the design, control, and intelligent enhancement of exoskeleton robots. By focusing on the intersection between physical flexibility and human-centered interaction, this work aims to promote safer, more adaptive, and context-aware robotic assistance. Compliant actuators, particularly SEAs, play a vital role in achieving both physical safety and smooth motion in wearable systems. Through the lens of exoskeletons—a class of robots that closely interact with the human body—this book surveys the state-of-the-art from low-level actuator principles to emerging AI-driven interaction strategies. The conceptual flow illustrated in Fig. 6.2 highlights the progressive and hierarchical structure of this book, reflecting how exoskeleton systems evolve from foundational hardware principles to advanced semantic intelligence. Chapter 1–3 form the low-level development stage, where the groundwork is laid through the study of compliant actuators and robot-level control strategies. These chapters establish the mechanical and interactive reliability necessary for safely and effectively interfacing with the human body. Building on this, Chaps. 4 and 5 mark a transition to mid-level development, focusing on interactive learning and trajectory planning. These techniques enable exoskeletons to adapt in real time based on user feedback and task context, enhancing personalization and task-specific adaptability. Finally, this chapter represents the high-level development stage, introducing the integration of semantic reasoning and task planning through large language models. This chapter projects a vision in which exoskeletons not only respond to predefined movements but also comprehend high-level tasks, interpret natural language instructions, and autonomously generate context-aware assistance strategies.

Taken together, these six chapters present a coherent framework for the comprehensive development of exoskeleton robots. The structure moves step-by-step from low-level compliant control to high-level intelligent behavior, reflecting the natural progression of robotic capability in human-centric applications. Each stage builds upon the former: low-level safety and control enable adaptive learning, which in turn enables semantic-level decision making. With the rapid development of foundation models like LLMs and Vision-Language-Action systems, exoskeletons are expected to go beyond their traditional functions. In the future, these systems may assist users in a wide range of everyday and unpredictable situations, such as helping with housework, supporting teamwork in the workplace, or adjusting to the specific needs of elderly care. Ultimately, exoskeletons could become intelligent and dependable partners that extend and enhance human abilities in daily life.

References

1. A. Brohan, Y. Chebotar, C. Finn, K. Hausman, A. Herzog, D. Ho, J. Ibarz, A. Irpan, E. Jang, R. Julian, et al., Do as i can, not as i say: Grounding language in robotic affordances, in *Conference on Robot Learning* (PMLR, 2023), pp. 287–318
2. B. Zitkovich, T. Yu, S. Xu, P. Xu, T. Xiao, F. Xia, J. Wu, P. Wohlhart, S. Welker, A. Wahid et al., Rt-2: Vision-language-action models transfer web knowledge to robotic control, in *Conference on Robot Learning* (PMLR, 2023), pp. 2165–2183
3. R. Parameswari, C. Di Natali, D.G. Caldwell, J. Ortiz, et al., Assessment and benchmarking of xonli: a natural language processing interface for industrial exoskeletons, in *2024 IEEE International Conference on Robotics and Automation (ICRA)* (IEEE, 2024), pp. 3333–3340
4. W. Chen, G. Li, M. Li, W. Wang, P. Li, X. Xue, X. Zhao, L. Liu, Llm-enabled incremental learning framework for hand exoskeleton control. IEEE Trans. Autom. Sci. Eng. (2024)
5. M.J. Kim, K. Pertsch, S. Karamcheti, T. Xiao, A. Balakrishna, S. Nair, R. Rafailov, E. Foster, G. Lam, P. Sanketi et al., Openvla: An open-source vision-language-action model (2024). arXiv:2406.09246
6. K. Black, N. Brown, D. Driess, A. Esmail, M. Equi, C. Finn, N. Fusai, L. Groom, K. Hausman, B. Ichter et al., *pi*_0: A vision-language-action flow model for general robot control (2024). arXiv:2410.24164
7. Y. Liu, X. Cao, T. Chen, Y. Jiang, J. You, M. Wu, X. Wang, M. Feng, Y. Jin, J. Chen, A survey of embodied ai in healthcare: Techniques, applications, and opportunities (2025). arXiv:2501.07468

Appendix

Dynamic Analysis of SEA

In this section, we analyze the dynamic characteristics of a Series Elastic Actuator and examine the performance enhancements introduced by the impedance controller. The control strategy, based on singular perturbation theory, relies on accurate knowledge of the system parameters. Under this assumption, we consider that the nonlinearities present in the model can be fully compensated, allowing the focus to be placed on a simplified linear SEA dynamic system, formulated as follows:

$$M\ddot{q} = K(\theta - q), \tag{A.1}$$

$$B\ddot{\theta} + K(\theta - q) = u. \tag{A.2}$$

where M and B are positive constants representing the mass at the joint and motor sides, respectively. The variable u denotes the system input, while θ and q represent the motor and joint positions. The parameter K corresponds to the stiffness of the series elastic element.

Applying the Laplace transform to the above equations yields the transfer function between the input u and the joint position q:

$$D(s) = \frac{Q(s)}{U(s)} = \frac{K}{BMs^4 + K(M + B)s^2} \tag{A.3}$$

Under this simplified dynamic model, the original impedance controller can be reformulated as a position controller, expressed by:

$$u = -K_v(\dot{\theta} - \dot{q}) - K_z z + (M + B)\ddot{q}_r, \tag{A.4}$$

$$z = \dot{q} - \dot{q}_r, \tag{A.5}$$

$$\dot{q}_r = \dot{q}_d - K_q(q - q_d), \tag{A.6}$$

X. Li and H. Yu, *Compliantly-Driven Robots with Application to Exoskeletons*,
https://doi.org/10.1007/978-981-96-9544-7

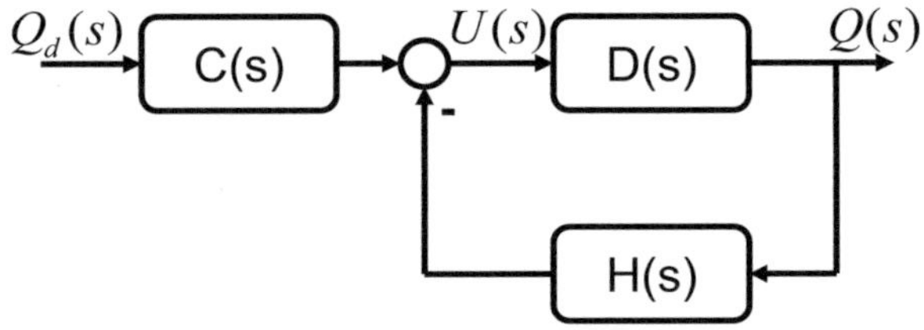

Fig. A.1 Block diagram representation of the SEA dynamic system in the frequency domain

Table A.1 Simulation parameters

Parameter	M	B	K	K_q	K_z	K_v
Value	1	0.01	100	100	1	10

where K_v, K_z, and K_q are positive controller gains. Here, z denotes the impedance error, while q_d and q_r represent the desired and reference trajectories, respectively. Transforming the control law into the frequency domain leads to the following expression for the control input $U(s)$:

$$\begin{aligned} U(s) =&(Ms^2 + (MK_q + K_z)s + K_zK_q)Q_d(s) \\ &- (\frac{K_vM}{K}s^3 + (MK_q + K_z)s + K_zK_q)Q(s), \\ =&C(s)Q_d(s) - H(s)Q(s), \end{aligned} \tag{A.7}$$

where $C(s)$ and $H(s)$ represent polynomials characterizing the feedforward and feedback paths, respectively.

The structure of the SEA system combined with the impedance controller is illustrated in Fig. A.1, showing the relationship between the system dynamics and the control inputs in the frequency domain. In the frequency domain, the open-loop and closed-loop transfer functions are given by:

$$L(s) =D(s)H(s), \tag{A.8}$$

$$G(s) =\frac{D(s)C(s)}{1 + D(s)H(s)}. \tag{A.9}$$

To further evaluate the dynamic behavior of the SEA system and validate the performance of the impedance controller, simulations were conducted based on the transfer functions (A.8) and (A.9). The specific simulation parameters used are summarized in Table A.1.

Figure A.2a presents the open-loop transfer function $L(s)$ of the SEA system. From the Bode plot, it can be observed that the system exhibits a phase margin of approximately 80° and a gain margin of approximately 20 dB. These margins indicate that the open-loop system maintains a reasonable level of robustness against model uncertainties and external disturbances. The system's characteristics reflect the

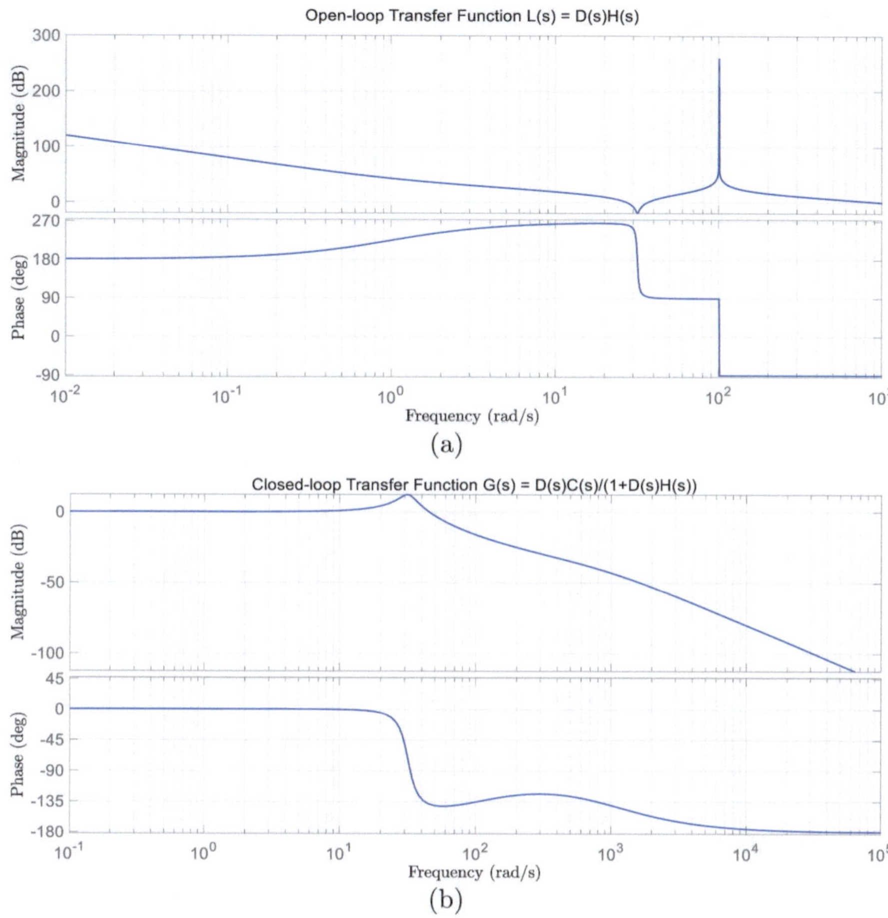

Fig. A.2 Transfer function representations of the SEA dynamic system: **a** Open-loop transfer function; **b** Closed-loop transfer function

intrinsic compliant nature of the SEA, with a relatively low gain crossover frequency, emphasizing safety and stability rather than aggressive dynamic response. Subsequently, Fig. A.2b illustrates the closed-loop transfer function $G(s)$. The closed-loop system demonstrates a bandwidth of approximately 50 rad/s, which determines the frequency range within which the system can effectively track the desired trajectory. The absence of excessive overshoot and the steady phase behavior confirm that the impedance controller successfully stabilizes the SEA dynamics while preserving the compliant interaction characteristics. These results collectively verify the effectiveness of the designed control strategy in regulating the SEA under nominal parameter settings.

In addition, a parametric study was performed to investigate the influence of the spring stiffness K on the closed-loop performance. By varying K from 100 N·m/rad to 600 N·m/rad under the same controller settings, it was found that the closed-loop bandwidth increased from 50.51 rad/s to 139.43 rad/s. This trend reflects that a higher spring stiffness leads to a stiffer and faster system response, enhancing the closed-loop tracking capability. However, it is important to note that excessive stiffness may reduce the system's compliance, potentially compromising safety in physical human-robot interactions. Therefore, selecting an appropriate stiffness value requires a careful trade-off between dynamic performance and interaction safety, depending on the specific application requirements of the exoskeleton.

Lyapunov Theory

Definition A.1 (*Equilibrium Point*) Consider a nonlinear system described by

$$\dot{x}_s = f_s(x_s, t), \tag{A.10}$$

where $x_s \in \mathbb{R}^n$ denotes the system state and $f_s(\cdot)$ is a nonlinear function. A point x_e is called an equilibrium if, whenever $x_s(t_0) = x_e$ for some initial time t_0, then $x_s(t) = x_e$ for all $t \geq t_0$. This implies that $f_s(x_e) = 0$.

Definition A.2 (*Stability of Equilibrium*) An equilibrium point $x_e = 0$ of the system is said to be stable if, for any $\varepsilon > 0$, there exists $\sigma > 0$ such that whenever $|x_s(t_0)| < \sigma$, it follows that $|x_s(t)| < \varepsilon$ for all $t \geq t_0$.

Definition A.3 (*Asymptotic Stability*) An equilibrium point $x_e = 0$ is asymptotically stable if it is stable and, in addition, there exists $\sigma > 0$ such that whenever $|x_s(t_0)| < \sigma$, It can be mathematically represented as

$$\lim_{t \to \infty} |x_s(t)| = 0. \tag{A.11}$$

Definition A.4 (*Positive Definite Functions*) A continuous function $V(x_s, t) : \mathbb{R}^n \times \mathbb{R}^+ \to \mathbb{R}$ is called locally positive definite if there exists $\varepsilon > 0$ such that:

- $V(0, t) = 0$ for all $t \geq 0$,
- $V(x_s, t) > 0$ for all $x_s \in B_\varepsilon$, where $B_\varepsilon = x_s \in \mathbb{R}^n : |x_s| < \varepsilon$.

If these conditions hold globally, then $V(x_s, t)$ is called a positive definite function. If $V(x_s, t) \geq 0$, it is referred to as positive semi-definite.

Theorem A.1 (Lyapunov Stability Theorem) *Let $x_e = 0$ be an equilibrium point of the system $\dot{x}_s = f_s(x_s, t)$. Assume that there exists a continuously differentiable function $V(x_s, t)$ such that:*

- *$V(x_s, t)$ is locally positive definite,*
- *The time derivative of V along system trajectories, given by*

$$\dot{V} = \frac{\partial V}{\partial x_s} f_s(x_s, t) + \frac{\partial V}{\partial t}, \tag{A.12}$$

  is locally negative semi-definite.

Then, the equilibrium point at the origin is locally stable. Moreover:

- *If $\dot{V}$ is locally negative definite, the origin is locally asymptotically stable.*
- *If V is positive definite and $\dot{V}$ is negative definite globally, the origin is globally asymptotically stable.*

Theorem A.2 (LaSalle's Invariance Principle) *If the Lyapunov function derivative $\dot{V}(x_s)$ is negative semi-definite rather than strictly negative definite, asymptotic stability cannot be immediately concluded from Lyapunov's theorem. In such cases, LaSalle's invariance principle can be applied to establish asymptotic stability by examining the largest invariant set where $\dot{V} = 0$.*

The Lyapunov framework provides a powerful method for verifying stability without explicitly solving the system's differential equations. It also offers systematic approaches for the design of stabilizing control laws for nonlinear systems.

Index

X. Li and H. Yu, *Compliantly-Driven Robots with Application to Exoskeletons*,
https://doi.org/10.1007/978-981-96-9544-7

MIX
Papier aus verantwortungsvollen Quellen
Paper from responsible sources
FSC® C105338

If you have any concerns about our products,
you can contact us on
ProductSafety@springernature.com

In case Publisher is established outside the EU,
the EU authorized representative is:
Springer Nature Customer Service Center GmbH
Europaplatz 3, 69115 Heidelberg, Germany

Printed by Libri Plureos GmbH
in Hamburg, Germany